The St. Petersburg School of Number Theory

HISTORY OF MATHEMATICS • VOLUME 26

The St. Petersburg School of Number Theory

B. N. Delone

Translated by Robert Burns

AMERICAN MATHEMATICAL SOCIETY • LONDON MATHEMATICAL SOCIETY

Б. Н. ДЕЛОНЕ

ПЕТЕРБУРГСКАЯ ШКОЛА ТЕОРИИ ЧИСЕЛ

Издательство Академии Наук СССР

Москва–Ленинград, 1947

Translated from the Russian by Robert Burns

2000 *Mathematics Subject Classification.* Primary 01A72, 11–03;
Secondary 01A55, 01A60.

For additional information and updates on this book, visit
www.ams.org/bookpages/hmath-26

Library of Congress Cataloging-in-Publication Data

Delone, B. N. (Boris Nikolaevich), 1890–1980.
[Peterburgskaia shkola teorii chisel. English]
The St. Petersburg school of number theory / B. N. Delone ; translated by Robert Burns.
p. cm. — (History of mathematics, ISSN 0899-2428 ; v. 26)
Includes bibliographical references.
ISBN 0-8218-3457-6 (acid-free paper)
1. Number theory. 2. Mathematics—Russia (Federation)—Saint Petersburg—History. 3. Mathematicians—Russia (Federation)—Saint Petersburg—History. I. Title: Saint Petersburg school of number theory. II. Title. III. Series.

QA241.D413 2005
512.7—dc22 2005048205

∞ The paper used in this book is acid-free and falls within the guidelines
established to ensure permanence and durability.
The London Mathematical Society is incorporated under Royal Charter
and is registered with the Charity Commissioners.
Visit the AMS home page at http://www.ams.org/

10 9 8 7 6 5 4 3 2 1 10 09 08 07 06 05

Contents

Foreword to the English Edition

"A series of Russian mathematicians—Chebyshev, Korkin, Zolotarev, Markov, Voronoĭ, and others—have worked in the theory of numbers. One can become acquainted with the content of the classical work of these notable mathematicians in B. N. Delone's book "The St. Petersburg School of Number Theory"." So begins the Preface to the book "Elements of Number Theory" by I. M. Vinogradov. Vinogradov does not inform us that he himself is the subject of the longest and most detailed chapter of Delone's book. These mathematicians, whose life and work in number theory form the substance of this work, are indeed a very distinguished group. Their work is of the highest quality and of lasting significance.

Delone's book was published, in Russian, in 1947. The present volume is a translation of the original into English. The text is essentially unaltered, but footnotes are added to clarify the exposition and in some cases to give more up to date references. For each of the six mathematicians considered, a brief biography is provided followed by an exposition of several of their most important contributions to number theory. In all cases, the mathematical exposition is much longer and more detailed than the biography. In spite of that, we get a good sense of their professional lives. We also get to see how their individual work is interrelated with the work of the others.

It should be emphasized that this book is strictly about number theory. For example, Chebyshev and Markov made very important contributions to the theory of probability, but no mathematical discussion of their work on probability is presented. Another point to make to the potential reader is that the mathematical discussion is not superficial. Delone himself was a distinguished member of the St. Petersburg school of number theory, and was well versed in the work of his predecessors. His exposition is at times taxing, but the effort expended in following the argument is amply rewarded. This is beautiful material.

One final comment before providing an overview of the mathematical content on this book. In 1947, the Cold War was just getting under way in earnest. Perhaps because of this historical fact, Delone's treatment is marred somewhat by an excessive nationalistic pride. While this is grating on modern sensibilities, it should not interfere with one's enjoyment while reading this book. After all, the mathematical works of these authors are well worth being proud of, and the Soviet era Russians are not the only ones who have been guilty of this fault.

The book begins with a consideration of P. L. Chebyshev and his contributions to number theory. Chebyshev was a first rate mathematician who contributed important results in a variety of fields. His main work in number theory concerns the distribution of prime numbers. For a positive real number x, Chebyshev denotes by $\phi(x)$ the number of positive prime numbers which are less than or equal to x. In modern notation his function is called $\pi(x)$, but we will follow Chebyshev. Legendre proposed the approximation

$$\phi(x) \approx \frac{x}{\ln(x) - 1.08366}.$$

This agrees well with tables of prime numbers, but in his first memoir on the subject, Chebyshev shows that it cannot be true. He shows that, in fact, the best approximation (in a certain precise sense) to $\phi(x)$ is

$$Li(x) = \int_2^x \frac{dt}{\ln(t)} \sim \frac{x}{\ln(x)}.$$

As usual, $f(t) \sim g(t)$ means that $f(t)/g(t)$ tends to 1 as t tends to infinity. In the proofs, Chebyshev uses properties of the zeta function, $\zeta(s)$, as defined by Euler. Namely, he uses the product formula and the fact that $(s-1)\zeta(s) \to 1$ as $s \to 1$ from above.

In his second memoir on the subject of prime numbers, Chebyshev goes much further and comes within striking distance of the prime number theorem as proposed by Gauss, namely,

$$\phi(x) \sim \int_2^x \frac{dt}{\ln(t)}.$$

In this work, Chebyshev introduces the auxilliary functions

$$\theta(x) = \sum_{p \leq x} \ln(p) \quad \text{and} \quad \psi(x) = \sum_{p^m \leq x} \ln(p),$$

which have played a key role in the literature on the distribution of primes ever since. Perhaps the main result of this paper is

$$.92129 \frac{x}{\ln(x)} < \phi(x) < 1.10555 \frac{x}{\ln(x)}.$$

A consequence of this is a proof of Bertrand's hypothesis, which states that for all $n > 1$, there is a prime number between n and $2n$.

Chebyshev's work on prime numbers was an absolutely major step forward in our knowledge of how prime numbers are distributed. In spite of this, it took almost fifty more years before the first proofs of the prime number theorem were given by J. Hadamard and, independently, by Ch.-J. de la Vallé Poussin.

We now pass on to the work of later members of the St. Petersburg circle. There is a problem in the theory of quadratic forms that provides a unifying thread in the work of Korkin, Zolotarev, Markov, and Voronoĭ, so it is a good idea to begin by explaining the problem. We will work in Euclidean space $\mathbb{R}^n$, which we can think of as being given by column vectors of real numbers with the usual inner product $(x, y) = \sum x_i y_i$. A quadratic form is a homogeneous polynomial of degree 2, usually written $Q(x) = \sum\sum a_{ij} x_i x_j$. We require the coefficients to be real numbers and the symmetry condition $a_{ij} = a_{ji}$. The quantity $D = \det(a_{ij})$ is called the discriminant of Q. If $Q(x) \geq 0$ for all x, with $Q(x) = 0$ if and only if $x = 0$, we say that Q is positive definite. Let $G = \mathbb{Z}^n \subset \mathbb{R}^n$ denote the integer lattice. Set $\mu(Q)$ equal to the minimum value of $Q(\gamma)$ as γ runs through the nonzero elements of G. C. Hermite was able to show that the quantity $\mu(Q)/\sqrt[n]{D}$ was bounded by some constant depending only on the dimension n. The smallest value of this constant is sometimes called Hermite's constant for dimension n, and the problem is to determine its value for each $n > 1$. The answer for $n = 2$ is $2/\sqrt{3}$. For $n = 3$, the answer is $\sqrt[3]{2}$. On the basis of this (and perhaps more), Hermite conjectured that the answer in general is $2/\sqrt[n]{n+1}$. This was disproved by Korkin and Zolotarev in a short paper (1872) where they show that for $n = 4$ the answer is $\sqrt[4]{4}$. Delone gives a fairly complete overview of this paper and its successors.

The problem referred to in the last paragraph can be reformulated as a problem about the smallest vector in a lattice. This problem, in turn, is closely related to the problem of optimal sphere packing in $\mathbb{R}^n$. In order to orient the reader to other parts of Delone's exposition, it is worthwhile to briefly explain these connections. Let $\{v_1, v_2, \ldots, v_n\}$ be a basis of $\mathbb{R}^n$. A lattice, L, in $\mathbb{R}^n$ is the free abelian group generated by a basis, i.e.

$$L = \mathbb{Z}v_1 + \mathbb{Z}v_2 + \cdots + \mathbb{Z}v_n.$$

Among all the nonzero lattice vectors λ, it is easy to see that there will be one (not necessarily unique), λ_o, such that $\|\lambda_o\| = \sqrt{(\lambda_o, \lambda_o)}$ is least. If $\lambda_o = \sum n_i v_i$, then $(\lambda_o, \lambda_o) = \sum\sum(v_i, v_j)n_i n_j$, so that $\|\lambda_o\|^2$ is the least value of the quadratic form $Q(x) = \sum\sum(v_i, v_j)x_i x_j$ evaluated at nonzero elements of $\mathbb{Z}^n$.

Conversely, if a positive definite quadratic form $Q(x) = \sum\sum a_{ij}x_i x_j$ is given, set $A = (a_{ij})$. We can find a nonsingular matrix B such that $B^t(a_{ij})B = I$, the identity matrix. Let $C = B^{-1}$. Then $A = C^t C$. Let v_i be the ith column vector of C. From $C^t C = I$ it follows immediately that $(v_i, v_j) = a_{ij}$. Let L be the lattice generated by the vectors $\{v_1, v_2, \ldots, v_n\}$. Then, the values of $Q(x)$ at elements of $\mathbb{Z}^n$ coincides with the values $\|\lambda\|^2$ as λ varies through L. We have shown that the problem of finding the smallest value of $Q(x)$ at integer points coincides with the problem of finding vectors of least length in a lattice.

The lattice sphere packing problem has to do with placing nonoverlapping spheres of fixed radius at each lattice point and determining the proportion of space taken up by the spheres. The idea is to find the best possible lattice packing. For a fixed lattice, the maximal radius of the spheres to be used is clearly equal to half the length of the smallest nonzero element of the lattice. People also consider sphere packings of space for which the centers of the spheres do not necessarily fill out a lattice. This problem may seem special, but it has wide applications throughout mathematics. It is interesting to note that it is mentioned by Hilbert in his famous list of problems (1899). It is discussed toward the end of problem 18. The most comprehensive modern source on sphere packing is the 1988 book by J. H. Conway and N. J. A. Sloane [**1**]. A shorter, and perhaps more readable, treatment is given in the book of C. Zong [**11**]. Both books refer to the work of Korkin, Zolotarev, and Voronoĭ.

In another direction, it is interesting to note that the work of Korkin and Zolotarev has found applications to lattice reduction algorithms which are of practical importance in cryptography. A famous 1990 paper [**7**] of J. C. Lagarias, H. W. Lenstra, and C. P. Schnorr is the genesis of the LLL-BKZ algorithm. The last two letters in this acronym stand for Korkin and Zolotarev, a tribute to how their fundamental work continues to find new applications.

Delone spends a lot of time on another aspect of Zolotarev's work. In two papers, one published in 1874, the other in 1880, Zolotarev establishes a general theory of algebraic integers and prime decomposition. The work is somewhat later than Dedekind's famous Supplement XI to Dirichlet's book "Vorlesungen über Zahlentheorie", in which he establishes a general theory of algebraic integers based on the new notion of ideals as certain subsets of the ring of integers. Zolotarev's ideas are more in keeping with Kummer's treatment of the ring of cyclotomic integers using the notion of ideal numbers. In his first paper on this subject, he assumes that the ring of integers is of the form $\mathbb{Z}[\xi]$. If $f(x) \in \mathbb{Z}[x]$ is the irreducible polynomial for ξ, the prime decomposition of a rational prime p is related to the decomposition of the polynomial $\bar{f}(x) \in \mathbb{Z}/p\mathbb{Z}\,[x]$. In general, the ring of integers in a number field is not of the form $\mathbb{Z}[\xi]$. Zolotarev overcomes this difficulty in his (posthumous) paper of 1880. It is a sad fact that in 1878, at the age of 31, Zolotarev was fatally injured in a train accident and died shortly thereafter.

Up to now, we have been concerned with positive definite quadratic forms. Markov discussed in detail the theory of indefinite binary quadratic forms. These behave quite differently, but Markov was able to give a rather complete theory which connects the theory of indefinite binary quadratic forms in a nontrivial way with the theory of continued fractions. As a by-product of his investigation he is able to give a procedure for finding all integer solutions to the diophantine equation $x^2+y^2+z^2 = 3xyz$. Today, this is called Markov's equation in his honor. Delone gives a careful discussion of Markov's work on binary quadratic forms. In addition he reformulates much of it in geometric terms. The reader will find this reformulation to be a valuable aid to understanding Markov's analysis.

Although it is not directly related to the work described in the present book, it is worth giving the statement of the Oppenheim conjecture, which concerns indefinite quadratic forms $Q(x)$ in three or more variables. This states that unless $Q(x)$ is proportional to a rational form, i.e. one with rational coefficients, the values of $Q(x)$ on the integer lattice $\mathbb{Z}^n$ are dense in $\mathbb{R}$. The Russian mathematician G. Margulis (now at Yale University) proved this conjecture in 1986. Although the statement of the conjecture is certainly in the spirit of the work of the classical mathematicians covered in the present book, the methods of proof are very different.

Voronoĭ made important contributions to many aspects of number theory. While still in graduate school he published a paper on Bernoulli numbers which simultaneously found new results about their arithmetic properties and gave a new method of deriving some of their known properties. These are based on some beautiful congruences which today bear Voronoĭ's name. A complete treatment is given in the text of Uspensky and Heaslet [**8**]. This approach is used in the book of Ireland and Rosen [**6**], which also shows how these congruences are related to the construction of p-adic L-functions which were first defined and explored by H. W. Leopoldt in 1975.

In another direction, Voronoĭ made an interesting contribution to analytic number theory. Let $\tau(n)$ denote the number of positive divisors of a positive integer n. Dirichlet was able to show that

$$\sum_{n,x} \tau(n) = n\ln(n) + (2\gamma - 1)n + O(\sqrt{n}).$$

Here, γ denotes Euler's constant. It has been a big problem over the years to improve the error term in this formula. Voronoĭ gave the first significant improvement by showing that the error term was $O(\sqrt[3]{n}\ln(n))$. One of the first achievements of I. M. Vinogradov, whom we will consider later, was to significantly generalize Voronoĭ's work. It is conjectured that the error term is $O(n^{1/4+\epsilon})$, which would follow from the Riemann hypothesis. On the other hand, it is known that $O(n^{1/4})$ won't do.

A further important contribution of Voronoĭ was his thorough investigation of cubic number fields. In all cases he was able to construct an explicit integral basis of the ring of integers of such a field and to discuss in detail the prime decomposition of rational primes. Perhaps his main result is to give an efficient algorithm for finding a set of fundamental units. Here, he takes as his model the continued fraction algorithm for finding the fundamental unit in a real quadratic number field. Others have given algorithms for finding independent units, but Voronoĭ's is both more efficient and has the advantage of always leading to a set of fundamental units. This work is covered at length in the present book. A more elaborate presentation is given in the book "The theory of irrationalities of the third degree" by B. N. Delone and D. K. Faddeev [**3**].

The last topic concerning Voronoĭ's contributions to number theory which is covered by Delone is his work on quadratic forms and lattices. Voronoĭ introduced important new concepts, e.g., perfect quadratic forms and Voronoĭ cells of a lattice. We will not go into the notion of a perfect quadratic form in this introduction, but will briefly describe what is meant by a Voronoĭ cell of a lattice. Let $L \subset \mathbb{R}^n$ be a lattice. For each $\lambda \in L$ let $V(\lambda)$ be the set of points x in $\mathbb{R}^n$ such that x is closer to λ than to any other lattice point in L. It is not hard to see that $V(\lambda)$ is a convex polytope. It is called a *Voronoĭ cell* of the lattice. The Voronoĭ cells are disjoint and the union of their closures fills all of $\mathbb{R}^n$. This decomposition is discussed in detail in the book by Conway and Sloane [**1**]. Another good place to read about Voronoĭ cells is the charming book by Conway entitled "The sensual (quadratic) form" [**2**].

Finally, we come to I. M. Vinogradov. Before surveying the chapter on Vinogradov, it should be pointed out that the mathematical exposition was written by B. A. Venkov. In the Preface, Delone refers to Venkov as his coauthor and extends his gratitude for writing the exposition of the work of Vinogradov and two parts of the work of Voronoĭ (on sums of divisors and on perfect quadratic forms). Nevertheless, Venkov's name does not appear on the title page.

The first contribution of Vinogradov which is discussed is his generalization of Voronoĭ's work on the divisor problem. The sum

$$\sum_{n=1}^{N} \tau(n)$$

which is to be estimated, is easily seen to be the number of integral lattice points in the first quadrant under the curve $xy = N$ with x-coordinate between 1 and N. What Vinogradov sets out to do is to estimate the number of integral lattice points in such a region where the top curve $y = N/x$ is replaced by a much more general function $y = f(x)$. With some restrictions on $f(x)$, Vinogradov is able to show that the number of lattice points is equal to the area of the region plus a well controlled error term. The reader should consult the text for details. In the case of Dirichlet's divisor problem, Vinogradov gets an error term of order $\sqrt[3]{N}[\ln(N)]^2$, which is only slightly worse than Voronoĭ's. On the other hand, his result is much more widely applicable.

While this early work clearly shows promise, the next two accomplishments show that the early promise was more than fulfilled by the mature mathematician. Vinogradov made signal contributions to two of the most venerable problems of number theory, Waring's problem and the Goldbach conjecture.

In 1770 E. Waring conjectured that for every $n \geq 2$ there is a number r such that every positive integer is the sum of at most r nth powers, i.e.

$$N = x_1^n + x_2^n + \cdots + x_r^n$$

is solvable in positive integers for every $N \geq 0$. The prototype for such a result is Lagrange's theorem that every positive integer is the sum of four squares.

In 1909, Hilbert proved Waring's conjecture in full generality. This was a great step forward, but there remained the question of finding a good bound on r as a function of n. Hilbert's proof produced values which were much too large. In the early 1920s, Hardy and Littlewood, using their newly invented "circle method" produced the first reasonable bounds. Let us define $G(n)$ to be the smallest integer such that every N with at most finitely many exceptions is the sum of $G(n)$ nth powers. By allowing for a finite number of exceptions, the theory is made much easier.

It is not hard to see that $n+1$ is the lower bound for $G(n)$ (see Chapter 18 of the book "Introduction to number theory" by H. K. Hua [**4**]). The upper bound obtained by Hardy and Littlewood is

$$G(n) \leq (n-2)2^{n-1} + 5.$$

In 1937, Vinogradov was able to prove that for $n \geq 16$

$$G(n) < 6n(\ln(n) + 1).$$

This is a remarkable result. One goes from a bound which is exponential in n to one that is almost linear in n. Vinogradov was not content, and later (1959) he showed that for large n one has

$$G(n) < n(2\ln(n) + 4\ln\ln(n) + 2\ln\ln\ln(n) + 13).$$

This upper bound was superseded, thirty years later, by a result of T. D. Wooley which shows that

$$G(n) < n(\ln(n) + \ln\ln(n) + O(1)).$$

In 1995, Wooley further improved the $O(1)$ term. This seems to be the best result at present. In less precise but simpler terms, we now know that for all large n, $n + 1 \leq G(n) < n^{1+\epsilon}$ for any $\epsilon > 0$.

The Goldbach conjecture states that every even number greater than four is the sum of two odd primes and that every odd number greater than seven is a sum of three odd primes. This famous conjecture was first stated in a letter from C. Goldbach to L. Euler in 1742. Once again, the first big step forward was due to Hardy and Littlewood. They showed that almost all even numbers are the sum of two odd primes. This means that $f(m)/m \to 0$ as $m \to \infty$, where $f(m)$ is the number of positive even numbers $\leq m$ which cannot be written as the sum of two odd primes. They also showed that every sufficiently big odd number is the sum of three odd primes provided that the generalized Riemann hypothesis is true. Unfortunately, to this day, neither the Riemann hypothesis nor the generalized Riemann hypothesis has been proved.

This was the situation in 1937 when Vinogradov proved, unconditionally, that every sufficiently large odd number is the sum of three odd primes. Yet another utterly remarkable result!

It is worth noting that neither Vinogradov nor anyone else has been able to resolve the Goldbach conjecture for even numbers. This remains inaccessible at the present time.

Vinogradov's success was due in large part to his ingenious methods for evaluating arithmetic sums, in particular, trigonometric sums. The last part of the chapter on his work is concerned with his estimates of Weyl sums (named after Hermann Weyl and introduced in Weyl's 1916 paper entitled "Über Gleichverteilung von Zahlen mod. Eins"). The sums in question look like

$$\sum_{n=1}^{N} e^{2\pi i f(n)},$$

where $f(x)$ is a polynomial with real coefficients and positive degree.

The reader who would like to see Vinogradov's method exposited in full generality should consult his book "The method of trigonometric sums in the theory of numbers" [**10**]. Another excellent resource is the recent book on analytic number theory by H. Iwaniec and E. Kowalski [**5**].

This brings us to the conclusion of this Foreword. Of course, the St. Petersburg school of number theory did not end in 1947 when the original of the present book was published. As Delone points out in his Preface (somewhat immodestly), "The next most important work of the St. Petersburg number-theoretic school, in the chronological order established above, belongs to the author of the present book, and to Venkov, Kuz'min, Tartakovskiĭ, and Linnik." While these are all estimable mathematicians, it is fair to say that if the volume were to continue beyond 1947, it would have to include Venkov and Linnik and perhaps the others. In any case, by composing the present volume, Delone did a great service to posterity. The new translation should have a wide readership among English speaking mathematicians with enough background to enjoy it.

Michael Rosen
Brown University

Bibliography

[1] J. H. Conway and N. J. A. Sloane, *Sphere packings, lattices, and groups*, Springer-Verlag, New York–Berlin–Heidelberg, 1988.
[2] J. H. Conway, *The sensual (quadratic) form*, Carus Math. Monograph No. 26, MAA, Washington, DC, 1997.
[3] B. N. Delone and D. K. Faddeev, *The theory of irrationalities of the third degree*, Amer. Math. Soc., Providence, RI, 1964.
[4] L. K. Hua, *Introduction to number theory*, Springer-Verlag, Berlin–New York, 1982.
[5] H. Iwaniec and E. Kowalski, *Analytic number theory*, Amer. Math. Soc., Colloquium Publ., Vol. 53, Providence, RI, 2004.
[6] K. Ireland and M. Rosen, *A classical introduction to modern number theory*, Grad. Texts in Mathematics, Col. 84, Springer-Verlag, New York–Berlin–Heidelberg, 1990.
[7] J. C. Lagarias, H. W. Lenstra Jr., and C. P. Schnorr, *Korkin–Zolotarev bases and successive minima of a lattice and its reciprocal lattice*, Combinatorica **10** (1990), 333–348.
[8] J. V. Uspensky and M. A. Heaslet, *Elementary number theory*, McGraw Hill, 1939.
[9] I. M. Vinogradov, *Elements of number theory*, Dover Publ., New York, 1954. (reprint, 2004)
[10] ———, *The method of trigonometric sums in the theory of numbers*, Interscience, London–New York.
[11] C. Zong, *Sphere packings*, Springer-Verlag, Berlin–Heidelberg–New York, 1999.

Translator's acknowledgments

I thank Steven Chen, Nina Ioukhoveli, Yuriĭ Kazmerchuk, Keith Matthews, and once again Abe Shenitzer, for help and encouragement while this translation was being carried out at York University, Toronto, and The University of Queensland, Brisbane.

2004 R. G. Burns

Preface

The work of Russian mathematicians in the theory of numbers constitutes a glorious contribution to Russian science. Although mention should perhaps be made in this connection of the brilliant work of the great Leonhard Euler, since he was a member of the Russian Academy of Sciences for a considerable portion of his life[1], in its proper sense the Russian school of number theory begins with Chebyshev. The St. Petersburg school was illumined by names such as that of Chebyshev himself, Korkin, Zolotarev, Markov and Voronoĭ[2]; and at the present time[3] the Soviet school can boast of several excellent number theorists, led by Academician Vinogradov.

The aim of the present book is to acquaint the lover of mathematics with the most important works of the above-named six preeminent members of the St. Petersburg school of number theory. For each of these six a short biography is given, followed by an exposition of two or three of the most significant of his number-theoretical contributions. Each such contribution is first expounded in the author's original terminology and notation, i.e., in the form of a summary, as it were, facilitating the reading of the original work itself, and this is then followed by a more or less broad commentary on it. Certain of the works in question, for instance those of Chebyshev on primes, have proved amenable to a relatively complete exposition, while others, more wide-ranging in nature, are dealt with much more briefly—for instance Zolotarev's dissertation on integral complex numbers.[4]

The next most important work of the St. Petersburg number-theoretic school, in the chronological order established above, belongs to the author of the present book, and to Venkov, Kuz'min, Tartakovskiĭ and Linnik. Outstanding contributions to number theory have also been made by Chebotarev, and by the Moscow mathematicians Khinchin, Shnirel'man and Gel'fond. However considerations of space have made it impossible to include expositions of the work of these authors, let alone the great many other lesser Russian contributions to number theory.

I wish to express my deepest gratitude to my coauthor, Professor Boris Alekseevich Venkov, for writing the exposition of the work of I. M. Vinogradov and of two of Voronoĭ's works—on the sum of the numbers of divisors and on complete forms. In addition I thank both him and Corresponding Member of the Academy P. O. Kuz'min for valuable advice concerning the compilation of material for this book.

1947 B. Delone

[1]1725–1741 and 1766–1783. *Trans.*

[2]These have stressed syllables as follows: Chebyshóv, Kórkin, Zolotaryóv, Márkov, Voronóĭ. *Trans.*

[3]In 1947. *Trans.*

[4]Algebraic integers. *Trans.*

Pafnutiĭ L'vovich Chebyshev (1821–1894)

Pafnutiĭ L'vovich CHEBYSHEV

Pafnutiĭ L'vovich Chebyshev was born on May 14 (26), 1821, on his father's property, which included the village Okatovo, situated in the Borovsk district, Kaluga province.[5] The Chebyshevs belonged to an old Russian noble line. For his time Chebyshev's father was a personage of substance. Pafnutiĭ L'vovich received his early education at home. His mother, Agrafena Ivanovna, taught him to read and write, while other subjects, including arithmetic and French, were taught to him by his cousin A. K. Sukhareva, a young woman of high education. In 1832 the whole Chebyshev family moved to Moscow in order to prepare Pafnutiĭ and his elder brother to enter the university.

To this end the best Moscow teachers were enlisted; in mathematics Chebyshev was coached by Pogorel'skiĭ, at that time well known in Moscow. In 1837 Chebyshev enrolled in the faculty of physics and mathematics of Moscow University, and on promotion to the second year was awarded a silver medal for his essay "Calculation of the roots of an equation". The seriousness of his work attracted the attention of Professor N. D. Brashman, whom he thenceforth held in the deepest respect. Chebyshev graduated from the university in 1841.

In 1840 famine raged over a large part of Russia, and the affairs of the Chebyshevs fell into disarray. In order to avoid going hungry the young Chebyshev, fresh from his university studies, was compelled either to serve somewhere or give lessons. Reasoning that such occupations would distract him from mathematics, he chose want. Around this time his first published articles appeared, and he completed his master's thesis "An experiment in elementary analysis in the theory of probability", which he defended at Moscow University in 1846.

In 1847 Chebyshev moved to St. Petersburg, having been awarded the post of adjunct professor at the university there subsequent to defending his dissertation "On integration with the aid of logarithms". This dissertation was submitted in handwritten form and only recently, in 1937, thanks to Academician Krylov, has it been published in "Izvestiya Akademii Nauk". On the strength of this work Chebyshev obtained the right to give lectures at the university. In 1849 he was awarded the doctoral degree with his third dissertation "The theory of congruences".

At almost the same time as Chebyshev there came to St. Petersburg University as ordinary professor the powerful mathematician Viktor Yakovlevich Bunyakovskiĭ, who also enjoyed the position of Ordinary Academician at the Academy of Sciences. At that time also the widely known mathematician Iosif Ivanovich Somov held the rank of extraordinary professor in the department of applied mathematics of St. Petersburg University. With these two Chebyshev became very close and to the

[5]This sketch reproduces in shortened form that composed by Academicians A. A. Markov and N. Ya. Sonin after the article by K. A. Posse in Vengerov's "Critical-Biographical Dictionary of Russian Writers and Scholars".

end of their lives remained on excellent terms with them. He was especially close to Bunyakovskiĭ, who, appreciating Chebyshev's great mathematical power, brought him into the Academy of Sciences initially as collaborator in the preparation of Euler's works for publication, and then as associate member. In 1853 Chebyshev was elected adjunct and finally in 1859 was made Ordinary Academician of the Academy of Sciences.

Over this period Chebyshev's means were scanty, so that he was forced to be frugal; and thus he remained till the end of his life. The only sort of expense he never spared was on mechanical devices of his own invention; on these he spent hundreds, perhaps thousands, of roubles. From early childhood Chebyshev had loved to devise various mechanisms, beginning with toys fashioned with a penknife, and culminating in the complicated calculating machine preserved in the Conservatoire des Arts et Métiers in Paris. Other mechanical gadgets built by his own hands are preserved to this day in the Academy of Sciences.

Chebyshev's circle of acquaintance was not large. He loved visiting Bunyakovskiĭ, at whose home many mathematicians, including M. B. Ostrogradskiĭ, gathered regularly. In his younger days at St. Petersburg University he often visited Somov to air his discoveries. By calling on Somov's vast erudition he was able to determine whether or not his latest discovery had already been made; and sometimes his results had indeed been anticipated. Chebyshev himself liked to do research independently of the relevant mathematical literature, especially that contemporaneous with him. Although thoroughly familiar with the work of Euler, Lagrange, Gauss and Abel, he assigned little importance to being up with current mathematical output. He used to assert that the excessive effort needed to read others' work affected the independence of his own ideas negatively.

Chebyshev loved to travel abroad. In the early part of his career such trips had a strictly scientific purpose. The report on his official trip abroad in 1852 contains a description of the trip, of visits to various factories in order to see how mechanics was applied, meetings of scientific societies, and discussions on scientific topics with famous scholars of various countries. Later in his career Chebyshev took advantage of these trips also to rest from his unceasing scientific work. France was his favorite place to visit and he often took the opportunity to lecture on his work there. In Paris, thanks to his excellent command of French, he felt himself at home. He was acquainted with many leading French mathematicians: Liouville, Bertrand, Serret, and Hermite, among others. When in Paris, in keeping with his characteristic thriftiness, Chebyshev usually stayed at the same very modest hotel (the Hôtel Corneille, opposite the Odéon). Only in his later years did he depart to some extent from such habits—even to the extent of giving lunches for his French friends.

Chebyshev was professor at the University of St. Petersburg over a period of 35 years, from 1847 to 1882. At different times he lectured on analytic geometry, higher algebra, number theory, the integral calculus, probability theory, the calculus of finite differences, elliptic functions, and the theory of definite integrals. His attitude to lecturing was pedantically strict. He almost never missed a lecture, was never late, and at the sound of the bell announcing the end of the lecture period, stopped talking immediately, sometimes even in the middle of a word. An unfinished proof would always be given from the very beginning in the next lecture, unless that followed immediately on the first. He would precede each complicated calculation with an explanation of its purpose and overall structure, and then write it out in detail, mostly in silence. The calculation would be performed quickly, but in such minute detail that it was very easy to follow. He would also during the lecture often step away from the blackboard to tell the audience his own views and of conversations he had had with other mathematicians concerning the topic at hand, and to explain the relative significance of and connections between various mathematical questions.

Chebyshev's merits as professor remained forever in the memories of those who had the enviable good fortune to study under him. He continued to guide them for some time after the completion of their degree. The first steps in the careers of those of his students wishing to devote themselves to mathematics were made under his close supervision. Once a week at a fixed time his home was open to anyone with something to communicate to the famous mathematician about their own work. In his discussions with students he would direct them to fruitful topics for independent investigation, and draw their attention to questions that were likely to yield worthwhile results. For instance the work of Lyapunov on the equilibrium trajectories of heavenly bodies began in this way.

Chebyshev died in 1894, in his 74th year. On the morning of 8th December, sitting at his writing desk, he suddenly felt unwell, and after a short agony died of a heart attack.

Chebyshev is one of the two supreme Russian mathematicians, the other, of course, being Lobachevskiĭ. Their work shares a common feature. It fell to each of them to make significant progress on problems that had resisted all efforts at solution for more than 2000 years — in Lobachevskiĭ's case the deepest question concerning the foundations of geometry, and in Chebyshev's the most difficult problem concerning the distribution of primes in the sequence of natural numbers.

There can be no doubt that the ingenious work of Lobachevskiĭ and Chebyshev gave in the past and continues to give Russian mathematicians the confidence needed to overcome daunting difficulties in their research and produce outstanding results. One can only hope that this wave of confidence, surging very satisfactorily at present[6], will continue into the future.

The fates of these two great men were however completely different.

Lobachevskiĭ lived in the provincial city Kazan. His ideas were too novel for his time. The foremost Russian mathematicians were to be found then in St. Petersburg, at the Academy of Sciences, and these, such as, for instance, Bunyakovskiĭ, did not believe in what Lobachevskiĭ was doing, and therefore failed to provide him with the support of their authority. As a result Lobachevskiĭ was not able to create a school.

On the other hand Chebyshev worked from the start in areas of interest to a great many of the most prominent mathematicians of the period: probability, the integration of algebraic differentials, and number theory. In each of these areas Chebyshev obtained first-rate results, and on the problem of the distribution of the primes produced results of world-wide importance, attracting general attention. Having secured for himself through this work a place in the front rank of

[6] In 1947. *Trans.*

contemporary mathematicians, he transferred his energies to his own, original topics, namely the problem of closest approximation of a given function on an interval by a polynomial function, and the theory of hinged lever mechanisms.

The variety and many-sidedness of his work astounds: Following on his memoir "An experiment in elementary analysis in the theory of probability", comes his dissertation "On integration with the aid of logarithms", and then his most celebrated works "On the determination of the number of primes not exceeding a given bound", and "On the prime numbers", and later the memoir "The theory of mechanisms known by the name of parallelograms", and so on. In this respect Chebyshev resembles the great classical mathematicians Euler and Lagrange. Here, for example, are a few randomly chosen works of Lagrange, in chronological order: "A new method for solving literal equations by means of series", "On the force of a compressed spring", "On Kepler's equation", "New arguments concerning tautochrones", "Proof of a theorem in arithmetic". Chebyshev's most fundamental work was in number theory, probability theory, the theory of best approximation of functions by polynomials, the integration of algebraic differentials, and the theory of mechanisms.

By the mid-1850s Chebyshev had already become generally acknowledged as a first-rank mathematician. He was as well-known overseas as in Russia, in Paris as in St. Petersburg, held the rank of adjunct in the Academy of Sciences, and of professor at St. Petersburg University. In 1859 he was appointed Academician. He had bestowed on him also the rare honor of being made an active foreign member of the Paris Academy of Sciences (membre associé). He remained to the end of his long life one of the world's most famous living mathematicians.

Naturally the St. Petersburg mathematicians, with easy access to Chebyshev, were deeply influenced by him. Around him the famous Chebyshev school, usually called the St. Petersburg school, formed itself, producing many remarkable scholars: Aleksandr Nikolaevich Korkin, Egor Ivanovich Zolotarev, Andreĭ Andreevich Markov, Georgiĭ Fedoseevich Voronoĭ, Aleksandr Mikhailovich Lyapunov, Vladimir Andreevich Steklov; and among the strongest mathematicians of our own day[7] belonging to this school, Sergeĭ Natanovich Bernshtein and Ivan Matveevich Vinogradov.

Chebyshev liked concrete, clearly stated problems. He considered a mathematical problem solved only when it had been brought into the form of a convenient formula or computational recipe, i.e., algorithm. This love of concreteness and algorithms was inherited by his students. Another marked preference of his—also absorbed by his students—was for basing theory on practice; he was of the opinion that "the closer a theory is to practice the more beneficial the results, and this is not only a matter of deriving practical applications, since in fact the sciences develop in this way out of practical considerations." For instance the theory of functions deviating least from zero was invented by Chebyshev in order to answer a question in the theory of mechanisms.

The methods Chebyshev used in solving problems were usually of the simplest, often, however, distinguished by a rare originality and subtlety of mind.

[7]In 1947. *Trans.*

Chebyshev's Articles on the Prime Numbers

On the Determination of the Number of Primes not Exceeding a Given Number

In the 1840s the Russian Academy of Sciences undertook to publish an edition of Euler's works under the title "Opera minora collecta". Among the first volumes to appear (in 1849) were the two entitled "Leonardi Euleri commentationes arithmeticae collectae". Academician Bunyakovskiĭ, the editor for the project, involved Chebyshev in the work just after his arrival in St. Petersburg from Moscow. Chebyshev had by then completed his research in probability and analysis, and the task of editing Euler's "Commentationes arithmeticae" led him to start thinking about number theory. He wrote "The theory of congruences", which he used as his doctoral dissertation, defended in 1849. For a long time this work was the sole Russian text on number theory. It contained three appendices: "On quadratic residues", "On the determination of primitive roots", and "On the number of primes not exceeding a given number". This third appendix contains Chebyshev's famous first memoir on the prime numbers.[8]

This first memoir begins by considering an empirical result appearing in the second volume of Legendre's "Theory of numbers". Legendre suggests the following formula as an approximation to the number of primes less than or equal to a given number x:

$$\frac{x}{\ln x - 1.08366}.$$

This formula fits well with the actual empirical values obtained from the table of primes available to Legendre. Chebyshev writes: "In spite of the apparent agreement of Legendre's formula with values calculated from the table of primes, we feel compelled to register doubt as to its real accuracy, and consequently are unable to consider deductions based on it as true. We are brought to this conclusion by a certain theorem concerning properties of the function giving the number of primes less than a given number, having many curious consequences."

The theorem in question is as follows.

THEOREM I. *Let $\phi(x)$ denote the number of primes less than x where x ranges over the integers from 2 to ∞, let n be any fixed nonnegative integer, and ρ a positive real variable. Then the sum*

$$\sum_{x=2}^{x=\infty} \left[\phi(x+1) - \phi(x) - \frac{1}{\ln x}\right] \frac{\ln^n x}{x^{1+\rho}},$$

[8]This work was also published in "Mémoires des savants étrangers de L'Académie Impériale des Sciences de St. Petersbourg" in 1848, and in 1852 in Liouville's journal under the title "Sur la fonction qui détermine la totalité de nombres premiers à une limite donnée".

considered as a function of ρ, approaches a finite limit as $\rho \to 0$.[9]

PROOF. Since $\phi(x+1)$ and $\phi(x)$ denote the numbers of primes less than $x+1$ and x respectively, the difference $\phi(x+1) - \phi(x)$ will be 1 when x is a prime and 0 when x is composite. Hence

$$\sum_{x=2}^{x=\infty} [\phi(x+1) - \phi(x)] \frac{\ln^n x}{x^{1+\rho}} = \sum_{\mu} \frac{\ln^n \mu}{\mu^{1+\rho}},$$

where μ ranges over the primes. Changing notation in the remaining term as follows:

$$\sum_{x=2}^{x=\infty} \frac{1}{\ln x} \frac{\ln^n x}{x^{1+\rho}} = \sum_{m} \frac{\ln^{n-1} m}{m^{1+\rho}},$$

where m ranges over all natural numbers from 2 to ∞, we now have the sum considered in the theorem in the form

$$\sum_{\mu} \frac{\ln^n \mu}{\mu^{1+\rho}} - \sum_{m} \frac{\ln^{n-1} m}{m^{1+\rho}}.$$

This in turn is equal to

$$\begin{aligned} &\frac{d^n}{d\rho^n}\left[\sum \ln\left(1 - \frac{1}{\mu^{1+\rho}}\right) + \sum \frac{1}{\mu^{1+\rho}}\right] \\ &\quad + \frac{d^n}{d\rho^n}\left[\ln \rho - \sum \ln\left(1 - \frac{1}{\mu^{1+\rho}}\right)\right] \\ &\quad + \frac{d^{n-1}}{d\rho^{n-1}}\left(\sum \frac{1}{m^{1+\rho}} - \frac{1}{\rho}\right). \end{aligned}$$

This is readily verified by first cancelling the two like terms and then carrying out the differentiations. Hence the theorem will follow if one can show that each of the three terms in this expression tends to a finite limit as ρ tends to 0.

For the third term Chebyshev begins with the equalities

$$\begin{aligned} \int_0^\infty \frac{e^{-x}}{e^x - 1} x^\rho dx &= \sum \frac{1}{m^{1+\rho}} \int_0^\infty e^{-x} x^\rho dx, \\ \int_0^\infty e^{-x} x^{-1+\rho} dx &= \frac{1}{\rho} \int_0^\infty e^{-x} x^\rho dx, \end{aligned}$$

the first of which is obtained by observing that

$$\frac{e^{-x}}{e^x - 1} = \sum_2^\infty e^{-mx},$$

and summing the equalities

$$\frac{1}{m^{1+\rho}} \int_0^\infty e^{-x} x^\rho dx = \int_0^\infty e^{-mx} x^\rho dx, \qquad m = 2, 3, 4, \ldots$$

(make the substitution $u := mx$ in the right-hand side), while the second expresses a well-known (and easily derived) property of the gamma function $\Gamma(\rho)$ (defined as

[9]That is, as $\rho \downarrow 0$. *Trans.*

the left-hand side of that equality). These yield immediately the following integral representation of the sum $\sum \frac{1}{m^{1+\rho}}$:

$$\sum \frac{1}{m^{1+\rho}} - \frac{1}{\rho} = \frac{\int_0^\infty \left(\frac{1}{e^x-1} - \frac{1}{x}\right) e^{-x} x^\rho dx}{\int_0^\infty e^{-x} x^\rho dx}.$$

From this equality (identity in ρ) it is easily seen[10] that the $(n-1)$st derivative of $\sum \frac{1}{m^{1+\rho}} - \frac{1}{\rho}$ with respect to ρ is a fractional expression with denominator $\left(\int_0^\infty e^{-x} x^\rho dx\right)^{n+1}$, and with numerator a polynomial expression with integer coefficients, in the following integrals:

$$\int_0^\infty \left(\frac{1}{e^x - 1} - \frac{1}{x}\right) e^{-x} x^\rho \, dx; \int_0^\infty \left(\frac{1}{e^x - 1} - \frac{1}{x}\right) e^{-x} x^\rho \ln x \, dx; \ldots;$$
$$\int_0^\infty \left(\frac{1}{e^x - 1} - \frac{1}{x}\right) e^{-x} x^\rho \ln^{n-1} x \, dx;$$
$$\int_0^\infty e^{-x} x^\rho \, dx; \int_0^\infty e^{-x} x^\rho \ln x \, dx; \ldots; \int_0^\infty e^{-x} x^\rho \ln^{n-1} x \, dx.$$

This fractional expression has, for every $n \geq 0$, a finite limit as $\rho \downarrow 0$ since $\int_0^\infty e^{-x} x^\rho dx = \Gamma(\rho + 1)$, and the gamma function $\Gamma(s)$ is certainly continuous at $s = 1$, while the other integrals similarly have finite limits as $\rho \downarrow 0$.[11] This completes the proof that the third term

$$\frac{d^{n-1}}{d\rho^{n-1}} \left(\sum \frac{1}{m^{1+\rho}} - \frac{1}{\rho}\right)$$

of the sum in question has a finite limit as $\rho \downarrow 0$.

For the second term of that sum Chebyshev uses Euler's identity

$$1 + \frac{1}{2^{1+\rho}} + \frac{1}{3^{1+\rho}} + \frac{1}{4^{1+\rho}} + \cdots = \left[\left(1 - \frac{1}{2^{1+\rho}}\right)\left(1 - \frac{1}{3^{1+\rho}}\right)\left(1 - \frac{1}{5^{1+\rho}}\right)\cdots\right]^{-1},$$

representing the series $1 + \sum \frac{1}{m^{1+\rho}}$, now called the *zeta-function* of $1 + \rho$, as the infinite product $\prod \left(1 - \frac{1}{\mu^{1+\rho}}\right)$, where, in contrast to the series, the product is taken over only the primes μ. Taking the logarithm of Euler's identity, Chebyshev obtains

$$\ln\left(1 + \sum \frac{1}{m^{1+\rho}}\right) = -\sum \ln\left(1 - \frac{1}{\mu^{1+\rho}}\right),$$

yielding

$$\ln \rho - \sum \ln\left(1 - \frac{1}{\mu^{1+\rho}}\right) = \ln\left[\left(1 + \frac{1}{m^{1+\rho}}\right) \cdot \rho\right],$$

whence

$$\ln \rho - \sum \ln\left(1 - \frac{1}{\mu^{1+\rho}}\right) = \ln\left[1 + \rho + \left(\sum \frac{1}{m^{1+\rho}} - \frac{1}{\rho}\right) \cdot \rho\right].$$

From this identity we see that the nth derivative with respect to ρ of

$$\ln \rho - \sum \ln\left(1 - \frac{1}{\mu^{1+\rho}}\right)$$

[10]Presumably differentiation under the integral sign is permissible here. *Trans.*

[11]In the Russian text this is claimed as "obvious". *Trans.*

(which nth derivative is what we are at present considering) can be represented as a fractional expression with denominator a positive integer power of

$$1+\rho+\left(\sum\frac{1}{m^{1+\rho}}-\frac{1}{\rho}\right)\cdot\rho,$$

and numerator a polynomial expression with integer coefficients, in ρ and the quantity

$$\sum\frac{1}{m^{1+\rho}}-\frac{1}{\rho}$$

and its derivatives with respect to ρ. That such a fractional expression approaches a finite limit as $\rho\downarrow 0$ can be seen as follows. First, the quantity

$$1+\rho+\left(\sum\frac{1}{m^{1+\rho}}-\frac{1}{\rho}\right)\cdot\rho,$$

a positive integer power of which comprises the denominator, approaches 1, since, as a particular case of what was shown above, $\sum\frac{1}{m^{1+\rho}}-\frac{1}{\rho}$ has a finite limit as $\rho\downarrow 0$. That the numerator tends to a finite limit as $\rho\downarrow 0$ follows from the fact established above that also all derivatives of $\sum\frac{1}{m^{1+\rho}}-\frac{1}{\rho}$ have finite limits as $\rho\downarrow 0$.

Finally we turn to the first term of the original three-term sum, namely the nth derivative with respect to ρ of the function

$$\sum\ln\left(1-\frac{1}{\mu^{1+\rho}}\right)+\sum\frac{1}{\mu^{1+\rho}}.$$

The first derivative is

$$\sum\frac{1}{\mu^{2+2\rho}}\cdot\frac{\ln\mu}{1-\frac{1}{\mu^{1+\rho}}},$$

whence it follows that each of the higher derivatives is a finite integral linear combination of series of the form

$$\sum\frac{1}{\mu^{2+2\rho}}\cdot\frac{\ln^p\mu}{1-\frac{1}{\mu^{1+\rho}}}\cdot\frac{1}{\mu^s\left(1-\frac{1}{\mu^{1+\rho}}\right)^r},$$

where p,r,s are nonnegative integers. In each such series (with $\rho\geq 0$), for any fixed $\epsilon>0$ the terms are all less than $\frac{1}{\mu^{2-\epsilon}}$ for sufficiently large μ. It follows that any such series represents a function of ρ right-continuous at 0, and so has a finite limit as $\rho\downarrow 0$.

This completes the proof of Theorem I. □

From Theorem I Chebyshev deduces the corollary that the function of ρ defined by the series

$$\sum_{x=2}^{x=\infty}\left[\phi(x+1)-\phi(x)-\int_x^{x+1}\frac{dt}{\ln t}\right]\frac{\ln^n x}{x^{1+\rho}}$$

also has a finite limit. For this purpose he first observes that the difference $\frac{1}{\ln x}-\int_x^{x+1}\frac{dt}{\ln t}$ is $o\left(\frac{1}{x}\right)$ as $x\to\infty$.[12] Hence

$$\left(\frac{1}{\ln x}-\int_x^{x+1}\frac{dt}{\ln t}\right)\frac{\ln^n x}{x^{1+\rho}}=o\left(\frac{1}{x^{2+\rho}}\right),$$

[12] By considering the region under the graph of $y=\frac{1}{\ln x}$ between x and $x+1$, one easily sees that the difference in question is bounded above by $\frac{1}{\ln x}-\frac{1}{\ln(x+1)}=\frac{\ln\left(1+\frac{1}{x}\right)}{\ln x\ln(x+1)}$, whence the desired conclusion follows from the expansion of $\ln\left(1+\frac{1}{x}\right)$ as a power series in $\frac{1}{x}$. *Trans.*

and consequently the series

$$\sum_{x=2}^{x=\infty} \left(\frac{1}{\ln x} - \int_x^{x+1} \frac{dt}{\ln t} \right) \frac{\ln^n x}{x^{1+\rho}}$$

is continuous from above at $\rho = 0$. Adding this to the series in Theorem I then yields the desired conclusion.

From this version of Theorem I Chebyshev next infers the important

THEOREM II. *For any positive real number α, no matter how small, and any n, no matter how large, the function $\phi(x)$ denoting the number of primes less than x, satisfies the inequality*

$$\phi(x) > \int_2^x \frac{dt}{\ln t} - \frac{\alpha x}{\ln^n x}$$

for infinitely many integer values of x, and also satisfies the inequality

$$\phi(x) < \int_2^x \frac{dt}{\ln t} + \frac{\alpha x}{\ln^n x}$$

for infinitely many integer values of x.

Since the proofs of the two inequalities are similar, Chebyshev confines himself to proving the second. Thus suppose that the inequality

$$\phi(x) < \int_2^x \frac{dt}{\ln t} + \frac{\alpha x}{\ln^n x} \tag{1}$$

holds for only a finite number of positive integers x, and let a be any integer larger than both e^n and the largest integer x for which the inequality (1) holds. Then for every integer $x > a$ we shall have

$$\phi(x) - \int_2^x \frac{dt}{\ln t} \geq \frac{\alpha x}{\ln^n x}, \qquad \frac{n}{\ln x} < 1. \tag{2}$$

However, then the series

$$\sum_{x=2}^{x=\infty} \left[\phi(x+1) - \phi(x) - \int_x^{x+1} \frac{dt}{\ln t} \right] \frac{\ln^n x}{x^{1+\rho}}$$

will tend to ∞ as $\rho \downarrow 0$. To see this, first consider this series as the limit as $s \to \infty$ of the finite sum

$$C + \sum_{x=a+1}^{x=s} \left[\phi(x+1) - \phi(x) - \int_x^{x+1} \frac{dt}{\ln t} \right] \frac{\ln^n x}{x^{1+\rho}}, \tag{3}$$

where C is the sum of its first $a-1$ terms, obviously a finite quantity for every $\rho \geq 0$. Chebyshev now applies to the sum in (3) Abel's transformation

$$\sum_{x=a+1}^{x=s} u_x(v_{x+1} - v_x) = u_s v_{s+1} - u_a v_{a+1} - \sum_{x=a+1}^{x=s} v_x(u_x - u_{x-1}),$$

with

$$v_x := \phi(x) - \int_2^x \frac{dt}{\ln t}, \qquad u_x := \frac{\ln^n x}{x^{1+\rho}}.$$

Taking into account the assumed inequality in (2) (i.e. $v_x \geq \frac{\alpha x}{\ln^n x}$), it is now not difficult to see that the sum (3), transformed in this way, tends to ∞ as $\rho \downarrow 0$. This contradiction shows that in fact the inequality (1) must hold for an infinity of positive integer values of x.

Having proved Theorem II in this way, Chebyshev goes on to deduce

THEOREM III. *The function* $\frac{x}{\phi(x)} - \ln x$ *cannot have limit other than* -1 *as* $x \to \infty$.

To see this, suppose $\frac{x}{\phi(x)} - \ln x$ has a limit L say, as $x \to \infty$. This means of course that given any positive number ϵ, no matter how small, for all sufficiently large x, say $x > N = N(\epsilon)$, we shall have

$$L - \epsilon < \frac{x}{\phi(x)} - \ln x < L + \epsilon. \tag{4}$$

Combining these two inequalities respectively with the inequalities

$$\phi(x) > \int_2^x \frac{dt}{\ln t} - \frac{\alpha x}{\ln^n x},$$
$$\phi(x) < \int_2^x \frac{dt}{\ln t} + \frac{\alpha x}{\ln^n x},$$

each holding, by Theorem II, for infinitely many (integer) values of x, and therefore for infinitely many values of $x > N$, we obtain the following two inequalities, each valid separately for infinitely many integer values of x:

$$L + 1 < \frac{x - (\ln x - 1)\left(\int_2^x \frac{dt}{\ln t} - \frac{\alpha x}{\ln^n x}\right)}{\int_2^x \frac{dt}{\ln t} - \frac{\alpha x}{\ln^n x}} + \epsilon,$$
$$L + 1 > \frac{x - (\ln x - 1)\left(\int_2^x \frac{dt}{\ln t} + \frac{\alpha x}{\ln^n x}\right)}{\int_2^x \frac{dt}{\ln t} + \frac{\alpha x}{\ln^n x}} - \epsilon.$$

By applying L'Hôpital's rule to each of the ratios in the right-hand sides of these inequalities, one finds that they each tend to 0 as $x \to \infty$.[13] Since these inequalities each hold for infinitely many (integer) values of x, it follows that the only possible value for the putative limit L is -1.

Having proved Theorem III, Chebyshev writes: "What we have proved concerning the limit of the values of $\frac{x}{\phi(x)} - \ln x$ as $x \to \infty$ contradicts the formula proposed by Legendre for approximately determining the number of primes less than a given integer. In his opinion, for large x the value of $\phi(x)$ should be given to sufficient accuracy by the equation

$$\phi(x) = \frac{x}{\ln x - 1.08366}.$$

However, this would imply that the limit of $\frac{x}{\phi(x)} - \ln x$ as $x \to \infty$ is -1.08366 rather than -1."

Chebyshev goes on to say: "Using Theorem II one can establish an upper bound on the accuracy to which $\phi(x)$, the number of primes less than x, can be

[13]For instance on differentiating top and bottom of the ratio in the second of these inequalities, one obtains, after rearranging,

$$\frac{1 - \frac{\ln x}{x}\int_2^x \frac{dt}{\ln t} - \frac{(n+2)\alpha}{\ln^{n-1} x} + \frac{\alpha}{\ln^{n-2} x} + \frac{n\alpha}{\ln^n x}}{1 - \frac{\alpha}{\ln^{n-1} x} + \frac{n\alpha}{\ln^n x}}.$$

A further application of L'Hôpital's rule to $\frac{\int_2^x \frac{dt}{\ln t}}{x/\ln x}$ shows that this ratio has limit 1 as $x \to \infty$, whence the desired conclusion. *Trans.*

approximated by any function $f(x)$ whatever." In fact Chebyshev compares the difference $f(x) - \phi(x)$ with the functions

$$\frac{x}{\ln x}, \quad \frac{x}{\ln^2 x}, \quad \frac{x}{\ln^3 x}, \ldots,$$

defining a function $A(x)$ to *be of order* $\frac{x}{\ln^n x}$ if $A(x)/\frac{x}{\ln^m x}$ tends to ∞ for $m > n$ and to 0 for $m < n$. Having made this definition he then proves

THEOREM IV. *If the expression*

$$\frac{\ln^n x}{x} \cdot \left(f(x) - \int_2^x \frac{dt}{\ln t} \right)$$

has either a positive finite limit or tends to ∞ *as* $x \to \infty$, *then* $f(x)$ *cannot approximate* $\phi(x)$ *to order* $\frac{x}{\ln^m x}$ *for* $m > n$, *i.e., the difference* $f(x) - \phi(x)$ *is of order at best* $\frac{x}{\ln^n x}$.

Under the stated assumption, there exists a fixed positive number l such that for all sufficiently large x we have

$$\frac{\ln^n x}{x} \left(f(x) - \int_2^x \frac{dt}{\ln x} \right) > l.$$

However, setting $\alpha = \frac{l}{2}$ in the second inequality of Theorem II, we have that

$$\phi(x) < \int_2^x \frac{dt}{\ln t} + \frac{lx/2}{\ln^n x}$$

for infinitely many positive values of x. Combining this with the preceding inequality, we infer that the inequality

$$\frac{\ln^n x}{x} (f(x) - \phi(x)) > \frac{l}{2}$$

holds for infinitely many positive integers x, so that

$$\frac{\ln^n x}{x} (f(x) - \phi(x))$$

cannot have limit zero as $x \to \infty$. Hence in accordance with the above definition, the difference $f(x) - \phi(x)$ cannot be of order $\frac{x}{\ln^m x}$ for $m > n$.

Taking $f(x) = \frac{x}{\ln x - 1.08366}$ in this theorem, and observing that

$$\frac{\ln^2 x}{x} \left(\frac{x}{\ln x - 1.08366} - \int_2^x \frac{dt}{\ln t} \right)$$

tends to 0.08366 as $x \to \infty$, [14] Chebyshev infers from Theorem IV that Legendre's suggested asymptotic formula cannot approximate $\phi(x)$ to order $\frac{x}{\ln^n x}$ with $n > 2$. It follows that in order for a formula of the form $\frac{x}{A \ln x + B}$ to approximate $\phi(x)$ to an order better than $\frac{x}{\ln^2 x}$ we must have $A = 1, B = -1$.

Finally Chebyshev formulates the following

[14] Applying L'Hôpital's rule to the ratio $\left(\frac{x}{\ln x + B} - \int_2^x \frac{dt}{\ln t} \right) / \frac{x}{\ln^2 x}$ eventually yields the limit $-B - 1$ as $x \to \infty$. *Trans.*

THEOREM V. *The only function $f(x)$ approximating $\phi(x)$, the number of primes less than x, to order $\frac{x}{\ln^{n+1} x}$, and algebraic in $x, \ln x, e^x$, is*

$$\frac{x}{\ln x} + \frac{1 \cdot x}{\ln^2 x} + \frac{1 \cdot 2 \cdot x}{\ln^3 x} + \cdots + \frac{1 \cdot 2 \cdot 3 \cdots (n-1) \cdot x}{\ln^n x}. \tag{5}$$

The proof depends on the one hand on Theorem IV, and on the other on the fact that a function algebraic in $x, \ln x, e^x$, has fixed sign for all sufficiently large x.

Observing that the expression (5) approximates the integral

$$\int_2^x \frac{dx}{\ln x}$$

to the order

$$\frac{x}{\ln^n x},$$

Chebyshev then concludes that this integral approximates $\phi(x)$ to the same order to which it is (approximable by a function) algebraic in $x, e^x, \ln x$.

Having thus demonstrated that Legendre's empirical approximation of $\phi(x)$ is invalid, Chebyshev goes on to show that the rest of Legendre's formulae are likewise in need of correction.

Commentary

What is the main result of Chebyshev's first memoir on the primes considered above? Clearly its essence is contained in Theorem II, according to which, no matter how small we choose $\alpha > 0$ and no matter how large we choose n, the function $\phi(x)$ giving the number of primes less than x grows at least as quickly as the function

$$\int_2^x \frac{dt}{\ln t} - \frac{\alpha x}{\ln^n x},$$

but less quickly than the function

$$\int_2^x \frac{dt}{\ln t} + \frac{\alpha x}{\ln^n x},$$

or, in other words, there is no value of x beyond which the graph of the function $y = \phi(x)$ lies completely below the graph of the function

$$y = \int_2^x \frac{dt}{\ln t} - \frac{\alpha x}{\ln^n x}$$

or above the graph of

$$y = \int_2^x \frac{dt}{\ln t} + \frac{\alpha x}{\ln^n x}.$$

This represents the first nontrivial result on the distribution of the primes since the time of Euclid. The remainder of the results contained in the memoir are essentially paraphrases of this one. In particular it follows immediately that the ratio of the arithmetic function $\phi(x)$ to the relatively simple function $\int_2^x \frac{dt}{\ln t}$ can have no limit other than 1.

Concerning the methods of the memoir, one can say that they all stem from Chebyshev's use of the function $\sum \frac{1}{m^{1+\rho}}$ of $1+\rho$, where the summation is over all natural numbers m, which was first introduced by Euler and which we now term the "zeta-function" in view of the standard notation $\zeta(1+\rho)$ for it.

To prove Theorem I Chebyshev applies a remarkable representation of the zeta-function as a ratio of two integrals of functions involving only the exponential function and powers of x, namely

$$\zeta(1+\rho)-\frac{1}{\rho}=\frac{\int_0^\infty\left(\frac{1}{e^x-1}-\frac{1}{x}\right)e^{-x}x^\rho dx}{\int_0^\infty e^{-x}x^\rho dx},$$

and also Euler's well-known identity

$$\sum\frac{1}{m^{1+\rho}}=\prod\frac{1}{\left(1-\frac{1}{\mu^{1+\rho}}\right)},$$

where as before m ranges over all natural numbers, but μ ranges over the primes. Setting $s=1+\rho$ in the zeta-function

$$\zeta(s)=\sum\frac{1}{m^s},$$

we may say, from the modern point of view, that Chebyshev singles out for attention the pole $s=1$ of $\zeta(s)$, and in considering small values of ρ is in effect investigating the behaviour of the zeta-function $\zeta(s)$ near this pole. He allows ρ to take only real values, i.e., considers the zeta-function $\zeta(s)$ only for real values of s. His proof involves the derivatives of all orders of $\zeta(s)$ and $\ln\zeta(s)$. The generalization of these methods, involving the consideration of $\zeta(s)$ not just on the real line, but for s taking values in the complex plane, fell subsequently to Riemann and his successors.

Of course it was Euler who first considered the zeta-function $\zeta(s)$, and discovered the famous identity

$$\sum\frac{1}{m^s}=\prod\frac{1}{\left(1-\frac{1}{\mu^s}\right)}.$$

However, the only application Euler gave of this result was to infer from the fact that the harmonic series diverges that the series of reciprocals of the primes also diverges, yielding a new proof of the infinitude of the primes. Thus to the present memoir of Chebyshev belongs the honour of containing the first nontrivial use of the zeta-function to obtain new results on the distribution of primes,[15] and of pointing the way to further research in this direction. In his memoir of 1859, appearing 10 years after Chebyshev's, Riemann initiated the later deep investigations of the zeta-function, now of a complex argument $s=\sigma+\tau i$, and stated the conjecture that all nontrivial zeroes of $\zeta(s)$ lie on the line $\sigma=\frac{1}{2}$, the famous "Riemann hypothesis", which remains unproven to this day. Riemann shows that if his conjecture turns out to be true, then many remarkable results about the primes would follow. In 1896, i.e., nearly 50 years after the appearance of Chebyshev's memoir, Hadamard, by exploiting certain general properties of analytic functions that he had discovered, was at last able to prove that the ratio

$$\frac{\phi(x)}{\int_2^x\frac{dt}{\ln t}}$$

[15]Mention might be made here of Dirichlet's use of the zeta-function in 1838 to prove his theorem on primes in arithmetic progressions. *Trans.*

does indeed tend to 1 as $x \to \infty$,[16] i.e., that the transcendental function $\int_2^x \frac{dt}{\ln t}$ represents the arithmetic function $\phi(x)$ asymptotically, yielding an estimate of the number of primes less than any given large positive number x. In the same year 1896, de la Vallée-Poussin gave a proof of this asymptotic formula yielding an estimate of the error term.

On the Prime Numbers

Chebyshev's second memoir on the primes, entitled "Mémoire sur les nombres premiers" (1850), is surprising on the one hand for the unusual simplicity of the methods employed and on the other for the exceptional subtlety of its arguments. The stimulus to the writing of this memoir was Bertrand's postulate, so-called. Not long before, Bertrand, in trying to prove a result about finite groups, had found that he needed the following lemma: *For every integer $n > 3$ there is always at least one prime number between n and $2n - 2$.* All his efforts to prove this lemma being in vain, Bertrand took it as a postulate. In this, the second of his articles on the primes, Chebyshev proves Bertrand's postulate — among other results.

Chebyshev begins his memoir thus: "All questions depending on the law of distribution of the prime numbers in the sequence

$$1, 2, 3, 4, 5, 6, 7, 8, 9, 10, 11, 12, \ldots,$$

present great difficulties. Conclusions about the primes derived from examination of tables of primes more often than not remain without rigorous proof." He then gives as examples Bertrand's postulate and several other very difficult questions also depending on the distribution of the primes in the sequence of all natural numbers. For instance, Euler, improving on the known result that the harmonic series

$$1 + \frac{1}{2} + \frac{1}{3} + \frac{1}{4} + \frac{1}{5} + \frac{1}{6} + \frac{1}{7} + \frac{1}{8} + \cdots$$

diverges, showed that the series of reciprocals of the primes

$$\frac{1}{2} + \frac{1}{3} + \frac{1}{5} + \frac{1}{7} + \frac{1}{11} + \frac{1}{13} + \frac{1}{17} + \cdots$$

also diverges, yet although the series

$$\frac{1}{2\ln 2} + \frac{1}{3\ln 3} + \frac{1}{4\ln 4} + \frac{1}{5\ln 5} + \frac{1}{6\ln 6} + \cdots$$

diverges, the series

$$\frac{1}{2\ln 2} + \frac{1}{3\ln 3} + \frac{1}{5\ln 5} + \frac{1}{7\ln 7} + \frac{1}{11\ln 11} + \cdots$$

converges. Chebyshev writes: "What sort of criterion might there be for convergence of series of terms with prime indices $2, 3, 5, 7, 11, \ldots$? And in the case of convergence how may one establish how closely the partial sums approximate the sum? This memoir contains the answer to these questions." Later he continues: "We also give a formula for computing the approximate number of primes not exceeding a given number, and establish the degree of error of this formula, something which has not hitherto been achieved." And indeed this last result is one of the first rank.

---

[16]Since also $\frac{\int_2^x \frac{dt}{\ln t}}{x/\ln x} \to 1$, this is equivalent to the "Prime Number Theorem" as usually stated. However, $\int_2^x \frac{dt}{\ln t}$ affords a considerably better estimate of $\phi(x)$. *Trans.*

In fact Chebyshev investigates the function $\theta(x)$ giving the sum of the logarithms of the primes less than or equal to x, or, in other words, the natural logarithm of the product of all primes not greater than x. The goal he sets himself is to find upper and lower bounds (depending on x) for $\theta(x)$, in order to understand how $\theta(x)$ grows with x, and hence how the product of the primes less than or equal to x grows with x.

For the product of all natural numbers less than or equal to x the corresponding question had been answered long before. Stirling had shown that

$$\ln\sqrt{2\pi}+a\ln a-a-\frac{1}{2}\ln a<\ln(1\cdot2\cdot3\cdot4\cdots a)<\ln\sqrt{2\pi}+a\ln a-a+\frac{1}{2}\ln a+\frac{1}{12a}.$$

Chebyshev's basic idea is to establish a connection between the values

$$\theta(x)=\ln(2\cdot3\cdot5\cdot7\cdot11\cdots)$$

of the function of interest to him, giving the natural logarithm of the product of all *primes* less than or equal to x, and the corresponding values of the simpler function

$$T(x)=\ln(1\cdot2\cdot3\cdot4\cdot5\cdot6\cdots)$$

giving the natural logarithm of the product of all *natural* numbers less than or equal to x, in order to exploit Stirling's formula.

Chebyshev begins with the simple identity

$$\begin{aligned}T(x)=\theta(x)&+\theta(\sqrt{x})+\theta(\sqrt[3]{x})+\theta(\sqrt[4]{x})+\cdots\\&+\theta\left(\frac{x}{2}\right)+\theta\left(\sqrt{\frac{x}{2}}\right)+\theta\left(\sqrt[3]{\frac{x}{2}}\right)+\theta\left(\sqrt[4]{\frac{x}{2}}\right)+\cdots\\&+\theta\left(\frac{x}{3}\right)+\theta\left(\sqrt{\frac{x}{3}}\right)+\theta\left(\sqrt[3]{\frac{x}{3}}\right)+\theta\left(\sqrt[4]{\frac{x}{3}}\right)+\cdots\\&+\theta\left(\frac{x}{4}\right)+\theta\left(\sqrt{\frac{x}{4}}\right)+\theta\left(\sqrt[3]{\frac{x}{4}}\right)+\theta\left(\sqrt[4]{\frac{x}{4}}\right)+\cdots\\&\dots\dots\dots\dots\dots\dots\dots\dots,\end{aligned}$$

where the sum on each line of this identity ends of course at the first zero term encountered; since 2 is the smallest prime, one has $\theta(z)=0$ for $z\le2$.

To verify this identity it suffices to observe first that both sides can be expressed as a sum of terms of the form $K\ln a$ for integral K and distinct primes a. Given a particular such term on the left-hand side $T(x)$ (take for example $a=3$), the integer K is the number of members of the sequence $1,2,3,4,\dots,[x]$ (where $[x]$ denotes as usual the largest natural number $\le x$) divisible by a plus the number of members of this sequence divisible by a^2, and so on. Now the number of numbers in the sequence $1,2,3,4,\dots,[x]$ divisible by a, is equal to the number of members of the sequence $x,\frac{x}{2},\frac{x}{3},\frac{x}{4},\dots$ greater than or equal to a, and the number of numbers in the former sequence divisible by a^2 is equal to the number of members of the sequence $\sqrt{x},\sqrt{\frac{x}{2}},\sqrt{\frac{x}{3}},\sqrt{\frac{x}{4}},\dots$ greater than or equal to a, and so on. Hence the two sides of the purported identity have the same terms of the form $K\ln a$, which establishes the identity.

Introducing the notation

$$\psi(x)=\theta(x)+\theta(\sqrt{x})+\theta(\sqrt[3]{x})+\theta(\sqrt[4]{x})+\cdots,$$

one may rewrite the above identity in the form

$$T(x) = \psi(x) + \psi\left(\frac{x}{2}\right) + \psi\left(\frac{x}{3}\right) + \psi\left(\frac{x}{4}\right) + \cdots .$$

Having established this identity, Chebyshev considers the following specially chosen linear combination:

$$T(x) + T\left(\frac{x}{30}\right) - T\left(\frac{x}{2}\right) - T\left(\frac{x}{3}\right) - T\left(\frac{x}{5}\right).$$

Rewritten in terms of ψ this becomes

$$\begin{aligned}
&\psi(x) + \psi\left(\frac{x}{2}\right) + \psi\left(\frac{x}{3}\right) + \psi\left(\frac{x}{4}\right) + \cdots \\
&+ \psi\left(\frac{x}{30}\right) + \psi\left(\frac{x}{2\cdot 30}\right) + \psi\left(\frac{x}{3\cdot 30}\right) + \psi\left(\frac{x}{4\cdot 30}\right) + \cdots \\
&- \psi\left(\frac{x}{2}\right) - \psi\left(\frac{x}{2\cdot 2}\right) - \psi\left(\frac{x}{3\cdot 2}\right) - \psi\left(\frac{x}{4\cdot 2}\right) - \cdots \\
&- \psi\left(\frac{x}{3}\right) - \psi\left(\frac{x}{2\cdot 3}\right) - \psi\left(\frac{x}{3\cdot 3}\right) - \psi\left(\frac{x}{4\cdot 3}\right) - \cdots \\
&- \psi\left(\frac{x}{5}\right) - \psi\left(\frac{x}{2\cdot 5}\right) - \psi\left(\frac{x}{3\cdot 5}\right) - \psi\left(\frac{x}{4\cdot 5}\right) - \cdots ,
\end{aligned}$$

which may in turn be rewritten in the form

$$A_1\psi(x) + A_2\psi\left(\frac{x}{2}\right) + A_3\psi\left(\frac{x}{3}\right) + A_4\psi\left(\frac{x}{4}\right) + \cdots + A_n\psi\left(\frac{x}{n}\right) + \cdots ,$$

where the coefficients $A_1, A_2, A_3, A_4, \ldots, A_n, \ldots$ are integers. Examination of these coefficients yields without difficulty that:

$$\begin{aligned}
A_n &= 1 \text{ if } n = 30m + k,\ k = 1, 7, 11, 13, 17, 19, 23, 29; \\
A_n &= 0 \text{ if } n = 30m + k,\ k = 2, 3, 4, 5, 8, 9, 14, 16, 21, 22, 25, 26, 27, 28; \\
A_n &= -1 \text{ if } n = 30m + k,\ k = 6, 10, 12, 15, 18, 20, 24; \\
A_n &= -1 \text{ if } n = 30m + 30.
\end{aligned}$$

In the first case, since these are precisely the values of n not divisible by $2, 3$ or 5, for each of them the formal term $\psi(x/n)$ will occur only in the first row, and so only once. In the second case the values of n listed are exactly those divisible by just one of $2, 3$ or 5, so that for each such n, in addition to the term $\psi(x/n)$ appearing in the first row, there will be just one further such term from one of the last three rows, however with a negative sign attached, and therefore cancelling with the identical term from the first row. The values of n in the third list are all those divisible by exactly two of $2, 3, 5$, so that in the last three rows there will be precisely two terms of the form $-\psi(x/n)$. Adding these to the term $\psi(x/n)$ in the first row yields $-\psi(x/n)$, i.e., $A_n = -1$, as claimed. In the fourth and final case, covering the remaining n, namely those divisible by 30, each of the five rows contains a single term of the form $\pm\psi(x/n)$, in the first and second with sign $+$, and in the last three with sign $-$.

Thus for

$$\begin{aligned}
n = 30m + k, \quad k = {}&1, 2, 3, 4, 5, 6, 7, 8, 9, 10, 11, 12, 13, 14, 15, \\
&16, 17, 18, 19, 20, 21, 22, 23, 24, 25, 26, 27, 28, 29, 30,
\end{aligned}$$

we have, in order,

$$A_n = 1,0,0,0,0,-1,1,0,0,-1,1,-1,1,0,-1,$$
$$0,1,-1,1,-1,0,0,1,-1,0,0,0,0,1,-1,$$

whence the original linear combination takes the form of the alternating series

$$\psi(x) - \psi\left(\frac{x}{6}\right) + \psi\left(\frac{x}{7}\right) - \psi\left(\frac{x}{10}\right) + \psi\left(\frac{x}{11}\right) - \psi\left(\frac{x}{12}\right) + \psi\left(\frac{x}{13}\right) - \psi\left(\frac{x}{15}\right) + \cdots .$$

Since this series is alternating and its terms are nonincreasing in absolute value, its sum is at most its first term and at least the sum of the first two terms, whence

$$\psi(x) - \psi\left(\frac{x}{6}\right) \le T(x) + T\left(\frac{x}{30}\right) - T\left(\frac{x}{2}\right) - T\left(\frac{x}{3}\right) - T\left(\frac{x}{5}\right) \le \psi(x).$$

Now from Stirling's bounds for $T(x)$ (see above) one obtains after some calculation (suppressed):

$$Ax - \frac{5}{2}\ln x - 1 < T(x) + T\left(\frac{x}{30}\right) - T\left(\frac{x}{2}\right) - T\left(\frac{x}{3}\right) - T\left(\frac{x}{5}\right) < Ax + \frac{5}{2}\ln x,$$

where

$$A = \ln\left(\frac{2^{1/2}3^{1/3}5^{1/5}}{30^{1/30}}\right) = 0.92129202\ldots .$$

This together with the previous bounds yields the inequalities

$$\psi(x) > Ax - \frac{5}{2}\ln x - 1, \qquad \psi(x) - \psi\left(\frac{x}{6}\right) < Ax + \frac{5}{2}\ln x.$$

The first of these gives a lower bound for $\psi(x)$, and from the second one an upper bound is obtained as follows.

Consider the function

$$f(x) = \frac{6}{5}Ax + \frac{5}{4\ln 6}\ln^2 x + \frac{5}{4}\ln x.$$

It is easy to check that

$$f(x) - f\left(\frac{x}{6}\right) = Ax + \frac{5}{2}\ln x.$$

Hence from the above upper bound for $\psi(x) - \psi\left(\frac{x}{6}\right)$ one obtains

$$\psi(x) - \psi\left(\frac{x}{6}\right) - f(x) + f\left(\frac{x}{6}\right) < 0,$$

or

$$\psi(x) - f(x) < \psi\left(\frac{x}{6}\right) - f\left(\frac{x}{6}\right).$$

Replacing x by $\frac{x}{6}, \frac{x}{6^2}, \ldots, \frac{x}{6^m}$ in succession in this inequality, one obtains

$$\psi(x) - f(x) < \psi\left(\frac{x}{6}\right) - f\left(\frac{x}{6}\right) < \psi\left(\frac{x}{6^2}\right) - f\left(\frac{x}{6^2}\right) < \cdots < \psi\left(\frac{x}{6^{m+1}}\right) - f\left(\frac{x}{6^{m+1}}\right).$$

If m is taken to be the largest integer for which $\frac{x}{6^m} \ge 1$, then $\frac{x}{6^{m+1}}$ will lie between 1 and $\frac{1}{6}$, whence

$$\psi\left(\frac{x}{6^{m+1}}\right) - f\left(\frac{x}{6^{m+1}}\right) < 1,$$

and therefore

$$\psi(x) - f(x) < 1.$$

Hence

$$\psi(x) < \frac{6}{5}Ax + \frac{5}{4\ln 6}\ln^2 x + \frac{5}{4}\ln x + 1.$$

Having thus bounded $\psi(x)$ from above and below, it is not difficult to infer bounds for $\theta(x)$. Taking into account the definition of $\psi(x)$ in terms of $\theta(x)$, $\theta(\sqrt{x})$, $\theta(\sqrt[3]{x}),\ldots$, one has

$$\psi(x)-\psi(\sqrt{x})=\theta(x)+\theta(\sqrt[3]{x})+\theta(\sqrt[5]{x})+\cdots,$$
$$\psi(x)-2\psi(\sqrt{x})=\theta(x)-[\theta(\sqrt{x})-\theta(\sqrt[3]{x})]-[\theta(\sqrt[4]{x})-\theta(\sqrt[5]{x})]-\cdots,$$

whence

$$\psi(x)-2\psi(\sqrt{x})\le\theta(x)\le\psi(x)-\psi(\sqrt{x}),$$

since the quantities

$$\theta(\sqrt[3]{x}),\theta(\sqrt[5]{x}),\ldots,\quad\text{and}\quad\theta(\sqrt{x})-\theta(\sqrt[3]{x}),\ \theta(\sqrt[4]{x})-\theta(\sqrt[5]{x}),\ldots$$

are, obviously, all nonnegative.

This and the bounds for $\psi(x)$ established above now yield

$$(6)\qquad \begin{aligned} Ax-\frac{12}{5}Ax^{\frac12}-\frac{5}{8\ln 6}\ln^2 x-\frac{15}{4}\ln x-3&<\theta(x)\\ &<\frac{6}{5}Ax-Ax^{\frac12}+\frac{5}{4\ln 6}\ln^2 x+\frac{5}{2}\ln x+2.\end{aligned}$$

These remarkable inequalities of Chebyshev play in connection with the product of the prime numbers less than x a role analogous to that played by Stirling's inequalities for the product of the natural numbers less than x. With these bounds Chebyshev is in a position to infer a whole series of results relating to the law of distribution of the primes.

As the first consequence of his bounds for $\theta(x)$ Chebyshev gives the following proof of Bertrand's postulate.

Suppose that there are exactly m primes contained (strictly) between the bounds l and L. The natural logarithm of the product of these primes will then lie between $m\ln l$ and $m\ln L$, i.e.,

$$m\ln l<\theta(L)-\theta(l)<m\ln L,$$

or

$$\frac{\theta(L)-\theta(l)}{\ln L}<m<\frac{\theta(L)-\theta(l)}{\ln l}.$$

Replacing $\theta(l)$ and $\theta(L)$ by the upper or lower bounds, as appropriate, established above, we obtain two functions of l and L between which m lies. From the lower of these bounds one infers that for the number of primes between l and L to exceed a given k, it suffices that

$$k=\frac{1}{\ln L}\left[A\left(L-\frac{6}{5}l\right)-\frac{12}{5}AL^{\frac12}-\frac{15}{8\ln 6}(\ln^2 L+2\ln^2 l)-\frac{5}{4}(3\ln L+2\ln l-5)\right],$$

or, equivalently, that

$$l=\frac{5}{6}L-2L^{\frac12}-\frac{25\ln^2 L}{16A\ln 6}-\frac{5}{6A}\left(\frac{25}{4}+k\right)\ln L-\frac{25}{6A}.$$

In particular for there to exist at least one prime between l and L, it suffices that the above equation hold with $k=0$, i.e., that

$$l=\frac{5}{6}L-2L^{\frac12}-\frac{25\ln^2 L}{16A\ln 6}-\frac{125\ln L}{24A}-\frac{25}{6A}.$$

It is not difficult to see that if $a > 160$ then between a and $2a-2$ there do indeed exist numbers l and L satisfying this equation, i.e., there exists a number L satisfying

$$a < \frac{5}{6}L - 2L^{\frac{1}{2}} - \frac{25\ln^2 L}{16A\ln 6} - \frac{125\ln L}{24A} - \frac{25}{6A} < L < 2a-2.$$

In fact one may take $L = 2a-3$, since the inequalities

$$a < \frac{5}{6}(2a-3) - 2\sqrt{2a-3} - \frac{25\ln^2(2a-3)}{16A\ln 6} - \frac{125\ln^2(2a-3)}{24A} - \frac{25}{6A} < 2a-3$$

hold for all $a > 160$, as is easily verified. Thus for every $a > 160$ there is at least one prime number lying between a and $2a-2$. For values of a less than or equal to 160, Bertrand's postulate can be verified directly from a table of primes.

The second corollary Chebyshev deduces from his bounds for $\theta(x)$ gives an exact condition for convergence of a series of the form

$$F(2) + F(3) + F(5) + F(7) + F(11) + F(13) + F(17) + \cdots,$$

i.e., whose terms are the values of a function $F(x)$ at the primes, where $F(x)$ is any function which for all sufficiently large x, i.e., for all x greater than some fixed number, remains positive and is such that $\frac{F(x)}{\ln x}$ is decreasing.

THEOREM. *If $F(x)$ is any function satisfying the above-mentioned conditions, then the series*

$$F(2) + F(3) + F(5) + F(7) + F(11) + F(13) + \cdots$$

converges if and only if the series

$$\frac{F(2)}{\ln 2} + \frac{F(3)}{\ln 3} + \frac{F(4)}{\ln 4} + \frac{F(5)}{\ln 5} + \frac{F(6)}{\ln 6} + \cdots$$

converges.

For the proof, Chebyshev first observes that if $\alpha, \beta, \gamma, \ldots, \rho$ are the primes lying between positive integers l and L, then the sum

$$U = F(\alpha) + F(\beta) + F(\gamma) + \cdots + F(\rho)$$

is equal to

$$\begin{aligned}&\frac{\theta(l)-\theta(l-1)}{\ln l}F(l) + \frac{\theta(l+1)-\theta(l)}{\ln(l+1)}F(l+1)\\ &\quad + \frac{\theta(l+2)-\theta(l+1)}{\ln(l+2)}F(l+2) + \cdots + \frac{\theta(L)-\theta(L-1)}{\ln L}F(L),\end{aligned}$$

since the function $\frac{\theta(x)-\theta(x-1)}{\ln x}$ of integral x takes the value 0 if x is composite, and 1 if x is prime. One next rearranges the terms in the latter sum to obtain

$$\begin{aligned}U = -\theta(l-1)\frac{F(l)}{\ln l} &+ \left[\frac{F(l)}{\ln l} - \frac{F(l+1)}{\ln(l+1)}\right]\theta(l)\\ &+ \left[\frac{F(l+1)}{\ln(l+1)} - \frac{F(l+2)}{\ln(l+2)}\right]\theta(l+1) + \cdots\\ &+ \left[\frac{F(L)}{\ln L} - \frac{F(L+1)}{\ln(L+1)}\right]\theta(L) + \frac{F(L+1)}{\ln(L+1)}\theta(L).\end{aligned}$$

Assuming that l is sufficiently large for the function $\frac{F(x)}{\ln x}$ to be positive and decreasing on the interval $l-1 \leq x \leq L+1$, the first term of this sum, involving $\theta(l-1)$, will be negative, while all later terms, involving $\theta(l), \theta(l+1), \ldots, \theta(L)$, will be positive. Hence denoting the upper bound for $\theta(x)$ in (6) by $\theta_I(x)$ and the lower by $\theta_{II}(x)$, i.e., writing (6) in the form $\theta_{II}(x) < \theta(x) < \theta_I(x)$, by appropriately replacing θ by θ_I or θ_{II} in the preceding expression for U, and rearranging, one obtains the following bounds for U:

$$\theta_{II}(l-1)\frac{F(l)}{\ln l} - \theta_I(l-1)\frac{F(l)}{\ln l} + \sum_{x=l}^{x=L} F(x)\frac{\theta_{II}(x)-\theta_{II}(x-1)}{\ln x} < U$$

$$< \theta_I(l-1)\frac{F(l)}{\ln l} - \theta_{II}(l-1)\frac{F(l)}{\ln l} + \sum_{x=l}^{x=L} F(x)\frac{\theta_I(x)-\theta_I(x-1)}{\ln x}.$$

The conclusion of the theorem now follows readily from these inequalities.

By way of an example of the application of this theorem Chebyshev deduces the convergence of the series

$$\frac{1}{2\ln 2} + \frac{1}{3\ln 3} + \frac{1}{5\ln 5} + \frac{1}{7\ln 7} + \frac{1}{11\ln 11} + \cdots.$$

Chebyshev next investigates the error in approximating the sum of a convergent series of the type considered in the theorem by its nth partial sum. For instance he shows that the sum of the preceding series is 1.63 to within 0.1, i.e., lies between 1.53 and 1.73.

In the final section of this article, Chebyshev exhibits two increasing functions of x, which for large x bound $\phi(x)$, the number of primes less than x, from either side. This is achieved by simply taking $F(x) \equiv 1$ and $l = 2$ in the above inequalities for U, which then becomes the number of primes less than or equal to L, bounded from below by

$$\frac{\theta_{II}(1)}{\ln 2} - \frac{\theta_I(1)}{\ln 2} + \sum_{x=2}^{x=L} \frac{\theta_{II}(x)-\theta_{II}(x-1)}{\ln x},$$

and from above by

$$\frac{\theta_I(1)}{\ln 2} - \frac{\theta_{II}(1)}{\ln 2} + \sum_{x=2}^{x=L} \frac{\theta_I(x)-\theta_I(x-1)}{\ln x}.$$

Using relatively simple calculations in terms of the explicit formulae for $\theta_I(x)$ and $\theta_{II}(x)$ in (6), it can be shown that the ratio of the difference of these bounds to their sum tends to $\frac{1}{11}$ as $L \to \infty$. Hence for all sufficiently large L this ratio will be less than $\frac{1}{10}$. Hence as an approximation to the number $\phi(L)$ of primes less than L, half the difference of the above two bounds will, for all sufficiently large L, differ from the actual value of $\phi(L)$ by less than $\frac{1}{10}$.

Commentary

The centre of gravity, so to speak, of Chebyshev's second memoir on the prime numbers that we have just examined, is, of course, the result obtained at the very end. At last an analytic function had been found (namely the average of the upper and lower bounds exhibited above) approximating the inscrutable function $\phi(x)$ giving the number of primes less than x for large x to within a determined error, namely less than $\frac{1}{10}$ on either side. It is not difficult to verify that instead of the function obtained by Chebyshev one may equally well take $\frac{x}{\ln x}$ or $\int_2^x \frac{dt}{\ln t}$, since the ratio of each of these functions to Chebyshev's function tends to 1 as $x \to \infty$. Thus for the first time it was shown to a good approximation how the number of primes less than x grows with x. A little further calculation shows, as one of several similar results, that

$$0.92129 < \frac{\phi(x)}{x/\ln x} < 1.10555.$$

These are results of fundamental importance.

Sylvester, in two articles of 1881 and 1892, extended Chebyshev's method, introducing in place of the expression

$$T(x) + T\left(\frac{x}{30}\right) - T\left(\frac{x}{2}\right) - T\left(\frac{x}{3}\right) - T\left(\frac{x}{5}\right),$$

significantly more complicated linear combinations of the $T(x/m)$ for various m. Using these he was able to obtain better bounds than Chebyshev's. Thus instead of the bounds 0.92129 and 1.10555 noted above, he obtained the bounds 0.95695 and 1.04423. The gap between the bounds was further narrowed by Schur in 1929, and by Breusch in 1932. However, hitherto no-one has been able to extend Chebyshev's method to a proof of the Prime Number Theorem, i.e. that $\phi(x)$ is asymptotic to $\frac{x}{\ln x}$.

Chebyshev's memoirs on the prime numbers constitute the first significant steps in the general theory of the primes since Euclid, containing as they do the first precise results in the general theory of the distribution of the primes. Problems in this area are distinguished by their extraordinary difficulty. Following on Chebyshev, Riemann and Hadamard, from towards the end of the 19th century until well into the the 20th several gifted researchers devoted much effort to such problems, and yet to this day a great many of them remain unsolved.

The first of Chebyshev's memoirs considered above involves the Eulerian zeta-function, an integral representation of it, estimates of its ordinary and logarithmic derivatives, and so on. The second memoir astounds by the elementary nature of the initial ideas. Both memoirs are striking for their freshness, subtlety, and for the indication they give of the intellectual strength of a youthful, first-rank mathematical talent.

It was with these two works of genius that the St. Petersburg school of number theory was inaugurated.

Aleksandr Nikolaevich Korkin (1837–1908)

Aleksandr Nikolaevich KORKIN

Aleksandr Nikolaevich Korkin was born on the 19th February, 1837, in a village six versts[17] from the town of Shuĭskoye, in the Tot'ma district of Vologda province. His father was a prosperous peasant who also engaged in trade. When Korkin was three, the family moved to Shuĭskoye. At the age of eight his upbringing was entrusted to a mathematics teacher at the Vologda provincial gymnasium, Aleksandr Ivanovich Ivanitskiĭ, a gifted teacher and former student of V. Ya. Bunyakovskiĭ. Ivanitskiĭ's wife, of good education, taught the young Korkin French and German. At ten he entered the Vologda provincial gymnasium. In 1849, following on the loss of nearly all of his wealth, Korkin's father died, leaving his widow and 12-year old son essentially without means. Korkin's mother, A. N. Korkina, lived to 80 without ever leaving Shuĭskoye, dying in 1888. Throughout his time as professor and noted scholar at St. Petersburg University, Aleksandr Nikolaevich visited his mother at Shuĭskoye every year during the vacation. In those days there was no railway, and it was necessary to travel hundreds of versts on the river Sukhona rowed or towed in a simple boat.

As a pupil at the gymnasium, Korkin's outstanding abilities were already in evidence. He used to do his homework while walking home from the gymnasium.

Having graduated at 16 with the gold medal, he was, on account of his age, unable to enter university as he had wished, and moved instead to Yaroslavl' to study at the Demidov *lycée*. However he soon returned to Shuĭskoye, remaining there till 1854, when he entered the mathematics stream of the physics and mathematics faculty of St. Petersburg University. At that time V. Ya. Bunyakovskiĭ, P. L. Chebyshev, I. I. Somich, A. N. Savich, and E. Kh. Lents were professors there. In 1856 Korkin presented an essay to the faculty on the set theme "On greatest and least quantities", for which he was awarded the gold medal. On the recommendation of Bunyakovskiĭ this work was published in "The Student Anthology" of 1857. During his student years Korkin lived on his stipend of 7 roubles a month, supplemented by earnings from private lessons.

On graduating from the university in 1858, Korkin served as mathematics teacher at the First Cadet Corpus. Having passed the examination for the Master's degree in 1860, in that same year he defended his thesis "On the definition of arbitrary functions in terms of integrals of linear partial differential equations". At that time V. Ya. Bunyakovskiĭ retired from the university, and a competition was announced for the resulting vacant position. Korkin was awarded the post, and directed to lecture on spherical trigonometry, analytic geometry, and the integral calculus.

[17] 1 verst $\equiv 3,500$ feet, or somewhat over a kilometre. *Trans.*

In 1862 the university was closed because of student unrest. Korkin was appointed to the Ministry for National Education at the same rank, and sent abroad on a two-year scientific leave.

In Paris he attended the lectures of Liouville, Lamé, and Bertrand; he also worked for a time in Berlin. Of all German mathematicians active at that time, Korkin rated Kummer highest. During this period Korkin composed his doctoral dissertation, and his interest in number theory began.

In 1868 Korkin defended his doctoral dissertation "On systems of first order partial differential equations and certain questions of mechanics". In the same year the council of the university appointed him professor in the department of mathematics, the posting being confirmed by the Ministry for National Education.

The years from 1871 to 1877 witnessed the continued collaboration between Korkin and Zolotarev on the theory of quadratic forms, earning them considerable repute.

Korkin lectured at St. Petersburg University for almost 50 years, from 1860 to 1908, the year of his death. Over this long period he had perforce to lecture on nearly every mathematical topic: spherical trigonometry; descriptive geometry; analytic geometry; higher algebra; differential calculus and its applications to geometry; integral calculus; differential equations; and the calculus of variations. In addition for more than 30 years he taught differential and integral calculus at the Nikolaevskaya Naval Academy; it was only in 1900 that he gave up this task, which was then assumed by his favorite student Alekseĭ Nikolaevich Krylov.

Korkin's lecturing style was one of supreme simplicity and clarity. He explained each and every result by means of examples, and went minutely through the steps of each argument. After each lecture he would dictate examples as homework exercises. Taking notes from his lectures was very easy; however, later on in his career Korkin came to the conclusion that he should dictate his lectures, attributed to his lack of faith in his hearers' ability to write them down well enough. Those attending his lectures usually mastered the subject-matter very thoroughly, and passed the examination with flying colours.

Through his ability to draw on a broad knowledge of various branches of mathematics, Korkin often set very difficult problems; these he proposed to the best of his students. When he observed exceptional ability in a student, he would often take him under his wing and help him in his studies. Almost till the end of his life his home was always open on "Korkin Saturdays", as they came to be called, to anyone wanting advice, or to talk mathematics.

Korkin's knowledge of the classical mathematical literature was both wide and deep. He knew in both depth and detail the works of Euler, Lagrange, Gauss, Legendre, Laplace, Monge, Fourier, Poisson, Abel, Jacobi, and Dirichlet. However, to the work of Weierstrass and Riemann he reacted negatively, calling it "decadent". Among his contemporaries, he ranked Hermite especially highly, and considered his memoir "Sur la fonction exponentielle", in which the transcendence of the number e is proved, a classic. Korkin was also very well acquainted with analytical mechanics, which he regarded as a part of mathematics.

Korkin's knowledge of French was excellent, and the majority of his articles were written in French. He also knew German well. His knowledge of Latin was such that he could read Horace freely. He was blessed with a wonderful memory; that one might forget what one had studied passed his comprehension. "No doubt they read badly and inattentively," he would say in such circumstances, "otherwise they wouldn't have been able to forget." He was also very familiar with the history of ancient Greece and Rome, and of the French revolution. In music his taste was for the classical style.

Korkin's personality was very well integrated and highly original. He paid no heed to the plaint "What will they say?" He never sought popularity. His character was very independent; he always expressed his views and opinions directly and distinctly, prepared however to change them in face of irrefutable evidence of their mistakenness. For instance he reacted very negatively to Zolotarev's Master's thesis, giving it an extremely bad report; however, later on he became his young colleague's best friend. The following anecdote has also come down to us: Once when Alekseĭ Nikolaevich Krylov, his former student, and at that time professor, praised Gräffe's method for computing the roots of equations, Korkin responded that the method "is good for nothing." However, when Krylov, using concrete examples suggested by Korkin himself, demonstrated the advantages of the method, Korkin completely revised his opinion and thenceforth stoutly defended it.

The Articles of Korkin and Zolotarev on the Minima of Positive Quadratic Forms

On Quaternary Positive Quadratic Forms

The joint articles of Korkin and Zolotarev on the minima of quadratic forms appeared in 1872, 1873, and 1877. These works are concerned with positive[18] quadratic forms

$$f = a_{11}x_1^2 + a_{22}x_2^2 + \cdots + a_{nn}x_n^2 + a_{12}x_1x_2 + \cdots$$
$$+ a_{n-1,n}x_{n-1}x_n + a_{n,n-1}x_nx_{n-1} = \sum a_{ik}x_ix_k,$$

with prescribed *determinant* (or *discriminant*)

$$D = \begin{vmatrix} a_{11} & a_{12} & \cdots & a_{1n} \\ a_{21} & a_{22} & \cdots & a_{2n} \\ \vdots & \vdots & \cdots & \vdots \\ a_{n1} & a_{n2} & \cdots & a_{nn} \end{vmatrix}$$

where the coefficients a_{ik} are arbitrary real numbers satisfying

$$a_{ik} = a_{ki}.$$

It is not difficult to show that if, for fixed coefficients a_{ik}, the variables $x_1, x_2, \ldots, x_n$ are restricted to (rational) integer values only other than $0, 0, \ldots, 0$, then among all the resulting values of the function f there will be a smallest one. This smallest value is called the *minimum* of the form. Since this minimum is determined once the coefficients a_{ik} are given, it is a function of those coefficients, whose domain is restricted only by the requirements that the form be positive and symmetric, and that the determinant D have the prescribed value. As the coefficients a_{ik} vary continuously in their domain, the minimum will also vary continuously. Among these forms, *equivalent* ones, i.e., those transforming into one another via a linear transformation of the variables with integer coefficients and determinant ± 1, will, of course, have the same minimum.

As the minimum varies continuously with the coefficients a_{ik}, it may peak at one maximum, or even at several (local) maxima, corresponding to inequivalent forms, the number of which will depend on n. Forms whose minimum is a relative maximum among these minima are called by Korkin and Zolotarev *extremal.*

Prior to the research of Korkin and Zolotarev, Hermite had worked on the problem of finding a useful upper estimate for the extremum, i.e., an upper bound for the minimum of any n-ary positive quadratic form of prescribed determinant

[18] A quadratic form is *positive* (or *positive definite*) if its values at all $(x_1, \ldots, x_n) \neq (0, \ldots, 0)$ are positive, or, equivalently, if its coefficient matrix $[a_{ik}]$ is positive definite. *Trans.*

D, as a function of n and D. "My first investigations," says Hermite, "of forms in n variables and of determinant D, yielded the upper bound $\left(\frac{4}{3}\right)^{\frac{1}{2}(n-1)} \cdot \sqrt[n]{D}$. However I suggest, although I am unable to prove it, that the numerical coefficient $\left(\frac{4}{3}\right)^{\frac{1}{2}(n-1)}$ can be replaced by $\frac{2}{\sqrt[n]{n+1}}$." Even the relatively crude upper bound for the minima obtained by Hermite allowed him to deduce several important results in number theory as applications of that estimate.

In their first elegant little paper of 1872 on quaternary positive quadratic forms ("Sur les formes quadratiques positives quaternaires") Korkin and Zolotarev write: "The search for the exact upper bound of the minima of positive quadratic forms of given determinant in integer-valued variables, presents great difficulties and constitutes one of the most important problems in the theory of such forms." In this short memoir Korkin and Zolotarev show that in the case of forms in four variables these minima are all less than or equal to $\sqrt[4]{4D}$, and furthermore that this value is actually attained as the minimum of the form

$$\sqrt[4]{4D}(x_1^2 + x_2^2 + x_3^2 + x_4^2 + x_1x_2 + x_1x_3 + x_1x_4),$$

which indeed has determinant D. This result was noteworthy in that for the first time the exact upper bound for the minima of positive quadratic forms in four variables had been revealed. At the same time it puts paid, already in the case $n = 4$, to Hermite's proposed upper bound of

$$\frac{2}{\sqrt[n]{n+1}}\sqrt[n]{D},$$

since $\sqrt[4]{4} > \frac{2}{\sqrt[4]{5}}$.

This article is short enough for us to reproduce it here in its entirety, translated essentially word for word from the French in which it was written.[19]

After the introductory remark quoted above, Korkin and Zolotarev continue:

In this note we will be concerned with quaternary forms, and as the first result of our research we derive an exact upper estimate for their minima. It is quite remarkable that this bound follows directly from the known exact bound for ternary forms.

Let

$$f = \sum_{i=1,k=1}^{i=4,k=4} a_{ik}x_ix_k$$

be a positive quaternary form of determinant D. We may assume that the coefficient a_{11} is the minimum of f, since there is always an equivalent form with this property.

This assumed, consider the form

$$F(X_1, X_2, X_3, X_4)$$

obtained from the form f by means of the linear transformation

$$\begin{aligned} x_1 &= X_1 + lX_2 + mX_3 + nX_4, \\ x_2 &= \qquad l_1X_2 + m_1X_3 + n_1X_4, \\ x_3 &= \qquad l_2X_2 + m_2X_3 + n_2X_4, \\ x_4 &= \qquad l_3X_2 + m_3X_3 + n_3X_4, \end{aligned}$$

[19]That is, from French, via Russian, into English. *Trans.*

where $l, m, n, l_1, m_1, n_1, \ldots$ are integers subject only to the condition

$$(1) \qquad \begin{vmatrix} l_1 & m_1 & n_1 \\ l_2 & m_2 & n_2 \\ l_3 & m_3 & n_3 \end{vmatrix} = \pm 1.$$

The common minimum a_{11} for f and F is again the coefficient of X_1^2. Setting $X_4 = 0$ in F and leaving the other three variables arbitrary, we denote by Δ the determinant of the resulting ternary form

$$F(X_1, X_2, X_3, 0).$$

It is easy to see that the minimum of this form is also a_{11}, so that, in view of known results on ternary forms, we must have

$$(2) \qquad a_{11} \leq \sqrt[3]{2\Delta}.$$

Now let y_1, y_2, y_3, y_4 be the variables in the form

$$\phi(y_1, y_2, y_3, y_4)$$

dual to the form f corresponding to x_1, x_2, x_3, x_4; thus $\phi(y_1, y_2, y_3, y_4)$ is the form obtained by transforming the product $D \cdot f(x_1, x_2, x_3, x_4)$ via the linear transformation

$$y_1 = \frac{1}{2}\frac{\partial f}{\partial x_1}, \quad y_2 = \frac{1}{2}\frac{\partial f}{\partial x_2}, \quad y_3 = \frac{1}{2}\frac{\partial f}{\partial x_3}, \quad y_4 = \frac{1}{2}\frac{\partial f}{\partial x_4}.$$

The number Δ is represented[20] by this form with

$$\begin{aligned} y_1 &= 0, \\ y_2 &= m_2 l_3 - m_3 l_2, \\ y_3 &= m_3 l_1 - m_1 l_3, \\ y_4 &= m_1 l_2 - m_2 l_1, \end{aligned}$$

where the integers $l_1, l_2, l_3, m_1, m_2, m_3$ are restricted only by the condition (1), which may always be secured by an appropriate choice of n_1, n_2, n_3 provided only that the integers

$$\begin{aligned} &m_2 l_3 - m_3 l_2, \\ &m_3 l_1 - m_1 l_3, \\ &m_1 l_2 - m_2 l_1 \end{aligned}$$

have highest common factor 1.

Choose $l_1, l_2, l_3, m_1, m_2, m_3$ so that the ternary form

$$\phi(0, y_2, y_3, y_4)$$

takes its minimum value at

$$\begin{aligned} y_2 &= m_2 l_3 - m_3 l_2, \\ y_3 &= m_3 l_1 - m_1 l_3, \\ y_4 &= m_1 l_2 - m_2 l_1. \end{aligned}$$

[20] "Mathematische Werke von Jacobi", 2 Band; Hermite, "Première lettre sur la théorie des nombres", p. 223.

That this is always possible follows from a known result,[21] and furthermore since integer values of y_2, y_3, y_4 yielding the minimum must have highest common factor 1, the only condition required of the integers $l_1, l_2, l_3, m_1, m_2, m_3$ is satisfied.

Hence the minimum of the form

$$\phi(0, y_2, y_3, y_4)$$

is just Δ, and, taking into account that the determinant of this form is $D^2 \cdot a_{11}$, we conclude that

$$\Delta \leq \sqrt[3]{2D^2 a_{11}}. \tag{3}$$

The inequalities (2) and (3) give

$$a_{11} \leq \sqrt[4]{4D}.$$

The upper bound $\sqrt[4]{4D}$ is exact, since it is the minimum of the form

$$\sqrt[4]{4D}(x_1^2 + x_2^2 + x_3^2 + x_4^2 + x_1x_2 + x_1x_3 + x_1x_4),$$

which has determinant D.

We have thus proved the following theorem:

The variables of any positive quaternary form of determinant D may always be assigned integer values so that the resulting value of the form is not greater than $\sqrt[4]{4D}$, and there are such forms having this quantity as minimum.

On Quadratic Forms

We now turn to a detailed examination of the long second memoir of Korkin and Zolotarev on quadratic forms, of 1873, entitled "Sur les formes quadratiques".

After defining an "extremal" form,[22] they give examples of various extremal forms, and then, noting again Hermite's conjecture that $\frac{2}{\sqrt[n]{n+1}}\sqrt[n]{D}$ is the exact upper bound for the minima of positive n-ary quadratic forms, they continue: "In attempting to verify this conjecture we have managed to prove that the quantity $\frac{2}{\sqrt[n]{n+1}}\sqrt[n]{D}$ is indeed the minimum of some extremal form, or, in other words, is the exact bound for the minima of a certain set of (equivalent) forms.[23] However from the above [i.e. from the examples of extremal forms given at the beginning of the article][24] we see that there are forms whose minima exceed Hermite's conjectured bound, so that that bound is not valid for all n-ary forms of determinant D. The correct exact bound for the set of all such forms will be the largest of the minima of all extremal forms contained in that set."

Next Korkin and Zolotarev digress on a connection with indefinite quadratic forms. They write: "The bounds that we have discussed above (for positive forms) are useful also in the theory of indefinite forms. Thus let Ω be any quantity not less than the exact bound on the minima of all n-ary positive forms with determinant D; it is easy to show that for any n-ary indefinite form with determinant $\pm D$,

[21] Gauss, "Disquisitiones arithmeticae", art. 279.

[22] As defined in their third memoir examined below, an *extremal form* is a positive quadratic form whose minimum decreases for every infinitesimal change (or perturbation) of its coefficients throughout which the determinant remains unchanged.

[23] Since equivalent forms have the same minimum, this does not seem to have the meaning perhaps intended, namely of a set of forms (or equivalence classes of forms) neighbouring on the extremal form in question. *Trans.*

[24] Or, in the case $n = 4$, from the first article considered above. *Trans.*

integer values may be assigned to the variables so that the absolute value of that form is also less than or equal to Ω. To see this consider the arbitrary indefinite form of determinant D, in canonical form

$$f = \pm A_1 X_1^2 \pm A_2 X_2^2 \pm \cdots \pm A_n X_n^2,$$

where $A_1, A_2, \ldots, A_n$ are positive constants,[25] and $X_1, X_2, \ldots, X_n$ are real linear combinations of the original variables $x_1, x_2, \ldots, x_n$. Consider the positive form

$$\phi = A_1 X_1^2 + A_2 X_2^2 + \cdots + A_n X_n^2,$$

which has determinant D, the absolute value of the determinant of the form f. If we assign the variables those values at which ϕ takes on its minimum value, then we obtain immediately

$$|f| \leq \phi \leq \Omega \qquad \text{at those values of the variables.}$$

In particular for indefinite binary forms with determinant $-D$, one obtains in this way the upper bound $\sqrt{\frac{4}{5}D}$.[26]

"However," continue Korkin and Zolotarev, "in relation to such bounds there is a large difference between indefinite and definite (i.e. positive) forms. In order to indicate this difference, at least for binary forms and the bound $\sqrt{\frac{4}{5}D}$, we note that if we exclude the particular form $\sqrt{\frac{4}{5}D}(x^2 + xy - y^2)$ and those equivalent to it, then the exact upper bound for all others with the same determinant D is $\sqrt{\frac{D}{2}}$." This aside concerning indefinite forms, given without proof, contains the germ of the remarkable researches of Markov on binary quadratic forms carried out 8 years after the present article of Korkin and Zolotarev,[27] and also of the great quantity of research on various kinds of indefinite forms, nonquadratic as well as quadratic, which today,[28] 70 years after Markov's work, are of such interest to English and American mathematicians.

In the rest of their memoir the authors confine themselves to the question of bounding the minima of positive quadratic forms only. To begin with they describe a special decomposition of an arbitrary positive form as a sum of squares, as follows. Let A be the minimum of the form. First transform the form to an equivalent one with A as the coefficient of x_1^2, a step known always to be possible. Next complete the square of the sum of all terms involving x_1, obtaining

$$A(x_1 + \alpha x_2 + \beta x_3 + \cdots + \gamma x_n)^2 + \text{terms in } \ x_2, x_3, \ldots, x_n.$$

The terms in $x_2, x_3, \ldots, x_n$ remaining outside this square will constitute a positive $(n-1)$-ary quadratic form, with minimum A', say. Now as before apply to $x_2, x_3, \ldots, x_n$ a linear transformation with integer coefficients and determinant ± 1, so that in the latter form the coefficient of the square x_2^2 of the new first variable is A', not forgetting to replace the old variables other than x_1 in the first square already obtained (with coefficient A) by their expressions in terms of the new ones.

[25]With their signs attached, they are the eigenvalues of the matrix of the original form in $x_1, x_2, \ldots, x_n$, so their product is $\pm D$. *Trans.*

[26]That is, the upper bound for the minima of the absolute values of such forms under substitution of integer values, not all zero, for the variables. *Trans.*

[27]See the chapter on Markov below. *Trans.*

[28]That is, in 1947. *Trans.*

Next complete the square of the terms of this transformed $(n-1)$-ary form involving x_2, and so on.

After at most n steps we will have the original form expressed as

$$\begin{aligned} &A(x_1 + \alpha x_2 + \beta x_3 + \cdots + \gamma x_n)^2 \\ &\quad + A'(x_2 + \delta x_3 + \cdots + \zeta x_n)^2 + \cdots \\ &\quad + A^{(n-2)}(x_{n-1} + \sigma x_n)^2 + A^{(n-1)} x_n^2, \end{aligned} \tag{1}$$

where A is the minimum of the whole form, A' is the minimum of the $(n-1)$-ary form obtained by deleting the square resulting from the first completion of the square, A'' is the minimum of the $(n-2)$-ary form obtained by deleting also the next completed square, and so on. Korkin and Zolotarev call this representation of a given positive form by an equivalent form (1), a *decomposition of the form along its successive minima.* A form may have several different such decompositions, but for present purposes any one will serve. For the sake of brevity we shall call A *the first coefficient of the decomposition*, A' the *second*, and so on. It is clear that if we delete some of the first squares from a decomposition of a form along its successive minima, or alternatively set some of the last variables equal to zero, then we shall obtain another form decomposed along its successive minima.

Consider the binary form

$$f = k(x + \lambda y)^2 + ly^2,$$

decomposed along its successive minima k and l. The value $k\lambda^2 + l$ obtained by setting $x = 0, y = 1$, cannot be less that the minimum k of the form f, i.e., $k\lambda^2 + l \geq k$, or $l \geq k(1 - \lambda^2)$. Now we may assume that $|\lambda| \leq \frac{1}{2}$, since in the contrary case, applying the transformation

$$x = X + tY, \quad y = Y,$$

where t is an integer, we obtain

$$f = k(X + tY + \lambda Y)^2 + lY^2,$$

and then t can be chosen so that $|t + \lambda| \leq \frac{1}{2}$. Hence $l \geq \frac{3}{4}k$, and equality may indeed occur here, as for instance in the decomposition along successive minima

$$\left(x + \frac{1}{2}y\right)^2 + \frac{3}{4}y = x^2 + xy + y^2.$$

We conclude therefore that:

For a decomposition along successive minima of a positive binary quadratic form, we always have $A' \geq \frac{3}{4}A$, with equality holding for the form $x^2 + xy + y^2$.

By applying the above-described procedures for "truncating" the form (1) from the left or right, we obtain the following positive binary forms, all decomposed along their successive minima:

$$\begin{aligned} &A(x_1 + \alpha x_2)^2 + A' x_2^2, \\ &A'(x_2 + \delta x_3)^2 + A'' x_3^2, \\ &\dots\dots\dots\dots\dots\dots\dots\dots \\ &A^{(n-1)}(x_{n-1} + \sigma x_n)^2 + A^{(n-1)} x_n^2. \end{aligned}$$

Hence in view of what has just been proved for binary forms, we must have

$$A' \geq \frac{3}{4}A, \ A'' \geq \frac{3}{4}A' \geq \left(\frac{3}{4}\right)^2 A, \ldots, A^{(n-1)} \geq \frac{3}{4}A^{(n-2)} \geq \cdots \geq \left(\frac{3}{4}\right)^{n-1} A.$$

Since $AA' \cdots A^{(n-1)} = D$, the determinant of the form (1), these inequalities yield $D \leq A \cdot (3/4)A \cdot (3/4)^2 A \cdots (3/4)^{n-1}A$, whence Hermite's inequality

$$A \leq \left(\frac{4}{3}\right)^{\frac{n-1}{2}} \sqrt[n]{D}.$$

Ternary forms can be analysed in detail via decompositions along their successive minima in analogous fashion to the binary case considered above. By means of a fairly laborious analysis Korkin and Zolotarev obtain the following result:

If A, A', A'' are the successive minima of a positive ternary quadratic form, then one always has

$$A' \geq \frac{3}{4}A, \quad A'' \geq \frac{2}{3}A.$$

Further investigation showed them that equality holds in the second inequality $A'' \geq \frac{2}{3}A$, only for the two equivalence classes of forms represented by

$$\sqrt[3]{2D}\left[\left(x + \frac{1}{2}y + \frac{1}{2}z\right)^2 + \frac{3}{4}\left(y + \frac{1}{3}z\right)^2 + \frac{2}{3}z^2\right]$$
$$= \sqrt[3]{2D}(x^2 + y^2 + z^2 + xy + xz + yz),$$

and

$$\frac{3}{2}\sqrt[3]{\frac{D}{2}}\left[\left(x + \frac{1}{3}y + \frac{1}{3}z\right)^2 + \frac{8}{9}\left(y - \frac{1}{2}z\right)^2 + \frac{2}{3}z^2\right]$$
$$= \frac{3}{2}\sqrt[3]{\frac{D}{2}}\left(x^2 + y^2 + z^2 + \frac{2}{3}xy + \frac{2}{3}xz - \frac{2}{3}yz\right).$$

Korkin and Zolotarev now "truncate" the form (1) from the right and left in such a way as to obtain ternary forms with respective successive minima

$$\begin{array}{ccc} A, & A', & A'', \\ A', & A'', & A''', \\ \multicolumn{3}{c}{\dotfill} \\ A^{(n-3)}, & A^{(n-2)}, & A^{(n-1)}. \end{array}$$

Applying the above general inequalities to these ternary forms, Korkin and Zolotarev infer the following inequalities:

$$A' \geq \frac{3}{4}A, \ \ A'' \geq \frac{2}{3}A, \ \ A''' \geq \frac{2}{3}A' \geq \frac{2}{3} \cdot \frac{3}{4}A, \ \ A^{iv} \geq \frac{2}{3}A'' \geq \left(\frac{2}{3}\right)^2 A, \ldots,$$

whence they obtain

$$\begin{aligned} A^{(2i-1)} &\geq \frac{3}{4}\left(\frac{2}{3}\right)^{i-1} A, \\ A^{(2i)} &\geq \left(\frac{2}{3}\right)^{i} A. \end{aligned} \tag{2}$$

Taking into account once again that $AA'A''\cdots A^{(n-1)} = D$, they now easily infer that:

$$
\begin{aligned}
(3) \qquad & A \le \frac{3^{(m-2)/2}}{2^{(m-3)/2}}\sqrt[n]{D} \qquad \text{for even } n = 2m; \\
& A \le \sqrt[n]{\frac{3^{m(m-1)}}{2^{m(m-2)}}}\sqrt[n]{D} \qquad \text{for odd } n = 2m+1.
\end{aligned}
$$

From (2) one has in particular

$$A' \ge \frac{3}{4}A, \quad A'' \ge \frac{2}{3}A, \quad A''' \ge \frac{1}{2}A, \quad A^{iv} \ge \frac{4}{9}A.$$

We have seen already that the first and second bounds here are actually attained for certain positive binary and ternary forms. It turns out that the third bound is attained for only one class of forms, namely that represented by the form

$$
\begin{aligned}
&\sqrt[4]{4D}\left[\left(x+\frac{1}{2}y+\frac{1}{2}z+\frac{1}{2}t\right)^2+\frac{3}{4}\left(y+\frac{1}{3}z+\frac{1}{3}t\right)^2+\frac{2}{3}\left(z-\frac{1}{2}t\right)^2+\frac{1}{2}t^2\right] \\
&\quad = \sqrt[4]{4D}(x^2+y^2+z^2+t^2+xy+xz+xt+yz+yt).
\end{aligned}
$$

Detailed further investigations by Korkin and Zolotarev lead to the result that the fourth bound above, namely $A^{iv} \ge (4/9)A$, is not attained for any form, i.e., there is no positive form for which the fifth coefficient A^{iv} of some decomposition along successive minima is exactly $\frac{4}{9}$ of its first. Thus the bounds given in (3) for the minimum of an n-ary positive form are exact only for $n = 2, 3, 4$.

On Positive Quadratic Forms

The third extensive memoir of Korkin and Zolotarev on quadratic forms, published in 1876, is entitled "Sur les formes quadratiques positives". In their introduction the authors write: "In the course of our investigations into quadratic forms we encountered and determined forms which we call 'extremal'. The problem of finding the exact upper bound for the minima of n-ary quadratic forms reduces to searching for an extremal form with largest minimum. However, it is easy to see that there are other fundamental questions about quadratic forms depending on a knowledge of such forms.

"For these reasons and also by way of completing the research we have already published, we propose expounding in the present memoir certain fundamental properties of these special forms, as well as exhibiting all extremal binary, ternary, quaternary, and quinary forms.

"Our memoir is divided into two parts. In the first we consider mainly forms in any number of variables. The second is devoted to quinary forms."

Korkin and Zolotarev continue by first reminding us that an *extremal* form is defined to be a positive quadratic form whose minimum decreases for all sufficiently small perturbations of the coefficients leaving the determinant of the form unchanged, and also that all forms equivalent to an extremal form are, of course, also extremal, so that they will regard an equivalence class of forms essentially as (represented by) a single form. They go on to remark that although their most characteristic property is indeed that used to define them, the direct search for such forms in a given number of variables presents great difficulties. This search involves a detailed investigation of the properties of the integer values of the variables at which forms attain their minimum. “However,” they write, “notwithstanding its difficulty, the problem of finding the extremal forms is of the very first importance in the theory of quadratic forms.”

It turns out that for any given concrete form it is easy to decide whether or not it is extremal. In this connection the authors consider the following three forms:

$$U_n = 2\sqrt[n]{\frac{D}{n+1}}(x_1^2 + x_2^2 + \cdots + x_n^2 + x_1x_2 + x_1x_3 + \cdots + x_{n-1}x_n)$$

(all products x_ix_k occurring);

$$V_n = \sqrt[n]{2^{n-2}D}(x_1^2 + x_2^2 + \cdots + x_n^2 + x_1x_3 + x_1x_4 + x_{n-1}x_n)$$

(all products x_ix_k occurring except x_1x_2);

$$Z = \sqrt[5]{\frac{2^9}{3^4}D}\left(x_1^2 + x_2^2 + x_3^2 + x_4^2 + x_5^2 - \frac{1}{2}x_1x_2 - \frac{1}{2}x_1x_3 - \frac{1}{2}x_1x_4 - \frac{1}{2}x_1x_5 + \frac{1}{2}x_2x_3 + \frac{1}{2}x_2x_4 - x_2x_5 + \frac{1}{2}x_3x_4 - x_3x_5 - x_4x_5\right).$$

An ordered n-tuple of (integer) values of the variables at which a form takes on its minimum value is called a *representation of the minimum.* Note that we shall not distinguish two representations of the minimum of a form if one is obtained from the other by multiplying all the values by -1.

It is not difficult to see that the possible representations of the minimum of the form U_n are of just the following two types:

1) the standard basis vectors, i.e. n-tuples where all entries but one are zero, and the single nonzero entry is 1;

2) those n-tuples where all entries but two are zero, and one of the two nonzero entries is 1, the other -1.

Thus altogether the form U_n has $\frac{n(n+1)}{2}$ representations.

Consider the form f obtained by perturbing the coefficients of U_n by infinitesimally small amounts $\alpha_{11}, \alpha_{22}, \ldots, \alpha_{nn}, \alpha_{12}, \alpha_{13}, \ldots, \alpha_{n-1,n}$. Equating the determinant $\det[a_{ik} + \alpha_{ik}]$ of f to that of U_n, namely $\det[a_{ik}]$ $(= D)$, we obtain after simplifying,

$$(1) \qquad n(\alpha_{11} + \alpha_{22} + \cdots + \alpha_{nn}) - (\alpha_{12} + \alpha_{13} + \cdots + \alpha_{n-1,n}) + \Omega = 0,$$

where Ω is a sum of terms of degree two or more in the infinitesimal quantities α_{ik}.

Now setting $x_i = 1$, and $x_j = 0$ for all $j \neq i$, in f, we obtain the value

$$2\sqrt[n]{\frac{D}{n+1}}(1 + \omega_{ii}),$$

where $\omega_{ii} = \alpha_{ii}$. Setting $x_i = 1, x_k = -1$ in f for any two distinct i, k, and all other variables zero, we obtain the value

$$2\sqrt[n]{\frac{D}{n+1}}(1 + \omega_{ik}),$$

where $\omega_{ik} = \omega_{ki} = \alpha_{ii} + \alpha_{kk} - \alpha_{ik}$. Denoting by $\sum \omega$ the sum of all the ω_{ii} and ω_{ik}, we therefore have

$$\sum \omega = n(\alpha_{11} + \alpha_{22} + \cdots + \alpha_{nn}) - (\alpha_{12} + \alpha_{13} + \cdots + \alpha_{n-1,n}),$$

so that the equation (1) may be rewritten as

$$\sum \omega + \Omega = 0. \tag{2}$$

Since Ω contains only terms of the second order of smallness in the α_{ik}, there must be among the ωs both positive and negative quantities. (They cannot all be zero, since then all αs would likewise be zero, which we do not, of course, allow.) However, if some ω is negative, then the value

$$2\sqrt[n]{\frac{D}{n+1}}(1+\omega)$$

obtained for f on substituting the corresponding representation of the minimum of U_n, will be less than that minimum, i.e., less than $2\sqrt[n]{\frac{D}{n+1}}$. Hence the minimum of F is less than that of U_n, whence the extremality of U_n.

The extremality of the forms V_n and Z is proved in very much the same way. Having established that the forms U_n, V_n, Z are extremal, the authors next prove the following important theorem.

THEOREM. *The number of representations of the minimum of an extremal form is at least $\frac{n(n+1)}{2}$, and the set of all such representations completely determines the coefficients of the form, provided its minimum is given.*

The proof is as follows. Let

$$(p_1, p_2, \dots, p_n), \ (q_1, q_2, \dots, q_n), \ \dots$$

be all of the representations of the minimum M of an arbitrary given extremal form f with determinant D, i.e.,

$$\sum p_i p_k a_{ik} = M, \quad \sum q_i q_k a_{ik} = M, \dots . \tag{3}$$

Assume that the totality of representations of the minimum does not determine the coefficients uniquely. This assumption entails that the linear system (3) in the unknowns a_{ik} has more than one solution and hence infinitely many. It is well known that this implies that the corresponding homogeneous system

$$\sum p_i p_k \xi_{ik} = 0, \quad \sum q_i q_k \xi_{ik} = 0, \dots$$

also has infinitely many solutions and therefore a solution where not all the ξ_{ik} are zero; let λ_{ik} be such a solution. We shall show that the existence of such a solution leads to a contradiction of the extremality of the form f. Since there will always be such a solution if the number of variables, namely $\frac{n(n+1)}{2}$, exceeds the number of equations, the first claim of the theorem will then also follow.

Consider the positive form $f + \epsilon \cdot \phi$, where ϵ is a number of vanishingly small absolute value, and ϕ is the (indefinite) form $\sum \lambda_{ik} x_i x_k$. The form $f+\epsilon\cdot\phi$ attains its minimum for exactly the same values of the variables as does f, and moreover since ϕ is zero at these values, its minimum is also M. We now show that for appropriate choice of ϵ the determinant of the form $f + \epsilon \cdot \phi$ is less than D. Denoting this determinant by D', we have

$$\begin{aligned} D' = D + \epsilon \biggl(&\frac{\partial D}{\partial a_{11}}\lambda_{11} + \frac{\partial D}{\partial a_{22}}\lambda_{22} + \dots + \frac{\partial D}{\partial a_{nn}}\lambda_{nn} \\ &+ \frac{\partial D}{\partial a_{12}}\lambda_{12} + \frac{\partial D}{\partial a_{13}}\lambda_{13} + \dots + \frac{\partial D}{\partial a_{n-1,n}}\lambda_{n-1,n} \biggr) + \Omega, \end{aligned}$$

where Ω is a sum of terms of higher degree in ϵ. If the quantity P contained in parentheses here is not zero, then by choosing the sign of ϵ appropriately, we will have $D' < D$. If, on the other hand, $P = 0$, we proceed as follows. Writing $\Omega = K\frac{\epsilon^2}{2} + \Omega_1$, where Ω_1 is a sum of terms of degree higher than two in ϵ, we consider the expression (for the moment making no assumptions about P)

$$P^2 - DK,$$

a combined invariant of the forms f and ϕ. Now apply to f a linear transformation from the variables $(x_1, x_2, \ldots, x_n)$ to new variables $(z_1, z_2, \ldots, z_n)$, in terms of which f is a sum of squares, $f = \sum \alpha_{ii} z_i^2$. Since f is positive, we must then have $\alpha_{ii} > 0$. Write

$$\phi = \sum \mu_{ik} z_i z_k$$

in terms of the new variables. It can now be calculated that

$$P^2 - DK = D^2 \left[\left(\frac{\mu_{11}}{\alpha_{11}}\right)^2 + \left(\frac{\mu_{22}}{\alpha_{22}}\right)^2 + \cdots + \left(\frac{\mu_{nn}}{\alpha_{nn}}\right)^2 + 2\frac{\mu_{12}^2}{\alpha_{11}\alpha_{22}} + \cdots + 2\frac{\mu_{n-1,n}^2}{\alpha_{n-1,n-1}\alpha_{nn}} \right],$$

whence it is immediate that $P^2 - DK > 0$, since all $\alpha_{ii} > 0$. Hence in the situation we wished to consider, namely $P = 0$, it follows that $K < 0$. However, since in this case

$$D' = D + K\frac{\epsilon^2}{2} + \Omega_1,$$

we shall again have $D' < D$.

Having established that $D' < D$, we may write $D' = D(1 - \eta)$, where $\eta > 0$. However, then the form

$$\frac{f + \epsilon\phi}{\sqrt[n]{1 - \eta}},$$

whose coefficients can be made as close as we please to those of f, will have determinant D, and minimum $\frac{M}{\sqrt[n]{1-\eta}} > M$, contradicting the assumption that f is extremal.

The following three conclusions summarize and supplement the foregoing:

1°. *Every extremal n-ary form has at least $\frac{n(n+1)}{2}$ representations of its minimum, and these, together with that minimum, determine the form uniquely.*

2°. *If f is an extremal form with minimum M, then the form $\frac{1}{M}f$ has rational coefficients, since the coefficients of the latter form are solutions of the system (3) with M replaced by 1, which is a linear system over the integers.*

3°. *The determination of the exact upper bound for the minima of positive n-ary forms reduces to finding an extremal form with largest minimum.*

Korkin and Zolotarev next undertake a detailed investigation of the table of representations of the minimum of a given positive form f. Let

$$(p_1, p_2, \ldots, p_n),\ (q_1, q_2, \ldots, q_n),\ \ldots$$

be the representations of the minimum of f. To begin with, Korkin and Zolotarev consider the nth order determinants that can be formed having as rows various of the representations of the minimum:

$$\begin{vmatrix} p_1 & p_2 & \cdots & p_n \\ q_1 & q_2 & \cdots & q_n \\ \vdots & \vdots & \cdots & \vdots \\ r_1 & r_2 & \cdots & r_n \end{vmatrix}.$$

They call these determinants *characteristic*; their method for finding extremal forms hinges of the study of these determinants, concerning which they prove the following two theorems.

I. *Every extremal form has a characteristic determinant not equal to zero.*

II. *There is an upper bound in terms of n alone for the absolute value of the characteristic determinants of every positive n-ary quadratic form having at least n representations of its minimum.*

Korkin and Zolotarev prove the first of these results as follows. Let f be a positive n-ary form of determinant D, with table of representations of its minimum

$$\begin{matrix} p_1 & p_2 & \cdots & p_n \\ q_1 & q_2 & \cdots & q_n \\ \vdots & \vdots & \cdots & \vdots \end{matrix},$$

and assume that all of the characteristic determinants of f are zero. This implies that there exist numbers $\xi_1, \xi_2, \dots, \xi_n$, not all zero, such that

$$\begin{gathered} p_1\xi_1 + p_2\xi_2 + \cdots + p_n\xi_n = 0, \\ q_1\xi_1 + q_2\xi_2 + \cdots + q_n\xi_n = 0, \\ \dots\dots\dots\dots\dots\dots\dots\dots\dots\dots \end{gathered}$$

Consider the form $f' = f - \epsilon(\xi_1 x_1 + \xi_2 x_2 + \cdots + \xi_n x_n)^2$, where ϵ is positive. It is clear that for sufficiently small ϵ the representations of the minimum of f' must be contained among those of f. However, since

$$\xi_1 x_1 + \xi_2 x_2 + \cdots + \xi_n x_n = 0$$

for all those representations, it follows that for all sufficiently small ϵ, the forms f and f' have the same minimum. Now it is not difficult to see that the determinant of the form f is

$$D - \epsilon F(\xi_1, \xi_2, \dots, \xi_n),$$

where F is also a positive form, in fact the dual of the form f. Hence the determinant of f' is less than D, whence it follows as before that the form f cannot be extremal. This contradiction completes the proof of the first theorem above.

The second theorem is proved as follows. Let $f(x_1, x_2, \dots, x_n)$ now denote an arbitrary positive n-ary form of determinant D, and let

$$(p_1, p_2, \dots, p_n), (q_1, q_2, \dots, q_n), \dots, (r_1, r_2, \dots, r_n),$$

be n distinct representations of its minimum M with characteristic determinant Δ. Apply the linear change of variables

$$\begin{aligned} x_1 &= p_1 z_1 + q_1 z_2 + \cdots + r_1 z_n, \\ x_2 &= p_2 z_1 + q_2 z_2 + \cdots + r_2 z_n, \\ &\vdots \qquad\quad \vdots \\ x_n &= p_n z_1 + q_n z_2 + \cdots + r_n z_n, \end{aligned}$$

and write $\phi = \sum a_{ik} z_i z_k$ for the transformed form. Since the form attains its minimum at the standard basic n-tuples

$$(z_1, z_2, \ldots, z_n) = (1, 0, \ldots, 0), (0, 1, 0, \ldots, 0), \ldots,$$

it follows that $a_{11} = a_{22} = \cdots = a_{nn} = M$. As in the second memoir above, we now decompose the form ϕ along its successive minima (*à la* Jacobi), obtaining

$$\phi = a_{11}(z_1 + \cdots)^2 + K(z_2 + \cdots)^2 + L(z_3 + \cdots)^2 + \cdots + N z_n^2,$$

where

$$K = \frac{\begin{vmatrix} a_{11} & a_{12} \\ a_{21} & a_{22} \end{vmatrix}}{a_{11}}, \quad L = \frac{\begin{vmatrix} a_{11} & a_{12} & a_{13} \\ a_{21} & a_{22} & a_{23} \\ a_{31} & a_{32} & a_{33} \end{vmatrix}}{\begin{vmatrix} a_{11} & a_{12} \\ a_{21} & a_{22} \end{vmatrix}}, \ldots,$$

$$N = \frac{\begin{vmatrix} a_{11} & a_{12} & \cdots & a_{1n} \\ a_{21} & a_{22} & \cdots & a_{2n} \\ \cdots & \cdots & \cdots & \cdots \\ a_{n1} & a_{n2} & \cdots & a_{nn} \end{vmatrix}}{\begin{vmatrix} a_{11} & a_{12} & \cdots & a_{1,n-1} \\ a_{21} & a_{22} & \cdots & a_{2,n-1} \\ \cdots & \cdots & \cdots & \cdots \\ a_{n-1,1} & a_{n-1,2} & \cdots & a_{n-1,n-1} \end{vmatrix}}.$$

It is immediate that $a_{11} \cdot K \cdot L \cdots N = D \cdot \Delta^2$. It is not difficult to show[29] that also

$$K \leq a_{22}, \ L \leq a_{33}, \ldots, N \leq a_{nn}.$$

Since all $a_{ii} = M$, it follows that $M^n \geq D\Delta^2$. On the other hand by the fundamental theorem on the bound for the minimum of a form (see for instance the first memoir above) we know that $M \leq \sqrt[n]{\mu D}$ for some number μ depending only on n. This and the previous inequality together give $|\Delta| \leq \sqrt{\mu}$, completing the proof.

With the μ given by Hermite's bound for the minimum (see the first memoir above), namely $\left(\frac{4}{3}\right)^{\frac{n(n-1)}{2}}$, we infer the following particular cases:

For $n = 2$, Hermite's bound has $\mu = 4/3$, whence for binary forms $|\Delta| = 1$.

For $n = 3$, Hermite's bound has $\mu = (4/3)^3$, whence for ternary forms, $|\Delta| = 0$ or 1.

For $n = 4$, Hermite has $\mu = (4/3)^6$, whence for quaternary forms, $|\Delta| = 0, 1$ or 2.

[29]See for instance *Uspekhi matematicheskikh nauk*, Vol. III, p. 27.

For $n = 5$, we have by (3) in the second memoir above that we may take $\mu = 9$, whence $|\Delta| \leq 3$. However we cannot have $|\Delta| = 3$, since in this case it would follow from the above two inequalities that $M = \sqrt[5]{9D}$, which is impossible in view of the final result of that memoir. Hence for quinary forms also $|\Delta| = 0, 1$, or 2.

Thus in searching for the extremal n-ary forms for $n = 2, 3, 4$, and 5, the first step is to find all tables of representations of the minimum (which may be assumed equal to 1) with at least $n(n+1)/2$ rows, such that the characteristic determinants are, in absolute value, all 0 or 1, or, for $n = 4, 5$, all $0, 1$ or 2.

A detailed investigation leads eventually to the following result. (We omit the proof, which is, although elementary, rather involved, occupying 12 pages.)

If a matrix with integer entries has n columns, at least $n(n+1)/2$ rows, has no row equal to $\pm$ any other, is such that every $n \times n$ determinant made up of n of its rows is equal to either 0 or ± 1, with at least one such determinant nonzero, then the number of rows must in fact be exactly $n(n+1)/2$, and the matrix may be transformed (in an appropriate manner) to an equivalent one of the form

$$\begin{bmatrix} 1 & 0 & 0 & \cdots & 0 \\ 0 & 1 & 0 & \cdots & 0 \\ \vdots & \vdots & \vdots & \cdots & \vdots \\ 1 & -1 & 0 & \cdots & 0 \\ \vdots & \vdots & \vdots & \cdots & \vdots \\ 0 & 0 & \cdots & 1 & -1 \end{bmatrix}.$$

Since the rows of this matrix are precisely the representations of the minimum of the form U_n, we have the following corollary.

The form U_n is the only n-ary extremal form all of whose characteristic determinants Δ are 0 or ± 1.

Taking into account the fact established above that for binary and ternary forms Δ is always 0 or ± 1, we infer that in each of these cases there is essentially just one extremal form, namely:

$$\sqrt{\frac{4}{3}D}(x_1^2 + x_2^2 + x_1x_2) \text{ for } n = 2;$$
$$\sqrt[3]{2D}(x_1^2 + x_2^2 + x_3^2 + x_1x_2 + x_1x_3 + x_2x_3) \text{ for } n = 3.$$

In the second chapter of their memoir Korkin and Zolotarev determine all extremal forms in 4 and 5 variables. They already know one such form in each case, namely U_4 and U_5; by the above these are precisely the extremal forms for these two values of n having all their characteristic determinants equal to 0 or ± 1. However since, as we have seen, an extremal quaternary or quinary form may also have characteristic determinants equal to ± 2, it remains to find those 4- and 5-column integer matrices with representations of the minimum as rows, where all characteristic determinants are $0, \pm 1$ or ± 2, with at least one such equal to ± 2.

As the first step they show that it is impossible for all the characteristic determinants of a quaternary or quinary extremal form to be equal to either 0 or ± 2.

For suppose we have such a quinary form f, and let

$$\begin{vmatrix} p_1 & p_2 & p_3 & p_4 & p_5 \\ q_1 & q_2 & q_3 & q_4 & q_5 \\ \vdots & \vdots & \vdots & \vdots & \vdots \\ r_1 & r_2 & r_3 & r_4 & r_5 \end{vmatrix}$$

be a characteristic determinant of f equal to ± 2. Make the change of variables

$$\begin{aligned} x_1 &= p_1 z_1 + q_1 z_2 + \cdots + r_1 z_5, \\ x_2 &= p_2 z_1 + q_2 z_2 + \cdots + r_2 z_5, \\ &\cdots\cdots\cdots\cdots\cdots\cdots\cdots \\ x_5 &= p_5 z_1 + q_5 z_2 + \cdots + r_5 z_5, \end{aligned}$$

and denote the resulting form by $\phi(z_1, z_2, z_3, z_4, z_5)$. If in the above equations we set $x_1 = s_1, x_2 = s_2, x_3 = s_3, x_4 = s_4, x_5 = s_5$, where $(s_1, s_2, s_3, s_4, s_5)$ is any representation of the minimum of f, we obtain a linear system in the unknowns z_1, z_2, z_3, z_4, z_5, which by Cramer's rule, in view of the assumption that all characteristic determinants of f are 0 or ± 2, has a solution with all $z_i = 0$ or ± 1, since the denominator in Cramer's rule will be ± 2, while the numerator, also a characteristic determinant of f, is 0 or ± 2. Hence ϕ has the same minimum as f. Now ϕ is also an extremal quinary form (with these solutions as the representations of its minimum), whence its characteristic determinants are $0, \pm 1$ or ± 2. However, since in the above transformation the coefficient matrix has determinant ± 2, it is not difficult to see that they can only take the values 0 or ± 1.[30] Hence by what has already been proved above, ϕ is just the form U_n, except that if the determinant of f is D, then that of ϕ is $4D$. Hence the minimum of ϕ is $2\sqrt[5]{\frac{4D}{6}} = \sqrt[5]{\frac{64}{3}D}$ (see the beginning of this memoir, where U_n is introduced). Since ϕ and f have the same minimum M, this contradicts the fact established in the second memoir (see the inequalities (3) there) that $M \leq \sqrt[5]{9D}$. This concludes the proof that an extremal quinary form cannot have all of its characteristic determinants equal to 0 or ± 2; the proof in the quaternary case is similar.

Next the authors show that if the table of representations of the minimum of a quaternary form has rows

$$\begin{matrix} 1 & 0 & 0 & 0 \\ 0 & 1 & 0 & 0 \\ 0 & 0 & 1 & 0 \\ \alpha_1 & \alpha_2 & \alpha_3 & 2 \end{matrix}\text{'}$$

then the form must be one of the following eight:

$$M\left[\left(x_1 \pm \frac{x_4}{2}\right)^2 + \left(x_2 \pm \frac{x_4}{2}\right)^2 + \left(x_3 \pm \frac{x_4}{2}\right)^2 + \frac{x_4^2}{4}\right].$$

From this they deduce that:

Up to equivalence, the only extremal quaternary form other than U_4 is

$$\sqrt[4]{4D}(x_1^2 + x_2^2 + x_3^2 + x_4^2 + x_1x_4 + x_2x_4 + x_3x_4).$$

---

[30]Regarding an arbitrary characteristic determinant of f as the (signed) Euclidean volume of the 5-dimensional parallelepiped determined by its row vectors, it follows that application of the above linear transformation with determinant ± 2 will cut that volume in half; hence the characteristic determinants of ϕ are half those of f, i.e., half of 0 or ± 2. *Trans.*

(Note that this is the form they discovered in their first memoir.)

The memoir continues with a detailed further investigation of the table of representations of the minimum of an extremal quinary form having among its characteristic determinants at least one equal to ± 2. Although the argumentation is elementary, it is very involved, occupying as it does 20 pages, and we shall forego reproducing it here. The upshot is as follows:

There are, up to equivalence, just three extremal quinary forms, namely:

$$\begin{aligned} U_5 = 2\sqrt[5]{\frac{D}{6}}(&x_1^2 + x_2^2 + x_3^2 + x_4^2 + x_5^2 + x_1x_2 + x_1x_3 + x_1x_4 + x_1x_5 \\ &+ x_2x_3 + x_2x_4 + x_2x_5 + x_3x_4 + x_3x_5 + x_4x_5); \\ Z = \sqrt[5]{\frac{2^9D}{3^4}}\Big(&x_1^2 + x_2^2 + x_3^2 + x_4^2 + x_5^2 - \frac{1}{2}x_1x_2 - \frac{1}{2}x_1x_3 - \frac{1}{2}x_1x_4 \\ &- \frac{1}{2}x_1x_5 + \frac{1}{2}x_2x_3 + \frac{1}{2}x_2x_4 - x_2x_5 + \frac{1}{2}x_3x_4 - x_3x_5 - x_4x_5\Big); \\ V_5 = \sqrt[5]{2^3D}(&x_1^2 + x_2^2 + x_3^2 + x_4^2 + x_5^2 + x_1x_3 + x_1x_4 + x_1x_5 \\ &+ x_2x_3 + x_2x_4 + x_2x_5 + x_3x_4 + x_3x_5 + x_4x_5). \end{aligned}$$

Furthermore, since the largest minimum of these three is $\sqrt[5]{8D}$ *(the minimum of* V_5*), the exact upper bound for the minima of positive quinary quadratic forms is* $\sqrt[5]{8D}$.

Thus in their third memoir Korkin and Zolotarev found all extremal forms in two, three, four, and five variables. These results astounded Hermite. It was apparent that the failure of persistent attempts to find the exact upper bound for the minima of positive n-ary quadratic forms was to be explained by the fact that the problem was of far greater complexity than it had seemed initially. It turns out that for each $n \geq 4$ there are several locally maximal minima, and furthermore that there is no simple rule for determining the largest among these (representing, of course, the desired exact upper bound), or even how many there are.

An extensive correspondence between Korkin and Zolotarev (64 letters) has been preserved, covering the period 1871–1878, when they were writing their joint memoirs on positive quadratic forms. This correspondence, comprising 176 pages, is contained in the second volume of Zolotarev's collected works, published by the Academy of Sciences in 1932. In these letters, which are devoted almost exclusively to their work on quadratic forms, one can trace in detail the birth and development of the ideas so completely and consummately expounded in their articles.

Commentary

The memoirs of Korkin and Zolotarev on quadratic forms examined above call for elucidation. It is in the application of the question concerning the minima of quadratic forms to geometry that the significance of this question becomes graphically clear. This we shall now demonstrate.

We first remind the reader that a *quadratic form* is a homogeneous polynomial of degree two. One considers binary forms

$$a_{11}x_1^2 + 2a_{12}x_1x_2 + a_{22}x_2^2,$$

ternary forms

$$a_{11}x_1^2 + a_{22}x_2^2 + a_{33}x_3^2 + 2a_{12}x_1x_2 + 2a_{13}x_1x_3 + 2a_{23}x_2x_3,$$

quaternary forms

$$\begin{aligned} a_{11}x_1^2 + a_{22}x_2^2 + a_{33}x_3^2 + a_{44}x_4^2 + 2a_{12}x_1x_2 + 2a_{13}x_1x_3 \\ + 2a_{14}x_1x_4 + 2a_{23}x_2x_3 + 2a_{24}x_2x_4 + 2a_{34}x_3x_4, \end{aligned}$$

and so on. The general n-ary form is often written as

$$f = \sum_{i,k} a_{ik}x_ix_k, \quad i, k = 1, 2, \dots, n,$$

where it is assumed that $a_{ik} = a_{ki}$, whence the 2s in the above general expressions. The *determinant* or *discriminant* of the form is the determinant

$$D = \begin{vmatrix} a_{11} & a_{12} & \cdots & a_{1n} \\ a_{21} & a_{22} & \cdots & a_{2n} \\ \vdots & \vdots & \cdots & \vdots \\ a_{n1} & a_{n2} & \cdots & a_{nn} \end{vmatrix}.$$

In the *algebraic* theory of quadratic forms the variables x_i are allowed to take on arbitrary real or complex values; in the *arithmetic* theory, on the other hand, they are restricted to (rational) integer values.

When the coefficients a_{ik} are real, the form may be (negative or positive) definite, indefinite, or degenerate. An n-ary quadratic form is called *positive definite* (or just *positive*) if it can be expressed as the sum of the squares of n linear forms

$$\begin{aligned} &\lambda_{11}x_1 + \lambda_{12}x_2 + \cdots + \lambda_{1n}x_n, \\ &\lambda_{21}x_1 + \lambda_{22}x_2 + \cdots + \lambda_{2n}x_n, \\ &\dots\dots\dots\dots\dots\dots\dots\dots \\ &\lambda_{n1}x_1 + \lambda_{n2}x_2 + \cdots + \lambda_{nn}x_n, \end{aligned}$$

with real coefficients λ_{ik}, and nonzero determinant

$$\begin{vmatrix} \lambda_{11} & \lambda_{12} & \cdots & \lambda_{1n} \\ \lambda_{21} & \lambda_{22} & \cdots & \lambda_{2n} \\ \vdots & \vdots & \cdots & \vdots \\ \lambda_{n1} & \lambda_{n2} & \cdots & \lambda_{nn} \end{vmatrix}.$$

Thus if a form is positive, it can be expressed as

$$\begin{aligned} f = (\lambda_{11}x_1 + \lambda_{12}x_2 + \cdots + \lambda_{1n}x_n)^2 + (\lambda_{21}x_1 + \lambda_{22}x_2 + \cdots + \lambda_{2n}x_n)^2 \\ + \cdots + (\lambda_{n1}x_1 + \lambda_{n2}x_2 + \cdots + \lambda_{nn}x_n)^2. \end{aligned}$$

As will appear below, a positive form can be decomposed as such a sum of squares of linear forms in infinitely many ways.

Now let λ denote a rectangular Cartesian coordinate system (in n-dimensional Euclidean space), and consider the vectors which relative to this system have the above λ_{ik} as coordinates:

$$\begin{aligned}\mathfrak{A}_1 &= (\lambda_{11}, \lambda_{21}, \dots, \lambda_{n1});\\ \mathfrak{A}_2 &= (\lambda_{12}, \lambda_{22}, \dots, \lambda_{n2});\\ &\dots\dots\dots\dots\\ \mathfrak{A}_n &= (\lambda_{1n}, \lambda_{2n}, \dots, \lambda_{nn}).\end{aligned}$$

The coefficients a_{ik} of the above form f are easily calculated in terms of the λ_{ik} by expanding and collecting like terms, yielding

$$a_{ii} = \lambda_{1i}^2 + \lambda_{2i}^2 + \cdots + \lambda_{ni}^2, \quad a_{ik} = \lambda_{1i}\lambda_{1k} + \lambda_{2i}\lambda_{2k} + \cdots + \lambda_{ni}\lambda_{nk},$$

or

$$a_{ik} = \mathfrak{A}_i \cdot \mathfrak{A}_k,$$

the dot (or Euclidean scalar) product of $\mathfrak{A}_i$ and $\mathfrak{A}_k$. Hence the form f can be expressed as

$$f = \sum a_{ik}x_ix_k = (x_1\mathfrak{A}_1 + x_2\mathfrak{A}_2 + \cdots + x_n\mathfrak{A}_n)^2,$$

where the square (*scalar square*) here denotes the dot product of the vector $x_1\mathfrak{A}_1 + x_2\mathfrak{A}_2 + \cdots + x_n\mathfrak{A}_n$ with itself. It is easy to verify that the discriminant (or determinant) D of the form coincides with the square of the determinant $\det[\lambda_{ik}]$.

The n-frame $(\mathfrak{A}_1, \mathfrak{A}_2, \dots, \mathfrak{A}_n)$ is called a *frame* of the positive form f. It is not difficult to show that different decompositions of f as a sum of squares will yield identical frames, rotated relative to one another,[31] however, about the origin O of the coordinate system λ.

Consider the point lattice made up of all points having integer coordinates x_i relative to the frame $(\mathfrak{A}_1, \mathfrak{A}_2, \dots, \mathfrak{A}_n)$. Forms equivalent to f, i.e., forms obtained from f by applying an integer linear change of coordinates with determinant ± 1, will then correspond to the other fundamental frames of the lattice. In this sense we may say that the equivalence classes of positive forms correspond one-to-one to the point lattices of n-space with fixed origin O.

Since the value of the form f at an integer n-tuple $(x_1, x_2, \dots, x_n)$ is the scalar square of the vector $x_1\mathfrak{A}_1 + x_2\mathfrak{A}_2 + \cdots + x_n\mathfrak{A}_n$, we see that geometrically it is just the square of the distance from the origin O to the point with coordinates $(x_1, x_2, \dots, x_n)$ relative to the frame $(\mathfrak{A}_1, \mathfrak{A}_2, \dots, \mathfrak{A}_n)$. Hence the minimum of the form f is just the square of the distance from the origin O to a point nearest to the origin (other than the origin itself) of the lattice corresponding to f. In other words the minimum M of the form is the square of the shortest distance between any two points of the corresponding lattice.

The problem solved by Korkin and Zolotarev for $n \leq 5$ is thus equivalent to the following geometrical one:

Among all point lattices with fundamental parallelepiped of given volume $\sqrt{D}$, find those with the property that any infinitesimal perturbation of their fundamental parallelepiped not changing its volume causes the minimum to decrease.

We shall call such lattices *extremal*, following Korkin's and Zolotarev's terminology for the corresponding forms. Thus among the extremal lattices of prescribed dimension n and fundamental parallelepiped of volume $\sqrt{D}$, those having the largest minimum will furnish the exact upper bound for the minima of positive n-ary quadratic forms of discriminant D.

---

[31]That is, obtained from one another by means of a Euclidean motion. *Trans.*

Minkowski suggested inverting the problem as follows:

Among all lattices of prescribed dimension n and least distance between their points, find those with fundamental parallelepiped of least volume.

If one imagines spheres of the same dimension n as a given such lattice, with their centres at the points of the lattice and radius equal to half the given least distance between points of the lattice, then in Minkowski's formulation the problem becomes that of finding the "densest lattice packing" of nonoverlapping identical spheres of given radius. Here by the *density* of a packing we mean the amount of n-space occupied by the (solid) spheres in proportion to the whole space. We shall call a lattice packing of identical spheres *incompressible* if it is not possible to make an infinitesimal change in the shape of the fundamental parallelepiped of the lattice with points the centres of the spheres, so that the packing becomes denser, while the spheres continue to be nonoverlapping. Among the incompressible sphere packings (with prescribed radius) there will be a *densest* (or *closest*). The quadratic forms corresponding to the fundamental frames of the lattices with points the centres of the spheres of the incompressible packings, are precisely those termed "extremal" by Korkin and Zolotarev. (The equivalence class of an extremal form may likewise be called extremal.)

Thus the problem tackled by Korkin and Zolotarev is equivalent to the following geometrical problem:

Find all incompressible lattice packings of identical n-spheres in n-space.

In this geometrical formulation the problem is much easier to visualize, though not necessarily easier to solve.

We shall now elucidate certain of the results of Korkin and Zolotarev from this geometrical point of view.

We begin with the short article of 1872 in which they found the exact upper bound for the minima of positive quaternary forms, or, in geometrical terms, the densest lattice packing of identical 4-spheres in 4-space. We will however not comment further on the main conclusion of that article, since its full geometrical implication can be arrived at by direct, and more natural, geometric means, leading to a deeper understanding of the geometry of the situation. The following exposition is taken from a note by the present author (B. N. Delone).

It might seem likely at first glance that the densest lattice packing of identical n-spheres in n-space will be that obtained by generalizing from the known densest sphere packings in 2- and 3-space, i.e., of discs in the plane and spheres in space. (In these two cases it can be shown that the densest lattice packings have the following lattices: In the case of the plane it is the lattice consisting of the vertices of the tesselation of the plane by equilateral triangles, and in the case of space the lattice generated by the edges of a regular tetrahedron emanating from O, i.e., in both cases it is the lattice generated by the regular simplex of appropriate dimension.) However if this were the case, then since the quadratic form whose n-frame consists of n edges of a regular n-simplex emanating from the origin O can be shown to be (a multiple of) the form

$$U_n = x_1^2 + x_2^2 + \cdots + x_n^2 + x_1x_2 + x_1x_3 + \cdots + x_{n-1}x_n,$$

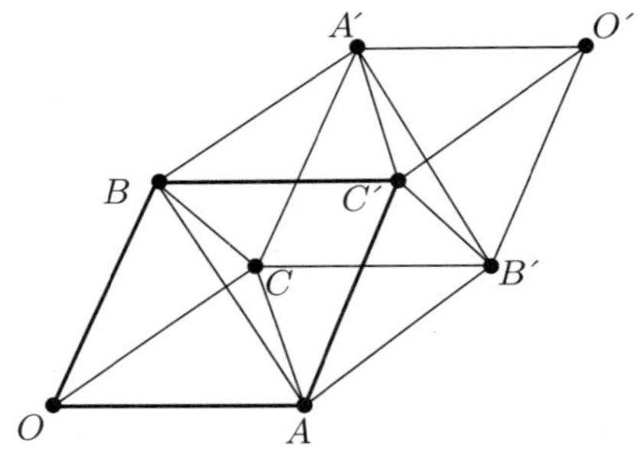

FIGURE 1

where the coefficients of all $x_i x_k$, $i \leq k$, are 1,[32] it would follow that the form U_n had the largest minimum of any positive n-ary form with the same determinant. This is in fact what Hermite had believed, until Korkin and Zolotarev showed in their first memoir above that this is not the case already for $n = 4$, since there is a positive quaternary form, namely V_4 (see the first or third memoir above) with larger minimum.

To show by direct geometric means that the sphere packing of Euclidean 4-space with lattice generated by the vectors represented by the edges of a regular 4-simplex emanating from the origin is not the densest, i.e., that there is another (in fact the one constructed will be that corresponding to the form V_4) which is denser, turns out to be rather simple. Consider the 3-dimensional sphere packing with lattice generated by the three edges emanating from O of a regular tetrahedron with one of its vertices at O. Let d denote the diameter of the spheres of the packing; of course d will then also be the length of an edge of the generating tetrahedron. The fundamental parallelepiped of the lattice is comprised of the generating tetrahedron $OABC$, the tetrahedron $O'A'B'C'$ opposite, and the octahedron $ABCA'B'C'$ in between, as shown in Figure 1. Now embed the 3-space of this regular sphere packing as a hyperplane of Euclidean 4-space, and consider the 4-spheres having the 3-spheres of the given packing as equators. These 4-spheres will then form a layer in 4-space, with centres lying in the original 3-dimensional hyperplane and coinciding with the points of the original 3-dimensional lattice. If we now place on this layer of 4-spheres an identical one (obtained as the image of the original under an appropriate translation) so that its spheres sit in the "pits" at the centres of the tetrahedra $OABC$, $O'A'B'C'$, and continue placing further layers in this way, we will finally obtain the lattice packing of 4-space by 4-spheres determined by the regular 4-simplex. However if we place each successive layer so that its 4-spheres sit

---

[32] If $\mathfrak{A}_1, \ldots, \mathfrak{A}_n$ are the edges of a regular n-simplex of side 1 attached to the origin O of Euclidean n-space, then $\mathfrak{A}_i \cdot \mathfrak{A}_i = 1$ and $\mathfrak{A}_i \cdot \mathfrak{A}_j = \cos 60^\circ = \frac{1}{2}$ for $i \neq j$, whence the associated form $(x_1\mathfrak{A}_1 + \cdots + x_n\mathfrak{A}_n)^2 = U_n$. For example for $n = 3$, one has

$$\begin{aligned} &x_1^2 + x_2^2 + x_3^2 + x_1x_2 + x_1x_3 + x_2x_3 \\ &= \left(x_1 + \frac{1}{2}x_2 + \frac{1}{2}x_3\right)^2 + \left(\frac{\sqrt{3}}{2}x_2 + \frac{\sqrt{3}}{6}x_3\right)^2 + \left(\frac{\sqrt{6}}{3}x_3\right)^2, \end{aligned}$$

whence

$$\mathfrak{A}_1 = (1,0,0), \quad \mathfrak{A}_2 = (1/2, \sqrt{3}/2, 0), \quad \mathfrak{A}_3 = (1/2, \sqrt{3}/6, \sqrt{6}/3),$$

which represent three of the edges of a regular tetrahedron attached to O. *Trans.*

instead in the "pits" at the centres of the octahedral portion (like $ABCA'B'C'$) of each fundamental parallelepiped of the preceding layer, then we obtain a different, and, as we shall now show, denser packing (in fact that corresponding to the form V_4). In the first case the height h_T of the centre of each sphere in the second layer (the one being placed), above the hyperplane containing the centres of the spheres of the first layer, is given by $h_T = \sqrt{d^2 - R_T^2}$,[33] where R_T denotes the radius of the sphere circumscribed about the tetrahedron $OABC$. Since $R_T = \frac{d\sqrt{6}}{4}$, we obtain $h_T = \frac{d\sqrt{10}}{4}$. On the other hand for the second type of placement of one layer on another, the height h_O of the centre of each sphere in the layer being placed, above the hyperplane through the centres of the spheres of the first layer, is $\sqrt{d^2 - R_O^2}$,[34] where R_O is the circumradius of the octahedron $ABCA'B'C'$. Since $R_O = \frac{d\sqrt{2}}{2}$, we have $h_O = \frac{d\sqrt{2}}{2}$. Hence $h_O < h_T$, so that, clearly, the second type of placement of successive layers of 4-spheres yields a lattice packing of 4-spheres denser than the first, determined by the lattice generated by the regular 4-simplex. It is important to note also that since $h_O > \frac{d}{2}$, the 4-spheres on either side of the "pit" in the middle of each octahedron will not even touch, let alone overlap, since they will lie wholly on opposite sides of the hyperplane determined by the centres of the spheres of the intermediate layer. Anticipating that this packing corresponds to the form V_4 of Korkin and Zolotarev, we shall denote it also by V_4 (and likewise for other sphere packings).

Our aim now is to prove geometrically that the above packing V_4 is indeed entitled to this label, i.e., that it is the densest lattice packing of 4-spheres in 4-space. However we first prove that the densest lattice packing of spheres in 3-space is that with fundamental parallelepiped as in Figure 1, i.e., generated by a regular tetrahedron, since the proof for the packing V_4 is in large part similar.

A. *The* 3-*dimensional case.*

1) If we have a lattice packing of identical spheres in space where neighbouring spheres do not touch, then obviously the packing is compressible, since we can compress the lattice uniformly in a direction parallel to one of the three generating vectors, towards any fixed sphere; this will compress the fundamental parallelepipeds in one direction while leaving the spheres nonoverlapping. Hence every lattice packing can be continuously compressed into one made up of rows of spheres with neighbouring spheres touching.

2) If there are neighbouring such rows that do not touch, then again we can compress the lattice along an axis parallel to another of the generating vectors of the lattice, towards any fixed row. Hence every lattice packing may be compressed continuously into one made up of parallel layers each of which consists of spheres with coplanar centres arranged in rows with neighbouring spheres touching and with neighbouring rows of spheres touching all along their length.[35]

3) Next, if there are neighbouring such layers not touching, then a third uniform compression parallel to the remaining generating vector of the lattice towards any fixed layer will bring neighbouring layers into contact. Hence every lattice sphere

[33] To see this choose a suitable plane section of the two layers, passing through the centre of the sphere of the second layer nestling above the pit in the middle of a tetrahedron of the first layer, and the centre of that tetrahedron, and then apply Pythagoras' theorem to an appropriate right triangle. *Trans.*

[34] See the preceding footnote, with "tetrahedron" replaced by "octahedron". *Trans.*

[35] Note that it is not assumed that each layer constitutes a planar closest packing. *Trans.*

packing in space can be continuously compressed into one made up of layers of spheres as in paragraph 2) with neighbouring layers touching.

4) Each layer then consists of a planar lattice with spheres centred at its points, and with fundamental region a rhombus with side equal to the diameter d of the spheres. Each point of the plane of this lattice is at a distance of at most $\frac{d\sqrt{2}}{2}$ from some point of the lattice, i.e., from the centre of some sphere of the layer, since in the "worst" case, namely when the rhombus is a square, the centre is exactly this distance from its corners. If we draw a circle of radius $\frac{d\sqrt{2}}{2}$ about each point of the lattice as centre, then in the case where the fundamental rhombus is square every point of the plane will lie in the interior of some circle except for the centres of the fundamental squares. If we then deform the square into a rhombus, two of the four circles that formerly intersected at the square's centre will now overlap, so that the the former centres of the squares become interior points of the these two circles.

5) It follows that two layers can be in contact only if they are adjacent, in the sense that there is no layer whose plane (i.e., the plane through the centres of its spheres) lies between their planes. For by paragraph 4) the height h of the centre of any sphere not in a given layer, above the plane of that given layer, is at least

$$h = \sqrt{d^2 - \left(\frac{d\sqrt{2}}{2}\right)^2} = \frac{d\sqrt{2}}{2},$$

whence $h > d/2$, so that no two layers can make contact through any of the gaps between the spheres of a layer lying between them.

6) Consider a layer I and another layer II lying directly on top of it. We now slide (or roll) the top layer II over layer I, attempting to seat the spheres of layer II as deeply as possible in the "pits" or interstices between the spheres of layer I. In order to make the process clearer, consider the spheres of layer I expanded about their centres to twice their radius. These spheres will of course now overlap one another, and by paragraph 4) there will no longer be any holes between them. However their tops will be visible from above; projected vertically onto the plane of the centres of layer I, each such top will become a Dirichlet region[36] P of the plane lattice consisting of the centres of the spheres of layer I (see Figure 2). Thus the action of sliding of layer II over layer I amounts in essence to that of sliding the plane of the centres of the spheres of layer II over the tops of these expanded spheres. In this process it suffices to consider the motion of just one such centre C, say, since the others can clearly be simultaneously situated analogously in view of the periodicity of the structure of the layers. Thus it suffices, in fact, to consider the motion of the projection $\bar{C}$ of the centre C on the plane of centres of layer I, inside a single polygon P.

7) If $\bar{C}$ lies in the interior of the polygon P, then we move it along a ray emanating from the centre of P until it reaches the boundary of P. Clearly, during this procedure the point C above $\bar{C}$ will move to a lower level (since the tops of the expanded spheres are convex upwards), and therefore the packing will become denser. If we next move $\bar{C}$ along an edge of P, then certainly in at least one direction along that edge $\bar{C}$ will move further away from the centre of P, so that

[36]The *Dirichlet region* around each point Q of the lattice is comprised of those points of the plane whose distance from Q is less than or equal to their distance from any other point of the lattice. *Trans.*

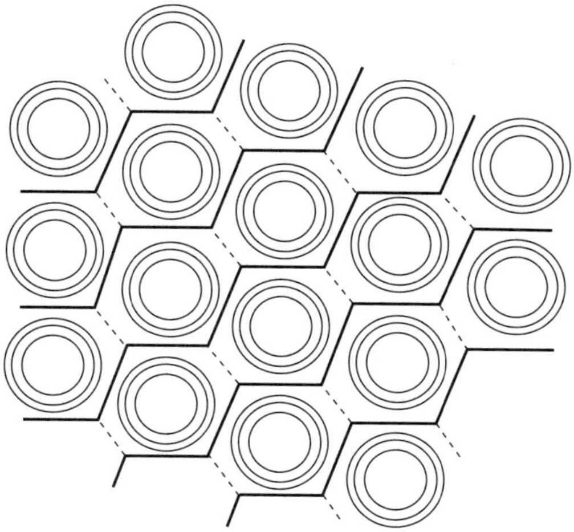

FIGURE 2

again the packing will become progressively denser till $\bar{C}$ arrives at a vertex of the polygon P.

8) The polygon P is in fact a hexagon (see for example the final appendix to Dirichlet's course in number theory)[37], with the two pairs of adjacent edges shown in bold in Figure 2 indicating where in layer I neighbouring spheres touch. (This becomes clear if one constructs the Dirichlet regions using Kelvin's method involving expanding circles.) We thus see that when $\bar{C}$ reaches a vertex of P, the sphere with centre C will be in contact with two (unexpanded) spheres of layer I touching each other. These three spheres together form one equilateral triangle of spheres of another layer (based, as are all layers, on the tesselation of its plane into equilateral triangles, and representing the densest planar packing), at an angle to layer I.

Thus by simply sliding layer II over layer I (and analogously for other pairs of layers) in a manner continuously increasing the density of the lattice packing, we arrive at a packing containing a layer based on the equilateral triangle of side d, i.e., most densely packed.

9) Take that layer now as the initial one (layer I). The hexagons will then be regular, and on successively positioning adjacent layers as above,[38] with the spheres of each layer nestled in the pits of its neighbours, we finally obtain the regular packing of 3-spheres, based on the regular tetrahedron.

10) To summarise: In paragraphs 1)–9) we have demonstrated that, starting from any lattice packing of identical nonintersecting spheres in space, we can, by means of a procedure throughout which the density of the packing increases, always arrive finally at the packing corresponding to the form U_3, i.e., the regular packing based on regular tetrahedra. Hence the latter packing is the unique incompressible packing, and therefore the densest.

[37]This is possibly: P. G. L. Dirichlet, "Vorlesungen über Zahlentheorie", with supplements by R. Dedekind and F. Vieweg; *Braunschweig*, Fourth Edition, 1894 (*Chelsea*, 1968). *Trans.*

[38]The layers parallel to this initial one must all be identical with it, since the packings are, throughout, lattice packings. *Trans.*

We now turn to

B. *The 4-dimensional case.*

1) Just as in paragraphs 1)–3) of the 3-dimensional case above, one shows that any lattice packing of Euclidean 4-space with identical nonintersecting 4-spheres can be continuously compressed so as to obtain in it a 3-dimensional layer of 4-spheres whose fundamental parallelepiped has edges of equal length, namely d, the diameter of the spheres. Then, as in paragraph 4) of the 3-dimensional case, it follows that every point of the 3-dimensional hyperplane determined by the centres of the spheres of this layer, lies not farther than $\frac{d\sqrt{3}}{2}$ (half the diagonal of a cube of side d) from one of those centres. Hence, much as in paragraph 5) of the 3-dimensional case, we see that two such 3-layers can at most just touch through the holes in a layer between them, since in the notation of that paragraph, $h \geq d/2$, with equality possible.

Remark. Beginning with 5-dimensional space, it can happen that h is less than $d/2$, so that layers on either side of an initial layer can prevent one another from fitting snugly into the "pits" of that intermediate layer. It is in fact just this possibility that lies at the root of the difficulties encountered in the theory of densest lattice sphere packings in higher-dimensional spaces.

2) Consider any such 3-layer, denoted by I, and an adjacent one II. If these did not touch, then we could increase the density of the packing by means of a uniform compression towards the 3-dimensional hyperplane of the centres of layer I. Assuming they do touch, as before we slide layer II over layer I so that the spheres of layer II sit in the "pits" of layer I, in order to increase the density of the packing. In order to make it clear just how this is to be achieved, we consider, as in paragraph 6) of the 3-dimensional case, auxiliary 4-spheres of double the radius with centres coinciding with those of the spheres of layer I (i.e., we blow the latter up to double the diameter), and project their tops, i.e., those portions not intersecting with neighbouring such spheres, perpendicularly onto the hyperplane of their centres, obtaining again the tesselation of that hyperplane into the polyhedral Dirichlet regions of the lattice consisting of those centres. Then, denoting by C the centre of any fixed sphere of layer II and by $\bar{C}$ its projection onto the aforementioned hyperplane, one shows, essentially as in paragraph 7) of the 3-dimensional case, that there is a way of sliding layer II over layer I so that the density of the original packing continuously increases, while $\bar{C}$ moves continuously within a Dirichlet region D to end at a vertex of D.

3) It is known that the region D can have at most 7 pairwise nonparallel faces. If one considers, via Kelvin's method of expanding spheres, the process of formation of the region D, it becomes clear that at three pairs of faces the region (or cell) D abuts with cells whose (unexpanded) spheres touch the (unexpanded) sphere associated with D, since in layer I each sphere is in contact with three pairs of spheres. Now in the case where D has 14 faces, each of its vertices is incident with six out of a possible seven nonparallel faces, and, in all other cases, with a full set of nonparallel faces. Hence the point $\bar{C}$, now at a vertex of D, will certainly lie on a face corresponding to two touching spheres. However, this means that the sphere of layer II whose centre C lies above $\bar{C}$, will now be in contact with these two spheres, so that as before we have a triple of spheres each in contact with the other two. As in paragraph 8) above, such a triple must form part of a 2-layer based on the equilateral triangle of side d, i.e., with the centres of its spheres forming a planar

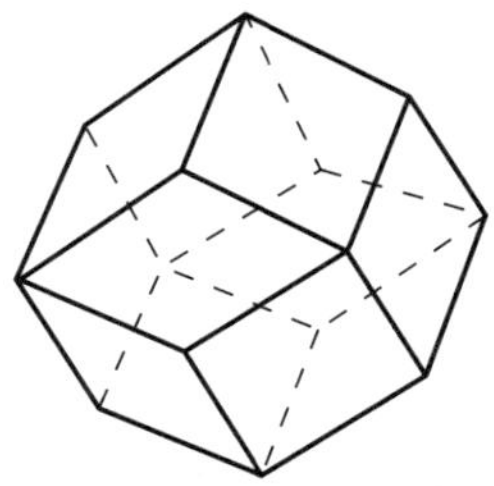

FIGURE 3

regular triangular lattice. Thus we have a regular 2-layer, based on the equilateral triangle, contained in the original 3-layer we were considering.

4) We now examine the 3-layer made up of the regular 2-layer just obtained together with the (identical) 2-layers parallel to it: this will be our new layer I. (Note that the 3-dimensional hyperplane determined by the centres of the spheres of this 3-layer will not necessarily be parallel to the original 3-layer I.) The (new) region D will have a subregion[39] corresponding to a regular 2-sublayer of layer I with faces just those faces of D arising from the spheres of that sublayer. Hence the spheres on opposite sides of these faces will touch, and the projection of this subregion on the plane of the sublayer will be a regular hexagon.

5) We now slide an adjacent layer, again denoted by II, over this new 3-layer I. Again we consider the centre C of an arbitrary sphere of layer II, and its projection $\bar{C}$ on the hyperplane determined by the centres of the spheres of layer I. As before we move C (uniformly with all of its "cognates", i.e., centres of the spheres of the layer II) so that $\bar{C}$ moves within D to a vertex, ensuring that throughout this process the density increases. However it turns out that by virtue of the fact that layers I and II are built up from regular 2-layers, all the vertices of D lie in the above-described subregion. (See Chapter III of "The mathematical fundamentals of the structural analysis of crystals", by Aleksandrov and Delone (Moscow–Leningrad, 1934 (Russian)).) It follows that when the projection $\bar{C}$ of the centre C of the 4-sphere of layer II is at any vertex of D, that 4-sphere is in contact with three 4-spheres of layer I in pairwise contact with one another, thus forming a regular tetrahedral structure. We conclude that any lattice packing of identical 4-spheres of diameter d in Euclidean 4-space can by means of compressions alone be transformed into a lattice packing containing four spheres whose centres are the vertices of a regular tetrahedron of side d, and therefore containing a 3-dimensional layer with fundamental parallelepiped determined by the regular tetrahedron of side d.

6) We now take the latter layer to be layer I. For this layer the region D turns out to be the rhombic dodecahedron shown in Figure 3. We carry out the usual procedure of sliding an adjacent such layer over this one (in such a way that the packing is progressively compressed) until each point $\bar{C}$ reaches a vertex of this dodecahedron.

However the vertices of the rhombic dodecahedron are of two types, namely of valency three and four. Those of valency three correspond to the centres of

[39] "Zone" in the original. *Trans.*

tetrahedra of the lattice I, and in this case we obtain the lattice packing of 4-spheres generated by the regular 4-simplex. On the other hand those of valency four correspond to the the centres of octahedra (see Figure 1), and here we obtain a completely well-defined but different lattice packing, shown at the beginning of this commentary to be denser.

7) Thus in paragraphs 1)–6) we have shown that an arbitrary lattice packing of 4-space with identical nonoverlapping 4-spheres can be continuously compressed into either that with lattice generated by the regular 4-simplex (corresponding to the quadratic form U_4) or else into the closest packing described above (corresponding to V_4).

8) Finally we remark that it can be shown geometrically that the packing corresponding to U_4 is incompressible. However, in fact this is better done analytically in the manner of Korkin and Zolotarev.

We now turn to the second memoir by Korkin and Zolotarev. Their approach there is based on the notion of the decomposition of a form along its successive minima. We shall now explain the geometric significance of such decompositions. As mentioned at the beginning of this commentary, such a decomposition of a given positive n-ary form corresponds to the choice of a special frame for the point lattice E of the form. We shall call the particular frame associated with the decomposition of a given form along its successive minima simply the *Korkin–Zolotarev frame* or, briefly, *KZ-frame*, of the form. This frame turns out to be the following one: The first vector is just a shortest among vectors from the origin to the points of the lattice E. The second vector is a shortest among those vectors of the lattice whose orthogonal projections onto the $(n-1)$-dimensional orthogonal complement of the first vector are shortest. The third is a shortest among those vectors of the lattice E with shortest orthogonal projections onto the $(n-2)$-dimensional orthogonal complement of the plane spanned by the first two, and so on. It is clear from this definition that a KZ-frame is indeed a generating frame for the lattice E. An alternative method of constructing a KZ-frame is as follows: Consider a sphere centred at the origin O, and let it expand until it first encounters a point A, say, of the lattice E. Next consider a cylinder with axis OA, and let its radius increase until it encounters a row of points of E parallel to OA. From this row of points choose one, B say, closest to O. Then let the 3-cylinder with axis the plane of OAB expand until it encounters a 2-dimensional sublattice of points of E, parallel to the plane of OAB, and choose a point C of that planar lattice closest to O, and so on. Then the ordered n-tuple of vectors $(\overrightarrow{OA}, \overrightarrow{OB}, \overrightarrow{OC}, \dots)$ will be a KZ-frame for the lattice E.

A Korkin–Zolotarev decomposition of a form f along its successive minima is essentially just a "Jacobi" decomposition of f. Of the various possible decompositions of a given positive quadratic form f as a sum of squares of linear forms (which, as noted above, correspond essentially to choosing different frames obtainable from each other by means of rotations—i.e., Euclidean motions—relative to a fixed rectangular system of coordinates $\xi_1, \xi_2, \dots, \xi_n$ in terms of which the frames are specified), the one in which the first squared term involves all n variables $x_1, x_2, \dots$, the second just the $n-1$ variables $x_2, x_3, \dots$, the third only the $n-2$ variables $x_3, x_4, \dots$, and so on (the $(n-1)$st involving only x_{n-1} and x_n, and the last only x_n), is called a *Jacobi* decomposition of f. If the indexing of the variables $x_1, x_2, \dots$ is fixed, then the Jacobi decomposition of the given form f is

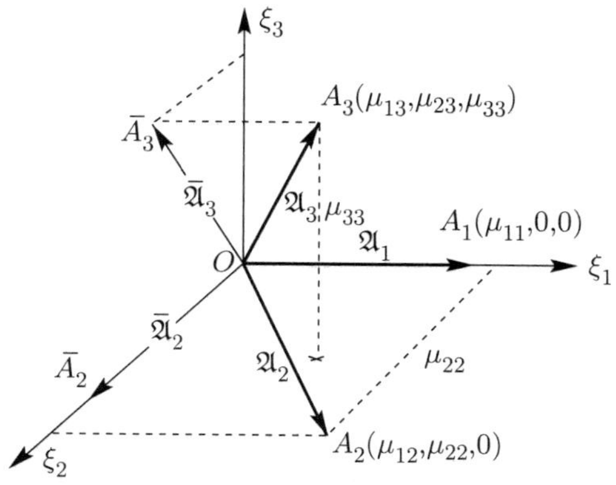

FIGURE 4

unique, say

$$f = (\mu_{11}x_1 + \mu_{12}x_2 + \cdots + \mu_{1n}x_n)^2 + (\mu_{22}x_2 + \mu_{23}x_3 + \cdots + \mu_{2n}x_n)^2 + \cdots \\ + (\mu_{n-1,n-1}x_{n-1} + \mu_{n-1,n}x_n)^2 + (\mu_{nn}x_n)^2.$$

The Jacobi decomposition of the form f corresponds to the rotation (i.e. Euclidean motion) of the frame of the lattice E of the form relative to the system of coordinates $\xi_1, \xi_2, \ldots, \xi_n$ in terms of which the frames are specified, so that the first vector $\mathfrak{A}_1$ of the frame is aligned with the ξ_1-axis, the second $\mathfrak{A}_2$ is in the $\xi_1\xi_2$-plane, the third $\mathfrak{A}_3$ in the 3-dimensional $\xi_1\xi_2\xi_3$-subspace, and so on. In fact, according to the definition given near the beginning of this commentary, they are as follows (see Figure 4):

$$\begin{aligned} \mathfrak{A}_1 &= (\mu_{11}, 0, \ldots, 0, 0), \\ \mathfrak{A}_2 &= (\mu_{12}, \mu_{22}, 0, \ldots, 0), \\ &\vdots \qquad \vdots \\ \mathfrak{A}_{n-1} &= (\mu_{1,n-1}, \mu_{2,n-1}, \ldots, \mu_{n-1,n-1}, 0), \\ \mathfrak{A}_n &= (\mu_{1,n}, \mu_{2,n}, \ldots, \mu_{n-1,n}, \mu_{nn}). \end{aligned}$$

Thus if we delete the first square in the above Jacobi decomposition of f, then what remains is an $(n-1)$-ary quadratic form in the variables $x_2, x_3, \ldots, x_n$, with frame comprised of the orthogonal projections $\bar{\mathfrak{A}}_2, \bar{\mathfrak{A}}_3, \ldots, \bar{\mathfrak{A}}_n$ of $\mathfrak{A}_2, \mathfrak{A}_3, \ldots, \mathfrak{A}_n$ onto the $(n-1)$-dimensional subspace determined by the ξ_i-axes, $i = 2, 3, \ldots, n$. If in addition we delete the second square, then the remaining quadratic form in the $n-2$ variables $x_3, x_4, \ldots, x_n$, has as frame the orthogonal projections $\bar{\bar{\mathfrak{A}}}_3, \bar{\bar{\mathfrak{A}}}_4, \ldots, \bar{\bar{\mathfrak{A}}}_n$ of $\mathfrak{A}_3, \mathfrak{A}_4, \ldots, \mathfrak{A}_n$ onto the $(n-2)$-dimensional subspace determined by the ξ_j-axes, $j = 3, 4, \ldots, n$, and so on.

It is clear from this that $|\mu_{11}|, |\mu_{22}|, \ldots, |\mu_{nn}|$ are the successive dimensions of the parallelepiped determined by the vectors $\mathfrak{A}_1, \mathfrak{A}_2, \ldots, \mathfrak{A}_n$, i.e., $|\mu_{nn}|$ is the height of the whole parallelepiped, $|\mu_{n-1,n-1}|$ the height of its base, and so on, until $|\mu_{11}|$, which is the length of an edge.

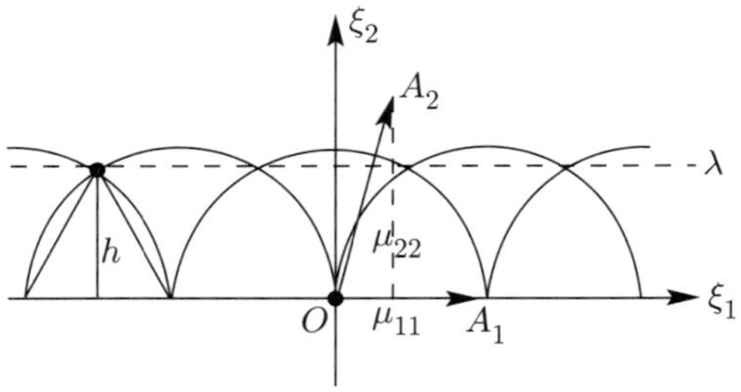

FIGURE 5

As noted above, given a fixed ordering of the variables, the corresponding Korkin–Zolotarev decomposition along the successive minima of a given positive form is essentially just the Jacobi decomposition. However, for Korkin and Zolotarev the point of this decomposition was not to simply furnish a convenient frame for the lattice E of the form, but its use as an aid in investigating the minima of positive forms. In terms of the Jacobi coefficients μ_{ik} the successive minima $A, A', A'', \ldots$ are respectively $\mu_{11}^2, \mu_{22}^2, \mu_{33}^2, \ldots$. It is geometrically obvious that the $(n-1)$-tuple $(\bar{\mathfrak{A}}_2, \bar{\mathfrak{A}}_3, \ldots, \bar{\mathfrak{A}}_n)$, the $(n-2)$-tuple $(\bar{\bar{\mathfrak{A}}}_3, \bar{\bar{\mathfrak{A}}}_4, \ldots, \bar{\bar{\mathfrak{A}}}_n)$, and so on, are on the one hand KZ-frames of the corresponding lattices of dimensions $(n-1), (n-2)$, and so on, while on the other they are situated *à la* Jacobi in the respective subspaces of $\mathbb{R}^n$ determined by the appropriate subsets of the ξ_i-axes.

If, working in the reverse order, we set $x_n = 0$, or $x_n = 0, x_{n-1} = 0$, etc., then the last vector $\mathfrak{A}_n$ falls out of reckoning, or the last two $\mathfrak{A}_n, \mathfrak{A}_{n-1}$, etc., while the frames $(\mathfrak{A}_1, \mathfrak{A}_2, \ldots, \mathfrak{A}_{n-1})$, or $(\mathfrak{A}_1, \mathfrak{A}_2, \ldots, \mathfrak{A}_{n-2})$, etc., remain KZ-frames situated *à la* Jacobi in the subspaces determined by the appropriate subsets of the ξ_i-axes.

The result of Korkin and Zolotarev that $A' \geq \frac{3}{4}A$, or, equivalently, $\mu_{22} \geq \frac{\sqrt{3}}{2}\mu_{11}$, is easy to see geometrically. In Figure 5, the two vectors $\overrightarrow{OA_1}, \overrightarrow{OA_2}$ are meant to be the first two vectors of the KZ-frame of f. By the definition of a KZ-frame given above, the point A_2 must lie above the interior of the strip between the ξ_1-axis and the line λ with height $h = \frac{\sqrt{3}}{2}\mu_{11}$, the height of the equilateral triangle of side μ_{11}.

The theorem of Korkin and Zolotarev that for any positive ternary form one has $A' \geq \frac{3}{4}A, A'' \geq \frac{2}{3}A$, for which they give a rather complicated proof, can likewise be proved geometrically. The result proved at the end of their memoir that the inequality $A^{\text{iv}} \geq \frac{4}{9}A$ is not exact for any positive form is however more subtle.

We now turn to the third memoir of Korkin and Zolotarev. The proof that the forms U_n, V_n, and Z are extremal is better carried out analytically as they do. However, there is a more natural geometrical explanation of their theorem that an n-ary extremal form has at least $\frac{n(n+1)}{2}$ representations of its minimum, and that these completely determine the coefficients of the form. (Recall that a *representation of the minimum* of a form is an ordered n-tuple of (integer) values of the variables at which the form attains its minimum.) Instead of considering spheres of radius $\frac{d}{2}$ centred at the points of whatever lattice is being investigated (where d is the least distance between points of the lattice), we now consider a single sphere of radius d centred at O. The problem then consists in finding those incompressible lattices with the centre O of this sphere as one of their points, but with no other points inside it, i.e., lattices whose fundamental parallelepipeds cannot be decreased in volume by any infinitesimal perturbation without causing at least one point of the lattice to enter the sphere. From this point of view the idea behind the theorem of Korkin and Zolotarev in question may be considered to be as follows: If the points of the lattice lying on the sphere do not fully determine the lattice, then it is natural to expect that the lattice may be altered ever so slightly so as to reduce the volume of the fundamental parallelepiped while keeping those points on the surface of the sphere and preventing any other points from entering the sphere. Since the presence of each point lying on the sphere (and therefore constituting a representation of the minimum of the corresponding form) imposes a single condition on the structure of the lattice, which is defined by generating vectors from O to nearby points, and since the n generating vectors of an n-dimensional lattice require $\frac{n(n+1)}{2}$ parameters for their specification up to a rotation, the theorem of Korkin and Zolotarev becomes natural.

The theorem stating that the characteristic determinants of a positive n-ary form are bounded in terms of n alone is easily proved by geometric means (see the exposition below of the work of Voronoĭ on perfect forms). Moreover it turns out to be easy to infer geometrically, as is done by Voronoĭ, the consequence that one can bound in terms of n also the number of rows of the matrix of representations of the minimum, and even the entries of the reduced matrix in the case of an extremal form. As a corollary one obtains the result that for each n there is a certain finite set of forms, which may in principle be written down, containing a full set of inequivalent extremal ones. And since, as Korkin and Zolotarev have shown at the beginning of their third memoir, it is an easy matter to determine if an arbitrary given form is extremal or not, it follows that one may find in a finite number of steps all extremal forms of given dimension n. It seems, however, that Korkin and Zolotarev failed to notice that the entries of the reduced matrix of representations of the minimum are bounded in terms of n; this was left to Voronoĭ to discover.

However, Korkin and Zolotarev's analysis of the table of representations of the minima of extremal quinary forms is very complex, and hitherto no one has been able to find by geometric means all extremal quinary forms.

In 1925 and 1926 the American mathematician Blichfeldt, using the methods of Korkin and Zolotarev, found the largest minima of forms in six and in seven variables. In more recent times[40] this has been done also for forms in eight and nine variables.

[40]That is, just prior to 1947. *Trans.*

Egor Ivanovich Zolotarev (1847–1878)

Egor Ivanovich ZOLOTAREV

E. I. Zolotarev was born in 1847 in St. Petersburg to Ivan Vasilievich and Agafya Izotovna Zolotarev. According to his passport lower middle class from Kiev, his father was a master watchmaker. At age ten E. I. was sent to a gymnasium, graduating in 1863 with the silver medal. At that time he was already attending lectures at Petersburg University on a voluntary basis, and a year later he was enrolled as a student in the mathematics section of the faculty of physics and mathematics. The brilliant lectures of Chebyshev and Korkin, together with their advice, exerted a profound influence on the development of Zolotarev's remarkable ability. In the autumn of 1867, having passed the examination and presented his essay "On the integration of the equations of motion of a top", he was awarded the degree of "kandidat".[41]

The next year he defended his dissertation *pro venia legendi* on the topic "On a certain question concerning minimum quantities". On the basis of this dissertation and two trial lectures, one on Abel's theorem and the other on the reduction of an elliptic function to normal form, Zolotarev was at the age of 21 permitted to give lectures as *Privatdozent.* He was directed to lecture on elliptic function theory to mathematics students and on differential calculus to physics students.

In the spring of 1869 he passed the examination for the degree of *Magister*, and towards the end of that year defended his dissertation "On an indeterminate equation of the third degree". His official opponents were Chebyshev and Sokhotskiĭ. At that defense Korkin voiced strong objections; however, already by the following year he and Zolotarev had formed a friendship, which continued uninterrupted until Zolotarev's tragic death. In 1870 they began their collaborative work on the minima of positive quadratic forms. At the same time Zolotarev was doing research on the theory of algebraic numbers, resulting in his dissertation "The theory of complex integers with an application to the integral calculus", at the defense of which Chebyshev and Korkin officiated as opponents. In 1876 he was appointed professor at the university, and in the same year, following on the death of Somov, adjunct of the Academy of Sciences in the department of applied mathematics.

Zolotarev made two official visits abroad, in 1872 and 1876, visiting Berlin, Paris, Heidelberg and Leipzig. He attended the lectures of Kummer and Weierstrass, and held discussions with Hermite, who knew and highly esteemed the research of Korkin and Zolotarev on quadratic forms. In a letter to Borchardt Hermite writes: "Deux géometres russes extrêmement distingués, M. Korkine et M. Zolotareff, ont récemment publié dans les Mathematische Annalen des recherches approfondies."

[41]The modern "kandidat" degree is more like the Ph.D. *Trans.*

On the 26th June, 1878, Zolotarev took the morning train from St. Petersburg travelling on the Warsaw line to the station Siverskaya, where relatives were staying at their summer cottage. At Aleksandrovskaya station he alighted and when the train started off again he fell under the engine. He was extricated from under the wheels with crushed left foot and right leg broken above the knee, and transported back to St. Petersburg to Aleksandrovskaya hospital, where he died after 12 days of agony.

Zolotarev lived for only 31 years, but in that short life-span he managed to write several articles of the first importance. It cannot be doubted that, had his life not been so soon cut short, he would have enriched science with many more remarkable works.

The extensive correspondence (64 letters) between Korkin and Zolotarev, dating chiefly from the period of their joint work on quadratic forms, may be of interest. We reproduce here some of this correspondence.

From a letter dated 7th December, 1871:

"Dear Egor Ivanovich! I have, it seems, found a general form giving the largest minimum

$$\sqrt[n]{2^{n-2}D}.$$

I need only to convince myself that it is [positive] definite. I am busy decomposing it as a sum of squares. I'll let you know when it's all ready. Your *A. Korkin*."

From letter No. 4, dated 8th December, 1871:

"Dear Egor Ivanovich! The form in n variables attaining the largest minimum $\sqrt[n]{2^{n-2}D}$ may be expressed very simply. It is as follows:

$$a(x_1^2 + x_2^2 + \cdots + x_n^2 + x_1x_3 + x_1x_4 + \cdots + x_2x_3 + \cdots + x_{n-1}x_n);$$

here all n squares are present, and of the products only x_1x_2 is absent. *A. Korkin*."

From letter No. 6, dated 13th January, 1872:

"Dear Aleksandr Nikolaevich! Wishing to find a general proof of the theorem of interest to us, I have simplified the proof in the case of four variables as far as I could... [The letter continues with a proof for $n = 4$ almost word for word as in their first memoir on quadratic forms of 1872, reproduced above.] Your devoted *Zolotarev*."

From letter No. 8, dated 22nd March, 1872:

"Dear Egor Ivanovich! It seems I have found a proof of the bound for the minimum for forms with five or more variables. At least I have the most reliable general methods for this purpose. I am writing this a quarter of an hour before I go to give a lecture. Why not drop by tomorrow, Thursday, in the evening, to talk about this? Your *A. Korkin*."

From letter No. 9, dated 6th April, 1872:

"Dear Egor Ivanovich! The theorem that I sent to you is not true, as may be seen by means of the simplest of examples. I thought it was correct, and felt pressed to send it to you because I had to hurry to my lecture. In general extremal forms are difficult things, and it would be worthwhile studying them more closely. When you think of visiting me, could you not let me know beforehand so that I stay at home? Your *A. Korkin*."

These are followed by long letters, some several pages in length, full of descriptions of their work on forms, both from Korkin to Zolotarev, and *vice versa.*

Letter No. 27 is of especial interest, since in it Zolotarev communicates to Korkin for the first time his research on algebraic numbers. We reproduce this letter here in its entirety.

"Dear Aleksandr Nikolaevich! I am writing to you without waiting for your answer to my last letter because tomorrow I am leaving again for Ropsha for two weeks. I had promised to tell you everything I have managed to achieve concerning the generalization of Chebyshev's methods to the case where the coefficients of the polynomial under the radical sign are arbitrary real numbers. I have, in fact, been able to carry out this generalization using certain new notions that I want to tell you about in a few words in this letter. There are only a few exceptional cases, which, however, it would be quite appropriate to call special.

I begin by reducing any differential of the form

$$\frac{(x+A)dx}{\sqrt{Rx}},$$

where Rx denotes an arbitrary 4th degree polynomial in x with real coefficients, to the form

$$(1)\qquad \frac{(x+A)dx}{\sqrt{x(x-1)(x-\alpha)(x-\beta)}},$$

where α and β are certain real parameters satisfying $1<\alpha<\beta$.

A necessary, though not sufficient, condition for such a differential to be integrable in terms of logarithms is that the parameters α and β satisfy an equation

$$(2)\qquad F(\alpha,\beta)=0,$$

where F denotes an entire function with integer coefficients. Thus if the parameters α and β are not related by at least one such equation, the given differential will not be integrable in terms of logarithms.

Suppose now that between α and β there is just one relation of the form (2). If the integral

$$\int \frac{(x+A)dx}{\sqrt{x(x-1)(x-\alpha)(x-\beta)}}$$

is integrable in logarithms, then it will be equal to

$$C\ln\left(P+Q\sqrt{x(x-1)(x-\alpha)(x-\beta)}\right),$$

where C is a constant, and P and Q are entire functions satisfying the equation

$$P^2-Q^2x(x-1)(x-\alpha)(x-\beta)=\text{constant}.$$

Of all solutions of this equation one may choose that with P of least degree n.[42] This assumed [of P] I exhibit an upper bound for n depending on the degree of the equation (2).

Finally, if α and β satisfy several equations of the form (2), then each of α and β must satisfy equations of the from

$$\phi(\alpha)=0,\qquad \phi_1(\beta)=0,$$

[42]So not just entire, but in fact a polynomial? *Trans.*

where ϕ and ϕ_1 are entire functions with integer coefficients. I show, and it is very easy to convince oneself of this, that then the parameters α and β can be expressed in terms of rational functions with rational coefficients depending on a single parameter λ satisfying an equation

$$F(\lambda) = 0 \tag{3}$$

with integer coefficients and leading coefficient 1. In this case, to integrate the differential (1) I use the same transformations as were used by Chebyshev when the coefficients of Rx are rational, except that I consider not the coefficients of the polynomials under the radical sign, but rather their roots. In order to establish the finiteness of the operations in this case I had to invent a new theory of complex numbers dependent on the roots of equation (3), which may be assumed irreducible. I ascribe much greater significance to this theory than to the generalization of Chebyshev's method.

Up to the present time only complex numbers dependent on roots of unity have been analysed. This was carried out by Kummer in his numerous impressive articles on this topic; however, the means used by Kummer to deal with these numbers hold out little hope of generalizing his theory to cover all irreducible equations with integer coefficients. Indeed he exploited the circumstance that the equations of subdivision of the circle can be reduced by means of properties of their periods, i.e., solved using equations of lower degrees.

My theory contains Kummer's as a special case yet is not at all more complicated than his, provided one excludes certain special functions $F(\lambda)$ requiring further generalizations.

I have not as yet investigated these special cases, and that is why I said that in my generalization of Chebyshev's method I have encountered certain exceptions.

This theory of complex numbers interests me greatly. It is so extensive that I cannot expound it all in a letter to you, and therefore am waiting till I see you to tell you about it in detail.

The properties of entire functions as expounded in the chapter 'Propriétés des fonctions entières d'une variable relativement à un module premier' of Serret's algebra text, are fundamental to my theory. I have no doubt that this theory of complex numbers will find applications other than just to Chebyshev's method.

I hope you spend the rest of the summer pleasantly and that your health is improved, and please give my best to your mother. Your *E. Zolotarev.*"

This correspondence shows very clearly the closeness in mathematical interests between Zolotarev and Korkin over the last 8 years of Zolotarev's life.

Zolotarev's Memoirs on the Theory of Ideal Numbers

The Theory of Complex Integers with an Application to the Integral Calculus

Doctoral Dissertation

During the period of his collaboration with Korkin on their brilliant research on quadratic forms, Zolotarev published a substantial opus on the theory of ideal numbers. As is clear from the letter reproduced above, he was led to these investigations by his work on a question concerning the integrability of elliptic differentials in logarithms. However, the centre of interest of this work is certainly not in this application, but in the theory of ideal numbers which it stimulated, and which Zolotarev created at the same time as, but independently of, Dedekind, so that in fact Dedekind and Zolotarev should both be considered the creators of the theory of ideal numbers. (A third independent exposition of the theory must be credited to Kronecker, and a fourth to Klein.)

Zolotarev bases the divisibility theory of integral complex numbers, or, as they are now called, algebraic integers, on the theory of congruences, or, more precisely, on the set of all polynomials with integer coefficients considered modulo a fixed prime p. This theory was first created by Évariste Galois, though with a somewhat different emphasis, in his article "Sur la théorie des nombres", *Bulletin de Ferussac*, 1830, and in that form is usually called "Galois' theory of imaginaries". Later this theory was expounded in detail in Serret's course in higher algebra. From the modern point of view it belongs to the theory of finite fields.

Zolotarev's dissertation on ideal numbers is 250 pages long. For this reason and also by the nature of the material, we are compelled to confine ourselves mainly to summarizing its content, omitting nearly all proofs, while nonetheless adhering to the same basic principles as before.

The dissertation is divided into four chapters: I. "On functional congruences" (32 pages); II. "On complex units" (44 pages); III. "Ideal factors of complex numbers" (40 pages); and IV. "Applications of the theory of complex numbers to the integral calculus" (82 pages). These chapters are far from being of equal weight.

The first half of Chapter I contains the general theory of the decomposition of an integer polynomial as a product of prime factors modulo a prime p, and the second half an application of this theory in certain special cases, namely to the so-called "cyclotomic polynomials", i.e., the polynomials $\frac{x^n-1}{x-1}$ where n is an arbitrary prime.

The theory expounded in detail by Zolotarev in the first half of this chapter was in fact widely known at that time by virtue of its treatment in Serret's algebra text. Zolotarev begins his exposition by recalling that two polynomials

$$A = a_0 + a_1 x + a_2 x^2 + \cdots + a_m x^m,$$
$$B = b_0 + b_1 x + b_2 x^2 + \cdots + b_m x^m,$$

with integer coefficients a_i, b_i, are said to be *congruent modulo* p, where p is a given prime, in symbols

$$A \equiv B(\text{mod } p),$$

if the coefficients of the same powers of x are congruent modulo p :

$$\left.\begin{array}{ccc} a_0 & \equiv & b_0 \\ a_1 & \equiv & b_1 \\ a_2 & \equiv & b_2 \\ \vdots & \vdots & \vdots \\ a_m & \equiv & b_m \end{array}\right\} (\text{mod } p).$$

If $A \equiv B, C \equiv D, E \equiv F(\text{mod } p)$, then

$$\left.\begin{array}{ccc} A + C + E & \equiv & B + D + F \\ A - C & \equiv & B - D \\ ACE & \equiv & BDF \end{array}\right\} (\text{mod } p)$$

and so on.

If A and B are two integer polynomials in x and one can find integer polynomials C and D such that

$$BC = A + pD,$$

or, equivalently,

$$BC \equiv A(\text{mod } p),$$

then we say that A is *divisible by* B *modulo* p, or that B is a *divisor modulo* p of A.

In order to actually divide A by B modulo p,[43] where $B \not\equiv 0(\text{mod } p)$, we first need to delete all terms of A and B whose coefficients are divisible by p. The remaining coefficients may then be replaced if so desired by other integers (for instance the appropriate residues) congruent to them modulo p. Suppose that A' and B' result from changing A and B in this way, and let ax^k and bx^l be the leading terms of A' and B' respectively. If $k < l$ then B does not divide A modulo p. Suppose $k \geq l$. Since b is not divisible by p, there exists an integer β such that $b\beta \equiv 1(\text{mod } p)$. Then $a\beta x^{k-l}B'$ has leading term $a\beta b x^k \equiv ax^k$, so that $A'' = A' - a\beta x^{k-l}$ is either 0 or has degree less than k. One now repeats this procedure with A'' in place of A, obtaining A''', and so on, until eventually one obtains either an $A^{(n)} \equiv 0(\text{mod } p)$, in which case B divides A modulo p, or else an $A^{(n)}$ which after deletion of all terms with coefficients divisible by p has degree less than B, in which case B does not divide A modulo p.

If,[44] starting with two integer polynomials A and $B\,(\not\equiv 0)$, one carries out successive divisions modulo p as in the Euclidean algorithm, then after finitely

[43]There follows a description of the usual long division algorithm for polynomials over the field of p elements $\mathbb{Z}/p\mathbb{Z}$—or for that matter over any field, i.e., for finding polynomials Q and R with $R = 0$ or $\deg R < \deg B$ such that $A = QB + R$. *Trans.*

[44]What follows is a brief exposition of the unique factorization property of $(\mathbb{Z}/p\mathbb{Z})[x]$ following Euclid's proof of unique factorization of the integers into primes. A proof abstracted from this one yields the same conclusion for the much larger class of "Euclidean domains", which include polynomial rings $F[x]$ over any field F. See any current textbook on modern algebra. *Trans.*

many steps one arrives at a last remainder $\not\equiv 0(\text{mod } p)$, which will be the greatest common divisor mod p of A and B, unique modulo p up to multiplication by an integer not divisible by p (and so unique if one demands that it be monic). If the greatest common divisor is congruent to a constant modulo p, then it is taken to be 1, and the polynomials are *relatively prime.* Next one infers the following results: If A and B are relatively prime modulo p, then there exist integer polynomials P and Q such that $PA+QB \equiv 1(\text{mod } p)$; and if A and B are relatively prime and A divides BC modulo p, then A divides C mod p. Definition: A polynomial of degree ≥ 1 modulo p, i.e., with the coefficient of some positive power of x not divisible by p, is called *irreducible* mod p if it is not divisible by any polynomial of lesser positive degree mod p.[45] For example x^2+1, x^2+2 are irreducibles ($\equiv$ primes in this context) modulo 7. Since the integers not divisible by p are invertible mod p, multiplication by them does not affect divisibility; two integer polynomials P and λP, where λ is any integer not divisible by p, are thus not distinguished as far as the present divisibility theory is concerned, nor if they are congruent mod p,[46] and we may always choose λ so that the leading coefficient is 1. This discussion culminates in the proof of the Unique Factorization Theorem for integer polynomials mod p: Every such polynomial $A \not\equiv 0(\text{mod } p)$ can be expressed uniquely (up to multiplication of the factors by appropriate such λ) as a product of finitely many primes mod p:

$$A = P^n P_1^{n_1} P_2^{n_2} \cdots (\text{mod } p),$$

where $P, P_1, P_2, \dots$ are primes mod p.

Next the following theorem is proved for such an integer polynomial A: If at least one of the exponents $n, n_1, n_2, \dots$ is greater than 1, then A and its first derivative have a common factor modulo p of degree ≥ 1, i.e., A and its derivative are not relatively prime. In fact if a prime factor, say P, goes into A to a power $n > 1$, then provided p does not divide n, P goes into the derivative to the power $n-1$, while if p divides n, the derivative of A is divisible by P^n (all modulo p, of course).

In Sections 9–13 of Chapter I, Zolotarev proves the theorem of Galois stating that the polynomial

$$x^{p^\nu} - x \ (\text{mod } p)$$

is congruent modulo p to the product of all the prime polynomials mod p of degrees dividing ν. It follows that if $[k]$ denotes the number of prime polynomials mod p of degree k, then

$$p^\nu = k[k] + \delta[\delta] + \delta'[\delta'] + \cdots + [1],$$

where $\delta, \delta', \delta'', \dots, 1$ are all the proper divisors of ν. Thus for $\nu = 1$ we have $p = [1]$, for $\nu = 2$, $p^2 = 2[2] + [1]$, whence $[2] = \frac{p^2-p}{2}$, and so on. Continuing in this way one derives inductively a formula for $[k]$.

In order to factorize an arbitrary given integer polynomial A as a product of primes modulo p, one may first use the above-mentioned result on the common

[45]It follows from the result just preceding this definition that if an irreducible polynomial divides a product BC modulo p, then it must divide at least one of B, C. Hence an irreducible polynomial mod p is even *prime* mod p. *Trans.*

[46]In modern parlance P and λP are called *associates* (mod p). *Trans.*

divisor of A and its derivative in order to obtain by long division mod p a decomposition of A of the form

$$A \equiv V_1^{n_1} V_2^{n_2} \cdots (\text{mod } p),$$

where $V_1, V_2, \ldots$ do not have multiple prime factors mod p. The factorization of A as a product of primes is thus reduced to the case where A has no multiple prime factors.

Assuming this of A, we then have that the greatest common divisor D_1 (mod p) of A and $x^p - x$ is the product of all distinct prime factors of A of degree 1, i.e., of the form $x + a$, $a = 0, 1, \ldots, p - 1$. Writing $A \equiv D_1 A_1$ (mod p), we then have that the greatest common divisor D_2 (mod p) of A_2 and $x^{p^2} - x$ is the product of all the distinct prime factors of A of degree 2 (which we may assume monic), and so on. Continuing in this way we ultimately compute the factorization

$$A = aD_1 D_2 \cdots D_m \, (\text{mod } p),$$

where a is an integer not divisible by p, and D_i is the product of all distinct (monic) prime factors of A of degree i. For the final step, namely the factorization of each D_i as a product of distinct primes mod p all of degree i, Zolotarev uses the method of "indeterminate coefficients".

In Chapter II Zolotarev expounds Dirichlet's theorem on the units of the ring of algebraic integers of the (unique) form[47]

$$\phi(\xi) = a_0 + a_1\xi + a_2\xi^2 + \cdots + a_{n-1}x^{n-1}, \tag{1}$$

where $a_0, a_1, a_2, \ldots$ are ordinary (rational) integers, and ξ is a root of a fixed nth degree polynomial $F(x)$ with integer coefficients and leading coefficient 1, which is irreducible over the rationals, i.e., not divisible by any rational polynomial of smaller positive degree. Zolotarev calls the set of all numbers of the form (1) *whole complex numbers depending on the root* ξ *of* $F(x)$.

The ultimate goal that Zolotarev sets himself is to investigate for any given irreducible such F, as Kummer had done earlier for the special case when $F(x) = 0$ is the cyclotomic equation, how the number (1) decomposes as a product of irreducible factors of the same form (1).

It is clear that as the first step in investigating the decomposition of the algebraic integers $\phi(\xi)$ as products of irreducibles, one needs to characterise the "compulsory" factors, so to speak, i.e., the *units* of R (the invertible elements of R), since these divide every algebraic integer. In contrast with the situation in the ring of rational integers, where there are just the two units 1 and -1, Dirichlet showed in 1846 that in the ring of all numbers of the form $\phi(\xi)$ there will be infinitely many units if the number h counting the number of real roots plus the number of complex conjugate pairs of roots of $F(x)$, exceeds 1. Now obviously h is equal to 1 in only two cases:

1) when the polynomial $F(x)$ has degree one; and

---

[47]In modern notation this ring is denoted by $\mathbb{Z}[\xi]$, the ring of integer polynomials in ξ, forming a subring of the ring of all algebraic integers, i.e., of all roots of integer monic polynomials. It can happen that there are algebraic integers in the number field $\mathbb{Q}[\xi]$ other than those in $\mathbb{Z}[\xi]$—see below. Dirichlet's theorem applies to the full ring of algebraic integers in $\mathbb{Q}[\xi]$, not just to the possibly smaller ring consisting of those numbers of the form (1); however, in the present context Zolotarev restricts himself to the situation where all the algebraic integers of $\mathbb{Q}[\xi]$ are of the form (1). *Trans.*

2) when $F(x)$ has degree two with nonreal roots.

Thus in all other cases the ring R of algebraic integers of the form $\phi(\xi)$ has infinitely many units. The whole of Chapter II is devoted to proving Dirichlet's theorem, which, stated in full, asserts that each unit of R can be expressed uniquely in the form

$$\epsilon(\xi) = \omega(\xi) \cdot \sigma_1(\xi)^{m_1} \cdot \sigma_2(\xi)^{m_2} \cdots \sigma_{h-1}(\xi)^{m_{h-1}},$$

where $\omega(\xi)$ is one of finitely many so-called "special" units, $\sigma_1(\xi), \sigma_2(\xi), \ldots, \sigma_{h-1}(\xi)$ form a system of "basic" units, and $m_1, m_2, \ldots, m_{h-1}$ are arbitrary rational integers. In his remarks concerning this theorem Zolotarev writes: "This theorem is also due to Dirichlet. In his superb note (in the *Monatsberichte* in 1846) he communicated the main steps in the proof of this theorem. In remarks prompted by a letter from Hermite, Jacobi says: 'Le théoréme complet, savoir: Qu'il y a effectivement, dans tous les cas, $m' - 1$ unités [Jacobi uses m' instead of h] complexes independantes par les produits des puissances desquelles on puisse répresenter tous les autres, est un des plus importants mais aussi un des plus épineux [i.e. thorny, or difficult] de la science des nombres'."

The proof of Dirichlet's theorem given by Zolotarev in Chapter II, is different from that of Dirichlet. Zolotarev extends Hermite's method involving the introduction of a continuously varying parameter, and also uses Hermite's upper bound for the minimum of a positive quadratic form.

In connection with Chapter II, we observe finally that if, as usual, one calls the rational number

$$N(\phi(\xi)) = \phi(\xi)\phi(\xi_1)\phi(\xi_2)\cdots\phi(\xi_{n-1}),$$

where $\xi, \xi_1, \xi_2, \ldots, \xi_{n-1}$ are the n roots of the polynomial $F(x)$, the *norm* of the algebraic integer $\phi(\xi)$, then the units of the ring R will be just those numbers with norm ± 1. Note that in fact the norm $N(\phi(\xi))$ is a rational integer since it is a symmetric function over $\mathbb{Z}$ of the roots of the integral monic polynomial F, and so is expressible as an integral linear combination of its coefficients.

Thus to summarize, in Chapters I and II Zolotarev provides an exposition of results on polynomials over $\mathbb{Z}/p\mathbb{Z}$, and complex integers known considerably earlier, but needed by him in the sequel.

In Chapter III, the centrepiece of the work, Zolotarev creates the general theory of ideal numbers. In the introduction to the chapter he writes: "I expound here a theory of complex numbers depending on a root of an irreducible equation

$$F(x) = 0.$$

... in order to avoid complicating the theory, I exclude certain special functions $F(x)$,[48] which we characterize below and which we shall consider on another occasion. This theory is based on the properties of functional congruences expounded in Chapter I of this essay."

He reminds us of the history of the question. He writes: "It is well known that Gauss introduced into the theory of numbers the complex integers of the form $a + bi$ (where a and b are rational integers), i.e., complex numbers depending on the equation

$$x^2 + 1 = 0.$$

[48]The completion of Zolotarev's theory, embracing also these special functions $F(x)$, is contained in his articles of 1878–1885, to be examined below. *Trans.*

This innovation was of such importance that it provided the key to the solution of many extremely difficult problems and also led to new and remarkable generalizations."

He writes further: "The complex numbers of the form $a + bi$, which include the ordinary whole numbers as a particular case, are themselves a particular case of [integral] complex numbers depending on the roots of an irreducible [monic] equation with integer coefficients. Perhaps the theory of such complex numbers failed to be created because no question had arisen to which it might be applied."

Zolotarev goes on the remind the reader of Kummer's divisibility theory for algebraic integers depending on the root of a cyclotomic equation, which he used to prove the nonexistence of nontrivial integer solutions of Fermat's equation

$$x^\lambda + y^\lambda = z^\lambda,$$

for many values of λ. He recalls also Kronecker's use of Kummer's theory to prove the new and important algebraic result that all abelian equations with Gaussian integers (so including the ordinary integers) as coefficients, are in fact equations of the division of the circle (i.e., cyclotomic equations) or of the lemniscate.

Following on this, in Section 44, headed "Statement of the main problem, to whose solution this chapter is devoted", Zolotarev writes:

"Turning now to the theory proper of ideal factors of complex numbers, I consider it appropriate to make a few remarks about the problems whose solutions we have in view. The chief among these is the following one:

To discover whether one complex number divides another[49] *without actually carrying out the division, but by examining each of these numbers separately.*

For the Gaussian integers $a + bi$, a, b integers, just as for the rational integers, the solution of this problem lies in the decompositions of the numbers in question as products of primes. However, it turns out to be impossible to solve the problem in this way for a general ring R of complex integers of the form $\phi(\xi)$ without introducing some kind of new conditions and changes of approach. In fact although each nonzero number in R can be expressed as a product of irreducibles from R, i.e., as a product (empty if the given number is a unit) of nonzero, nonunit numbers divisible only by units and associates, it is possible for such a number to have two or more essentially different such decompositions, i.e., not transformable into one another merely by inserting units of R. For example if we take $F(x) = x^2 + 11$, that is, $\xi = \sqrt{-11}$, then we have

$$15 = 3 \cdot 5 = (2 + \sqrt{-11})(2 - \sqrt{-11}),$$

and it can be shown that in the ring $R = \mathbb{Z}[\sqrt{-11}]$ (i.e., of all numbers of the form $a + b\sqrt{-11}$, a, b ordinary integers) the numbers $3, 5, 2 + \sqrt{-11}, 2 - \sqrt{-11}$ are all (nonassociated) irreducibles, i.e., none of them is expressible as a product of two elements of R neither of which is a unit.[50]

"In the following sections," continues Zolotarev, "we shall expound a method of solution of the above-mentioned problem for [integral] complex numbers; we shall

[49]Presumably here, as nearly everywhere in the present chapter, "number" means "integer". *Trans.*

[50]However, the full ring of algebraic integers of $\mathbb{Q}[\sqrt{-11}]$, i.e., of roots of monic integer polynomials lying in this field, consists of all integral linear combinations of 1 and $\frac{1+\sqrt{-11}}{2}$, and this ring is in fact a Euclidean domain. With the advantage of hindsight, one might suggest as a more appropriate example $\mathbb{Q}[\sqrt{-5}]$, where the full ring of algebraic integers is indeed $\mathbb{Z}[\sqrt{-5}]$,

see that it yields for general [integral] complex numbers the same conclusions as does decomposition into products of primes for integers of the form $a + bi$ (and for rational integers), and may therefore serve for solving many other problems." In particular the theory of ideal numbers provides the answer to the question as to how many essentially distinct factorizations there are of a nonzero, nonunit number in R as a product of irreducibles from R, i.e., distinct in the sense that the irreducible factors of the decompositions cannot be paired off in such a way that paired irreducibles are mutual associates.

Sections 45–53, occupying 12 pages, constitute the kernel of the dissertation. We shall now examine these in order.

Section 45 is especially important as there the numbers of R are classified as "simple" or "compound" (corresponding to "prime" and "composite"). Zolotarev begins by defining an ordinary prime number p to be *simple* if $F(x)$ is a prime polynomial mod p, otherwise *compound*. This terminology is motivated by the following result:

If a product of numbers from R $(= \mathbb{Z}[\xi])$ is divisible by an ordinary prime number p with the property that $F(x)$ is a prime polynomial modulo p, then at least one of the factors is divisible by p, i.e., when divided by p yields a number from R.

Since the proof is straightforward, we reproduce it here. Suppose p divides $\phi(\xi) \cdot \psi(\xi) \cdot \chi(\xi) \cdots$ with quotient $B(\xi)$, where $\phi(\xi), \psi(\xi), \chi(\xi), \ldots$, and $B(\xi)$ lie in R. This means that there is an integer polynomial $c(x)$ such that

$$\phi(x) \cdot \psi(x) \cdot \chi(x) \cdots = p \cdot B(x) + F(x) \cdot c(x),$$ [51]

so that the product $\phi(x) \cdot \psi(x) \cdot \chi(x) \cdots$ is divisible by $F(x)$ modulo p, and then since $F(x)$ is prime mod p, it follows that it must divide (mod p) at least one of the factors $\phi(x), \psi(x), \chi(x), \ldots$. If for instance $F(x)$ divides $\phi(x) \mod p$, then $\phi(x) = F(x)G(x) + pT(x)$, for some integer polynomials $G(x), T(x)$, whence $\phi(\xi) = pT(\xi)$, since $F(\xi) = 0$.

Consequently, simple primes p are divisible in R only by their associates and units; for if $p = \phi(\xi) \cdot \psi(\xi)$, then p divides at least one of the two latter factors, say $\phi(\xi)$, i.e., $\phi(\xi) = p \cdot \phi_1(\xi)$, $\phi_1(\xi) \in R$, so that $1 = \phi_1(\xi) \cdot \psi(\xi)$, whence $\psi(\xi)$ and $\phi_1(\xi)$ are units.

Remark. On the other hand, if p is compound, i.e., if $F(x)$ is not prime mod p, then of course $F(x)$ has the form

$$F(x) = V^m \cdot V_1^{m_1} \cdots V_s^{m_s} + p \cdot F_1(x), \tag{2}$$ [52]

where $V, V_1, \ldots, V_s$ are prime mod p, $m \geq 1$ and $m_i \geq 1$, $i = 1, 2, \ldots, s$, and $m + m_1 + \cdots + m_s \geq 2$ (since $F(x)$ is not prime mod p). Zolotarev shows earlier on that if any of the exponents $m, m_1, \ldots, m_s$ is greater than 1, then p divides the discriminant D of the polynomial $F(x)$. He excludes from his treatment all such F, i.e., with discriminant divisible by primes p, with the additional property that $F_1(x)$ is divisible mod p by at least one of $V, V_1, \ldots, V_s$ appearing with exponent greater than 1. These are the "exceptional cases" mentioned earlier. Zolotarev

---

which is not a unique factorization domain since for example $2 \cdot 3 = (1 + \sqrt{-5})(1 - \sqrt{-5})$. (See any modern text on algebraic number theory, and also the later papers by Zolotarev examined below, and the commentary.) *Trans.*

[51] Since, in modern notation, $\mathbb{Z}[\xi] \cong \mathbb{Z}[x]/F(x)\mathbb{Z}[x]$. *Trans.*

[52] Here, as noted earlier, the polynomials $V(x), V_1(x), \ldots, V_s(x)$ are considered defined only mod p and to within multiplication by integers λ not divisible by p. *Trans.*

shows that if p divides D but is not exceptional in this sense, i.e., is not among the excluded prime divisors of the discriminant D, then in fact none of $V, V_1, \ldots, V_s$ will divide mod p the polynomial $F_1(x)$.

With respect to a rational prime p not excluded in this way, but for which $F(x)$ is not prime mod p, so that we have the situation of equation (2) above, with none of $V, V_1, \ldots, V_s$ dividing $F_1(x)$, i.e., a compound p of R amenable to Zolotarev's treatment, Zolotarev says that p is made up of m identical irreducible factors *belonging to* $V(x)$, m_1 irreducible factors *belonging to* $V_1(x)$, and so on.[53] Zolotarev calls these irreducible factors *ideal.* Given any algebraic integer $\phi(\xi)$ from R, Zolotarev defines the terminology $\phi(\xi)$ *is divisible by an ideal factor of* p *belonging to* V to mean that $V(x)$ divides $\phi(x)$ mod p, and similarly for $V_1(x), V_2(x), \ldots$.

Section 46, which is devoted to the definition of the degree to which an ideal factor V say, of p, occurs in a given algebraic integer $\phi(\xi)$ of R, is of equal importance. Denote by $W, W_1, \ldots, W_s$ respectively the products

$$V_1^{m_1} \cdot V_2^{m_2} \cdots V_s^{m_s},\ V^m \cdot V_2^{m_2} \cdots V_s^{m_s}, \ldots, V^m \cdot V_1^{m_1} \cdots V_{s-1}^{m_{s-1}}.$$

In Zolotarev's terminology, an algebraic integer of the form $\phi(\xi)$ *involves an irreducible ideal factor of* p *belonging to* V λ *times* if , writing $\lambda = km + r$, where k is the quotient and r, $0 \leq r < m$, the remainder after division of m by λ, we have

$$\phi(x) \cdot V^{m-r} \cdot W^{k+1} \equiv 0 \mod (p^{k+1}, F(x)),^{54}$$

but

$$\phi(x) \cdot V^{2m-r-1} \cdot W^{k+2} \not\equiv 0 \mod (p^{k+2}, F(x)).$$

For example if $m = 1$, as must occur for any p not dividing the discriminant of $F(x)$, this congruence and noncongruence become:

$$\phi(x) \cdot V \cdot W^{\lambda+1} \equiv 0 \mod (p^{\lambda+1}, F(x));$$

$$\phi(x) \cdot V \cdot W^{\lambda+2} \not\equiv 0 \mod (p^{\lambda+2}, F(x)).$$

Although this definition is somewhat cumbersome, using it Zolotarev is able in Sections 47–49 to prove the following results, clarifying the role ideal numbers play in the theory of complex integers.

In Section 47: *If a complex integer of the form* $\phi(\xi)$ *involves an ideal factor of* p *belonging to* V *exactly* λ *times, and another such complex integer* $\psi(\xi)$ *does not involve* V, *then the product* $\phi(\xi)\psi(\xi)$ *involves that ideal factor the same number* λ *times.*

In Section 48: *The degree to which a product* $\phi(\xi) \cdot \psi(\xi) \cdots$ *of several complex integers from* R *involves an ideal factor is the sum of the degrees to which it is involved in each factor.*

In Section 49: *If* $\phi(\xi)$ *involves the ideal factors* $V, V_1, \ldots, V_s$ *to degrees at least* $\lambda m, \lambda m_1, \ldots, \lambda m_s$ *(where these positive integers are as above), then* p^λ *divides* $\phi(\xi)$.

In Section 50: *If the polynomials* $V(x), V_1(x), \ldots, V_s(x)$ *have degrees* $\nu, \nu_1, \ldots, \nu_s$ *respectively, and if* $\phi(\xi)$ *involves ideal factors of* p *belonging to* $V(x), V_1(x), \ldots, V_s(x)$ *to the powers* $\lambda, \lambda_1, \ldots, \lambda_s$ *exactly, then the norm* $N(\phi(\xi))$ *is divisible by* p *to the power* $\lambda\nu + \lambda_1\nu_1 + \cdots + \lambda_s\nu_s$ *exactly.*

---

[53]Substitution of ξ for x in (2), yields, taking into account that $F(\xi) = 0$, that $pF_1(\xi) = -V(\xi)^m V_1(\xi)^{m_1} \cdots V_s(\xi)^{m_s}$, where none of $V(\xi), V_1(\xi), \ldots, V_s(\xi)$ divides $F_1(\xi)$. *Trans.*

[54]Here, from a slightly more modern point of view, $(p^{k+1}, F(x))$ denotes the set (ideal) of all integer polynomials of the form $A(x)p^{k+1} + B(x)F(x)$, A, B arbitrary integer polynomials.

From this it follows in particular that each complex integer of the form $\phi(\xi)$ can involve only those ideal factors belonging to some rational prime dividing its norm, and moreover only a finite number of times. Hence we can determine for a given complex integer $\phi(\xi)$ which irreducible ideal factors it has.

In Section 52, by combining his results to this point, Zolotarev shows that for one complex integer $\phi(\xi)$ from R to divide another $\psi(\xi)$, it is necessary and sufficient that all ideal factors involved in $\phi(\xi)$ are included among those involved in $\psi(\xi)$, and to at most the same degree. In particular taking $\psi(\xi) = \phi(\xi)$ it follows that there is only one decomposition of $\phi(\xi)$ into irreducible ideal factors.

In Section 53 Zolotarev remarks that decompositions of complex integers from R into irreducible ideal factors enjoy the same properties as decompositions of ordinary integers as products of ordinary primes. We call two complex integers from R *relatively prime* when they have no irreducible ideal factor in common. Then if $\phi(\xi)$ and $\psi(\xi)$ are relatively prime and $\psi(\xi)$ divides a product $\phi(\xi)\cdot\phi_1(\xi)$, it must divide $\phi_1(\xi)$. It follows that if $\psi(\xi)$ is relatively prime to each of $\phi(\xi), \phi_1(\xi), \phi_2(\xi), \ldots$, then it is relatively prime to their product $\phi(\xi)\cdot\phi_1(\xi)\cdot\phi_2(\xi)\cdots$, and so on.[55]

In Sections 54 and 55 Zolotarev considers particular examples of such rings of complex integers. In Section 54 he considers the Gaussian integers $a + bi$, i.e., the case $F(x) = x^2 + 1$. As is well known (and easy to show) the congruence $x^2 + 1 \equiv 0 \pmod p$ has solutions $(\bmod\ p)$ precisely if $p = 2$ or p has the form $4k + 1$, i.e., only for these p does $x^2 + 1$ factor mod p into two degree-one factors, while for p of the form $4k + 3$, it is prime mod p. Hence ordinary primes remain primes in the ring of Gaussian integers if they are of the form $4k + 3$, but 2 and ordinary primes of the form $4k + 1$ split into products of nonreal algebraic integers. In fact $2 = (1 + i)(1 - i)$, and a prime p of the form $4k + 1$, being expressible by Fermat's theorem in the form $a^2 + b^2$ for integers a, b, satisfies $p = (a+bi)(a-bi)$, and it is not difficult to check that none of $1 \pm i, a \pm bi$ is a unit, in fact that they are irreducible in the ring of Gaussian integers. It follows from the theorem quoted above from Section 52,[56] that in this ring any other decomposition $p = (a' + b'i)(a' - b'i)$ will differ from the former one only by insertion of units (easily shown to be $\pm 1, \pm i$), so that the introduction of ideal factors is not needed in this case.

In Section 55 Zolotarev deduces from his general theory Kummer's theory of ideal numbers in the particular case of the cyclotomic polynomial, i.e., for $F(x) = \frac{x^n - 1}{x - 1}$, where n is prime, i.e., the theory Kummer applied in his investigations of Fermat's equation $x^\lambda + y^\lambda = z^\lambda$.

In the remaining sections of Chapter III Zolotarev considers a division of ideal numbers into classes.

Thus, to summarize, in Chapter III Zolotarev constructs a complete arithmetic for the ring of complex integers of the form

$$\phi(\xi) = a_0 + a_1\xi + a_2\xi^2 + \cdots + a_{n-1}\xi^{n-1},$$

[55] In view of these results, we may call the ideal irreducible factors ideal *prime* factors. *Trans.*

[56] Taking for example $p = 13$, equation (2) becomes

$$x^2 + 1 = (x - 5)(x - 8) + 13(x - 3),$$

among infinitely many possibilities, so here $V = x - 5, V_1 = x - 8$. However, the ideal $(i - 5, p)$ of the ring of Gaussian integers is a principal prime ideal generated by $2 + 3i$, and similarly $(i - 8, p) = (2 - 3i)$. One might consider the more instructive case $F(x) = x^2 + 11$ in order to contrast it with this. *Trans.*

where $a_0, a_1, a_2, \ldots, a_{n-1}$ are rational integers, and ξ is a root of an irreducible (over $\mathbb{Q}$) nth degree polynomial $F(x)$ with integer coefficients and leading coefficient 1, which is furthermore not "special", i.e., for no prime divisor p of its discriminant D is it the case that the polynomial F_1 of equation (2) is divisible mod p by any ideal factor involved in p to a power greater than 1.

It is now known that "special" polynomials are encountered rarely. All the same the exclusion of these polynomials from the theory renders it incomplete. Moreover it has to be said that the ring R of algebraic integers of the form $\phi(\xi)$ considered by Zolotarev, need not in general be the largest discrete subring of the algebraic number field $\mathbb{Q}[\xi]$ determined by a root ξ of the prime integer monic polynomial $F(x)$, and indeed it can happen that the largest discrete subring is not definable in this way in terms of any such polynomial. This situation links Zolotarev's theory more closely to the particular polynomial F rather than the number field as a whole.[57]

In addition to these shortcomings there is another, common to both Zolotarev's and Kummer's treatments: While Zolotarev defines clearly what he means by saying that a complex integer $\phi(\xi)$ of the ring R under consideration is divisible to some power by some prime ideal factor of an ordinary prime p, at the same time this factor appears only in the form of a pair $(V(x), p)$ where $V(x)$ is a prime factor of the polynomial $F(x)$ modulo p (and defined only up to multiplication by ordinary integers not divisible by p).

According to Zolotarev's definition (see earlier) the divisibility of $\phi(\xi)$ by the prime ideal factor V to the power $\lambda = km + r$, is determined by whether the polynomial $\phi(x)V(x)^{m-r}W(x)^{k+1}$ has the form $p^{k+1}A(x) + F(x) \cdot B(x)$, where $W(x)$ is as defined earlier and $A(x), B(x)$ are arbitrary integer polynomials. If Zolotarev had introduced ideals, as was in fact done by Dedekind, even in the form of a modulus $p \cdot A(x) + V(x) \cdot B(x)$, he would have been able to express more clearly what he meant by the product of two or more prime ideal numbers, which in fact he only indicates glancingly at the end of Section 53.

All the same the clear-cut goal which Zolotarev sets himself at the beginning of that chapter, namely: *To determine whether one complex integer $\phi(\xi)$ divides another $\psi(\xi)$ without actually carrying out the division*, is fully attained in the form of the theorem on the uniqueness of decomposition into prime ideal factors, with all its important consequences.

Zolotarev's articles of 1878 and 1885

The condition imposed on the polynomial $F(x)$ in his doctoral dissertation that it not be "special" is removed by Zolotarev in his articles of 1878 and 1885. The short (7-page) 1878 article, published in French in the *Bulletin of the St. Petersburg Academy of Sciences*, contains only a lemma fundamental to Zolotarev's strategy for circumventing the difficulties associated with "exceptional" primes. In the preface to that paper he writes: "In this note I set myself the goal of proving

[57]In his articles of 1878 and 1885 Zolotarev removes the restriction on F and also considers the largest discrete ring in a given number field, i.e., the ring of all algebraic integers, as they are now called, of the field—see below.

one fundamental theorem relating to that case, but shall not derive all of its consequences." In the long (60-page) 1885 article[58] published in French in the *Journal de Mathématiques Pures et Appliquées* entitled "Sur la théorie des nombres complexes", the first 17 pages (72–88)[59] reproduce almost word for word Chapter III of his doctoral dissertation, i.e., recapitulate his general theory of ideal numbers in the case when the fundamental polynomial $F(x)$ is not special. The following 16 pages (89–104) are devoted to the derivation of Kummer's theory as a special case of the general theory, as in the doctoral dissertation, i.e., the theory of ideal numbers of the cyclotomic number field, which falls into the case of nonspecial $F(x)$. Finally, the last 25 pages (104–129) contain an exposition of a completely general theory, valid for arbitrary polynomials $F(x)$, including special ones. We now examine this theory,

In Section 22 Zolotarev writes: "Hitherto we have excluded from our considerations equations

$$F(x) = x^n + a_1x^{n-1} + a_2x^{n-2} + \cdots + a_n = 0, \tag{1}$$

with the polynomial $F(x)$ having, relative to a prime modulus p, the form

$$F(x) = V^m \cdot V_1^{m_1} \cdots V_s^{m_s} + p \cdot G(x), \tag{2}$$

where the polynomial $G(x)$ is divisible mod p by one of the factors V, say, with exponent m greater than 1. We know that such "exceptional" p (for a given polynomial $F(x)$) are only finite in number (most often there are none). As far as other prime numbers are concerned, their decomposition into complex factors depending on equation (2) is carried out in the manner described above."

Zolotarev begins by showing that in the presence of "exceptional" primes p the lemma established in Section 12 need no longer hold. That lemma asserts the following: *In the case where there are no exceptional primes associated with $F(x)$, if the ratio $\frac{\phi(\xi)}{\psi(\xi)}$ (ξ a root of $F(x)$) of two complex integers (i.e., integers in the former sense that the polynomials $\phi(x)$ and $\psi(x)$ have rational integral coefficients) is not a complex integer, then it cannot be the root of an equation of the form*

$$z^k + q_1z^{k-1} + \cdots + q_k = 0,$$

with $q_1, q_2, \ldots, q_k$ rational integers.

In fact if $m > 1$, then for example the number

$$\frac{V^{m-1}(\xi)V_1^{m_1}(\xi) \cdots V_s^{m_s}(\xi)}{p}, \tag{3}$$

although not a complex integer in the former sense,[60] yet if p is "special"[61], turns out to be the root of a monic polynomial with rational integral coefficients.

[58]Since Zolotarev died in 1878, this work must have been submitted for publication at the latest in that year.

[59]The numbers in parentheses are those of the corresponding pages of Zolotarev's "Complete Collected Works", published by the Academy of Sciences, Leningrad, 1932.

[60]Since $F(x)$ is monic of degree n, we shall have $m \deg V + m_1 \deg V_1 + \cdots + m_s \deg V_s = n$, provided $V(x), V_1(x), \ldots, V_s(x)$ are chosen to have all their coefficients not divisible by p. However, then the number in (3) is a polynomial expression of degree $< n$ in ξ (and therefore unique to that number) with all of its nonzero coefficients rational but not integral. *Trans.*

[61]Note that here and elsewhere in this chapter "special" $\equiv$ "exceptional". *Trans.*

"In view of this situation, which does in fact sometimes occur when there are exceptional p," writes Zolotarev, "we need to generalize the concept of an integral complex number." He suggests that any number of the form

$$y = a + b\xi + \cdots + l\xi^{n-1},$$

where ξ is a root of $F(x)$ and the coefficients $a, b, \ldots, l$ are arbitrary *rational* numbers (whole numbers or fractions), be called a (complex) integer if it is the root of an integer monic polynomial.[62] In the remainder of his exposition the term "integral complex number" is to be understood in this way.

Zolotarev begins Section 23 with the observation that an integral complex number depending on a root ξ of the nth degree monic irreducible integer polynomial $F(x)$ must be expressible (uniquely) in the form

$$\eta = \frac{\alpha + \beta\xi + \gamma\xi^2 + \cdots + \lambda\xi^{n-1}}{N},$$

where $\alpha, \beta, \gamma. \ldots, \lambda, N$ are rational integers with greatest common divisor 1. He next argues that in view of the above-mentioned lemma concerning the case where there are no exceptional primes, the denominator N can have as its prime divisors only exceptional primes p. For if N had a prime divisor p that was not exceptional, then in view of the fact that η is a root of an integer monic polynomial, certainly the number

$$\frac{\alpha + \beta\xi + \gamma\xi^2 + \cdots + \lambda\xi^{n-1}}{p}$$

must also be a root of such a polynomial. However, it can be shown that since here p is not exceptional, all of the coefficients $\alpha, \beta, \gamma, \ldots, \lambda$ must then be divisible by p, and then since p divides N, we arrive at a contradiction of the relative primality of $\alpha, \beta, \gamma, \ldots, \lambda, N$.

Zolotarev next proves that the discriminant of $F(x)$ must be divisible by N^2. Combining this with the above, we infer that every prime that is exceptional relative to the polynomial $F(x)$ must divide the discriminant D of $F(x)$, and that the power to which each such p divides such a denominator N is at most half of that to which it divides the discriminant D.

In the long Section 25 Zolotarev classifies integral complex numbers depending on a given polynomial $F(x)$, relative to a given exceptional prime p for that F. He begins by considering all integral complex numbers of the forms

$$\alpha + \beta\xi + \gamma\xi^2 + \cdots + \lambda\xi^{n-1},$$
$$\frac{\alpha' + \beta'\xi + \gamma'\xi^2 + \cdots + \lambda'\xi^{n-1}}{p},$$
$$\frac{\alpha'' + \beta''\xi + \gamma''\xi^2 + \cdots + \lambda''\xi^{n-1}}{p^2},$$
$$\cdots\cdots\cdots\cdots\cdots\cdots$$

Obviously every integral complex number of the second type can be written in the form

$$\frac{\phi(\xi)}{p} + a + b\xi + c\xi^2 + \cdots + l\xi^{n-1},$$

where $a, b, c, \ldots, l$ are rational integers and $\phi(x)$ is an integer polynomial with nonnegative coefficients all less than p. Zolotarev now centres attention on those

[62] This is the modern definition of an "algebraic integer". *Trans.*

integral complex numbers of the form $\frac{\phi(\xi)}{p}$ where the degree of $\phi(x)$ is least. One can then construct from this complex integer one of the same least degree where the leading coefficient of $\phi(x)$ is 1. To see this write

$$\phi(x) = A + Bx + Cx^2 + \cdots + Lx^\mu,$$

and assume L is different from 1 (and satisfies $0 < L < p$). Taking g to be any integer satisfying $gL \equiv 1 \pmod p$, we can express $g\frac{\phi(x)}{p}$ in the form

$$g\frac{\phi(x)}{p} = \frac{\alpha + \beta x + \gamma x^2 + \cdots + x^\mu}{p} + \alpha' + \beta' x + \cdots + \lambda' x^{n-1},$$

thereby obtaining a complex integer of the form

$$\frac{\phi(\xi)}{p} = \frac{\alpha + \beta\xi + \gamma\xi^2 + \cdots + \xi^\mu}{p}, \tag{4}$$

as desired.

It is immediate that there is only one complex integer of this form, since if there were another different from this one, say

$$\frac{\phi_1(\xi)}{p} = \frac{\alpha_1 + \beta_1\xi + \gamma_1\xi^2 + \cdots + \xi^\mu}{p},$$

then from the difference $\frac{\phi(\xi) - \phi_1(\xi)}{p}$ we could construct a complex integer of the same form but lesser degree, contradicting the minimality of the degree of $\phi(x)$.

It is next shown that any complex integer of the form

$$\frac{a + b\xi + c\xi^2 + \cdots + l\xi^{n-1}}{p}$$

can be expressed in the form $A(\xi)\frac{\phi(\xi)}{p} + B(\xi)$, where $A(x)$ and $B(x)$ are integer polynomials, and $\phi(x)$ is the unique monic least-degree polynomial figuring in (4). To see this we first divide $a + bx + cx^2 + \cdots + lx^{n-1}$ by $\phi(x)$, obtaining

$$a + bx + cx^2 + \cdots + lx^{n-1} = A(x)\phi(x) + R(x),$$

where $R(x)$ is zero or has degree less than that of $\phi(x)$, and, since $\phi(x)$ is monic, $A(x)$ and $R(x)$ are both integer polynomials. Then since

$$\frac{R}{p} = \frac{a + bx + cx^2 + \cdots + lx^{n-1}}{p} - A\frac{\phi(x)}{p},$$

we infer that $\frac{R(\xi)}{p}$ is a complex integer. Since $R(x)$ has degree less than that of $\phi(x)$, it must be the case that its coefficients are all divisible by p. Hence $R(x)/p = B(x)$, say, is an integer polynomial. It follows that

$$\frac{a + b\xi + c\xi^2 + \cdots + l\xi^{n-1}}{p} = A(\xi)\frac{\phi(\xi)}{p} + B(\xi).$$

It can be shown similarly that any complex integer of the form

$$\frac{a + b\xi + c\xi^2 + \cdots + l\xi^{n-1}}{p^2} = A(\xi)\frac{\phi_1(\xi)}{p^2} + B(\xi)\frac{\phi(\xi)}{p} + C(\xi),$$

where $A(x), B(x), C(x)$ are integer polynomials, and $\phi_1(x)$ is an integer polynomial of least degree among those for which their quotient by p^2 yields a complex integer, and which is in addition monic with nonnegative coefficients all less than p^2.

Continuing in this way for complex integers of the forms

$$\frac{a+b\xi+c\xi^2+\cdots+l\xi^{n-1}}{p^3},\ldots,\frac{a+b\xi+c\xi^2+\cdots+l\xi^{n-1}}{p^m},$$

we obtain similar expressions involving (unique) complex integers

$$\frac{\phi_2(\xi)}{p^3},\ldots,\frac{\phi_{m-1}(\xi)}{p^m}$$

analogous to $\frac{\phi(\xi)}{p}$ and $\frac{\phi_1(\xi)}{p^2}$. Denoting by $\mu, \mu_1, \mu_2, \ldots, \mu_{m-1}$ the respective degrees of $\phi(x), \phi_1(x), \phi_2(x), \ldots, \phi_{m-1}(x)$, we have

$$\mu \leq \mu_1 \leq \mu_2 \leq \cdots \leq \mu_{m-1},$$

since the integrality of $\frac{\phi_i(\xi)}{p^{i+1}}$ implies that of $\frac{\phi_i(\xi)}{p^i}$ *a fortiori*, so that $\phi_i(x)$ cannot have degree less than that of $\phi_{i-1}(x)$.

If $\mu_i = \mu_{i+1}$ then we may replace $\phi_i(x)$ by $\phi_{i+1}(x)$, since $\frac{\phi_{i+1}(\xi)}{p^{i+1}} = p \cdot \frac{\phi_{i+1}(\xi)}{p^{i+2}}$ is a complex integer and $\phi_{i+1}(x)$ has the same degree as $\phi_i(x)$. Hence if

$$\begin{aligned}\mu &= \mu_1 = \mu_2 = \cdots = \mu_k\\ &< \mu_{k+1} = \mu_{k+2} = \cdots = \mu_{k_1}\\ &< \cdots\\ &< \mu_{k_h+1} = \mu_{k_h+2} = \cdots = \mu_{m-1},\end{aligned}$$

then each complex integer depending on the polynomial $F(x)$ (and p) is congruent modulo p to one of the form

$$(5)\qquad A(\xi)\frac{\phi_{m-1}(\xi)}{p^m} + B(\xi)\frac{\phi_{k_h}(\xi)}{p^{k_h+1}} + C(\xi)\frac{\phi_{k_{h-1}}(\xi)}{p^{k_{h-1}+1}} + \cdots + K(\xi)\frac{\phi_k(\xi)}{p^{k+1}} + L(\xi),$$

where $A(x), B(x), C(x), \ldots, L(x)$ are integer polynomials of degrees

$$n-1-\mu_{m-1},\ \mu_{m-1}-\mu_{k_h}-1,\ \mu_{k_h}-\mu_{k_{h-1}}-1,\ \ldots,\ \mu_k-1$$

respectively. Clearly any number of this form is a complex integer. It is easy to verify that if two such complex integers are congruent modulo p, then

$$A \equiv A',\ B \equiv B',\ C \equiv C',\ \ldots,\ L \equiv L' \pmod{p},$$

i.e., the coefficients of $A(x) - A'(x), B(x) - B'(x), \ldots, L(x) - L'(x)$ are all divisible by p.

It follows that the number of complex integers depending on the polynomial $F(x)$ and p, pairwise incongruent mod p,[63] and not divisible by p, is exactly $p^n - 1$, since the number of coefficients of the polynomials

$$A(x), B(x), C(x), \ldots, L(x)$$

is exactly

$$(n-\mu_{m-1}) + (\mu_{m-1}-\mu_{k_h}) + (\mu_{k_h}-\mu_{k_{h-1}}) + \cdots + \mu_k = n,$$

and there are p choices mod p for each such coefficient.

Note also that given $F(x)$ and p, the polynomials $\frac{\phi_{m-1}(x)}{p^m}, \frac{\phi_{k_h}(x)}{p^{k_h+1}}, \ldots,$ figuring in the formula (5) can be computed in finitely many steps. In Section 26, Zolotarev shows that once these are known, then it is easy to determine whether a complex number given in the form $\frac{f(\xi)}{\phi(\xi)}$, where $f(\xi)$ and $\phi(\xi)$ are complex integers, is again

[63]That is, the number of congruence classes mod p. *Trans.*

a complex integer. First the ratio $\frac{f(\xi)}{\phi(\xi)}$ is brought into the form $y = \frac{\psi(\xi)}{N}$,[64] where $\psi(\xi)$ is a complex integer and N a rational integer such that its highest common factor with the coefficients of $\psi(x)$ is 1. If the given ratio is a complex integer, then in the prime decomposition $N = p^m \cdot p_1^{m_1} \cdots p_s^{m_s}$, the primes $p, p_1, \ldots, p_s$ must all be exceptional relative to $F(x)$. It is obvious that if y is a complex integer, then so are $\frac{\psi(\xi)}{p^m}, \frac{\psi(\xi)}{p_1^{m_1}}, \ldots, \frac{\psi(\xi)}{p_s^{m_s}}$, but how to prove the converse, i.e., that if these are integral then y must be also? (The integrality of the numbers $\frac{\psi(\xi)}{p^m}, \ldots$ is easy to show using formula (5).)

In Sections 27–29 Zolotarev proves several theorems needed for his culminating theorem on the unique decomposition of complex integers into prime ideal factors.

He begins with some definitions. Let

$$\alpha_1, \alpha_2, \ldots, \alpha_\sigma, \tag{6}$$

where $\sigma = p^n - 1$, be a full set of incongruent representatives mod p of the complex integers not divisible by p in the ring of all complex integers depending on the given polynomial $F(x)$ (as defined in Section 25). A number β in this ring will be said to *have no common divisor with* p if none of the products $\alpha_i\beta$ is divisible by p. Otherwise β and p are, naturally, said to *have a common divisor*, possibly ideal.

Two complex integers β and γ are then defined as *not having a divisor of p as a common factor* if for all α_i from (6), $\alpha_i\beta, \alpha_i\gamma$ are not both divisible by p.

Zolotarev then proves the following results:

The norm $N(\beta)$ of a complex integer depending on $F(x)$ is not divisible by p if and only if β and p do not have a common divisor.

If two complex integers A and B do not have a divisor of p as a common factor, and similarly for A and C, then also A and BC do not have a common factor that is a divisor of p.

If A does not have a common divisor with p, then there is an α_i in (6) (= $M(\xi)$, say) such that

$$A(x)M(x) \equiv 1 \pmod{p}.$$

If A and B do not have a divisor of p as a common factor, then there exist complex integers M and N such that

$$A(x)M(x) - B(x)N(x) \equiv 1 \pmod{p}.$$

Section 30 is devoted to overcoming the main difficulty. Zolotarev begins by defining an ordinary (rational) prime p to be *irreducible (in the ring R of complex integers depending on $F(x)$)* if each of the complex integers in (6) is relatively prime to p, i.e., has no divisor in common with p.

To see how other rational primes decompose into prime ideal factors, Zolotarev chooses the complex integers α_i in (6) corresponding to a given such p (reducible in R) in a special manner. Let α be one of the α_i of (6) whose norm is divisible by p. Among all numbers of the ring R of the form $\alpha + p\eta$ congruent to α modulo p, where η is an element of R (any such complex integer may be taken in place of α in the list (6)), Zolotarev chooses one whose norm is divisible by the smallest power of p, i.e., chooses η so that this condition is satisfied. From this point on he assumes that the numbers in the list (6) having a common factor with p all satisfy this condition. For such a list (6) he then proves the following theorem:

[64]That is, one "rationalizes the denominator." *Trans.*

Any member α of the list (6) (modified as described above) with norm divisible by p^μ satisfies an equation

$$\alpha^n + b_1\alpha^{n-1} + b_2\alpha^{n-2} + \cdots + b_n = 0,$$

where the coefficient b_n is divisible by p^μ (by assumption),[65] *and the other coefficients $b_{n-1}, b_{n-2}, \ldots$ are divisible by $p^{\mu-1}, p^{\mu-2}, \ldots$ respectively.*

This theorem was the centrepiece of Zolotarev's 1878 paper, and is crucial to his method for accommodating exceptional primes p in the theory. However, its proof is difficult and we shall not reproduce it here.

From the proof of this theorem one obtains the following two corollaries:

I. *For each number α in the list (6) there is a rational integer H not divisible by p with the property that α divides Hp.*

II. *If β is any number from R (the ring of all complex integers depending on $F(x)$) not figuring in the list (6) and not divisible by p, then in the list (6) there is a number α which divides $H\beta$ for some rational integer H not divisible by p.*

In Section 31 Zolotarev makes the following definition: A complex integer β from R is said to *involve all (ideal prime) factors of p belonging to α*, if there exists a number A in R relatively prime to p such that α divides $A\beta$. Using the above pivotal theorem, he now proves the following result:

If the product $\beta\gamma$ of two numbers from the ring R involves all factors of p belonging to α, and if α and β have no common factor belonging to p, then the number γ involves all factors of p belonging to α.

Section 32 is devoted to the proof of the following result:

If numbers β and γ from the ring R have common divisors with p, and if neither involves all factors of p belonging to the other, then there exists a number α in R such that both β and γ involve all factors of p belonging to α and furthermore such that p divides $N(\alpha)$ to a lesser power than it divides $N(\beta)$ or $N(\gamma)$.

In Section 33 Zolotarev returns to the numbers in the list (6), assumed specially chosen as described above. He defines a number α from that list as *involving only one prime factor of p* if every number from R having a factor in common with α and p involves all (ideal prime) factors of p belonging to α. By the corollary II above, in order to determine whether an arbitrary number β involves only one prime factor of p,[66] it suffices to compare it with the numbers in the list (6). Using the last result quoted above concerning norms, one can show that in the list (6) there is a number α involving only one prime factor of p. To see this, consider any number β in that list not relatively prime to p, and let γ be another number in the list (6) having a factor in common with p belonging to β, but not involving all factors of p belonging to β. Then by the last quoted theorem above, there is in (6) a number α such that β involves all factors of p belonging to α, and the power of p dividing $N(\alpha)$ is less than that to which p divides $N(\beta)$. If this α involves more than one factor of p, then as before one can find in the list (6) a number α' such that α involves all factors of p belonging to α', and p divides $N(\alpha')$ to a lesser

[65] For a number β in $\mathbb{Q}[\xi]$, with minimal polynomial of degree m, the *norm* of β is the product of the m roots of its minimal polynomial, raised to the power n/m (an integer). Hence if β is an algebraic integer, its norm is the constant term of its (monic) minimal polynomial, raised to the power n/m. *Trans.*

[66] That is, corollary II may be used to extend the definition of the phrase "involves only one prime factor of p" to arbitrary numbers from R. *Trans.*

power than it divides $N(\alpha)$, and so on. Continuing in this way, after finitely many steps we reach a number in the list (6) involving only one (ideal prime) factor of p.

This argument shows that:

Every complex integer from R not relatively prime to p involves at least one ideal prime factor of p belonging to a number in the list (6) not involving any other factors of p.

Now let

$$\beta_1, \beta_2, \ldots, \beta_k \tag{6'}$$

be the numbers in (6) involving only one ideal prime factor of p. In the case where two of these are not relatively prime mod p, Zolotarev defines them as *involving the same ideal prime factor of p*. Let

$$\nu_1, \nu_2, \ldots, \nu_s \tag{6''}$$

be a maximal subset of numbers from the list (6′) involving distinct ideal factors of p. Zolotarev defines a number γ from the ring R as *involving the ideal prime factor of p belonging to ν_1* (say) *to the power m*, if it involves all ideal prime factors of p belonging to ν_1^m, but not all such factors belonging to ν_1^{m+1}. By means of an argument using norms, he then shows that if p^λ and p^μ are the largest powers of p dividing $N(\alpha)$ and $N(\nu_1)$ respectively, then $\lambda \geq m\mu$.

In Section 34 Zolotarev proves that:

The product $\beta\gamma$ of two numbers from the ring R involves an ideal prime factor of p belonging to ν_1 to the power equal to the sum of those to which that factor is involved in β and γ.

In conclusion Zolotarev remarks: "From the theorems thus established it is easy to infer all of the fundamental theorems concerning divisibility of complex numbers."

Commentary

For nonspecialists the very question of what motivates the articles by Zolotarev considered above requires elucidation. All of the theory of the ordinary rational integers is related to a large extent to questions of their divisibility. The ordinary (rational) integers are obtained very simply by means of addition and subtraction, beginning with the integer which, as far as addition and subtraction are concerned, is simplest, namely 1. In other words, with respect to addition the set of rational integers forms an infinite cyclic group generated by 1.

However the behaviour of the integers with respect to multiplication is incomparably more complex. Here of course they no longer form a group. What is fundamental in relation to multiplication is the fact that in the sequence of natural numbers (i.e., positive integers) there are the so-called "prime" numbers $2, 3, 5, 7, 11, 13, 17, 19, 23, \ldots$, and so on, to infinity, which are distributed amongst the natural numbers in an apparently very irregular manner. Their importance lies in the fact that every nonzero rational integer is expressible as a product of primes and possibly 1 or -1, moreover in essentially only one way.

This fundamental result has many consequences, used in the theory of numbers at every move, such as, for example, the property that if a product $a \cdot b$ is an nth power and a and b are relatively prime, i.e., have greatest common divisor 1, then each of a and b must also be nth powers, and so on. Of course it is by no means the case that all number-theoretic problems are so closely linked with questions of divisibilty. For instance in the articles by Korkin and Zolotarev on quadratic forms examined in the previous chapter the main interest is not centred in considerations of divisibility, but lies elsewhere, namely in the restriction of the variables of quadratic forms to integer values. All the same, in every situation where the accent is on integrality, divisibility questions inevitably play a role in some way or other.

Nowadays in algebra we call a set on which there are given two binary operations, termed "addition" and "multiplication", a *(commutative) ring* when the following conditions hold: application of either operation to any two elements of the set always yields as result another element of the set; with respect to the addition operation the set forms an abelian group, and with respect to the multiplication a commutative semigroup, i.e., $ab = ba$ and $a(bc) = (ab)c$ for all elements a, b, c; and finally the two operations should be linked by the distributive law $a(b + c) = ab + ac$.[67] The totality of rational integers under ordinary addition and multiplication is the prototypical ring, and in a sense the simplest. If a ring has the property that each of its nonzero elements can be decomposed (essentially) uniquely as a product of finitely many prime elements, then a theory analogous to ordinary number theory will be valid for it.

After the ring of integers, the simplest ring for which there does indeed exist such a theory is the ring of *Gaussian integers*, consisting of all complex numbers of the form $a+bi$, where $i = \sqrt{-1}$, the usual imaginary unit, and a and b are arbitrary rational integers. In 1824 Gauss showed that in this ring every nonzero element can be expressed as a product of finitely many prime elements,[68] in only one way up to the insertion of units, which here are just $1, -1, i, -i$. However, in other rings, even some apparently similar to the ring of Gaussian integers, for example those of all complex numbers of the form $a+b\sqrt{-5}$, or of the form $a+b\sqrt{-11}$,[69] not to mention more complicated rings, the situation is otherwise, although many such rings do possess "irreducible" elements (i.e., factoring only as a product of an associate and units) into products of which their nonzero elements decompose—although often in many essentially different ways.

One is thus faced with the following alternatives: To eschew considering such maverick rings from an arithmetic point of view altogether, or to study them more closely in order to penetrate to the underlying reason for their failure to have the unique factorization property. Many theorems concerning the rational integers require for their proofs the investigation of such rings, since they in some fashion reflect the situation of interest in the rational integers. Consider for example the question of the possible divisors of a sum of two squares $a^2 + b^2$ of integers. Since $a^2 + b^2 = (a + bi)(a - bi)$, this leads directly to the ring of Gaussian integers. (Put differently, the norm $z\bar{z}$ of the complex number $z = a + bi$ is $a^2 + b^2$.) Thus one

[67]If there is an identity element, i.e., a "1", with respect to the multiplication—which will always be the case here—the ring is called *unital. Trans.*

[68]A *prime* in a unital commutative ring is now defined to be a nonzero, nonunit element u with the property that if u divides vw for any elements v, w, then u must divide at least one of v and w, i.e., either $v = uz$ or $w = uz'$ for some z, z' in the ring. A *unit* is a divisor of 1, i.e., an invertible element of the multiplicative semigroup of the ring. *Trans.*

[69]One might say that the latter ring fails to enjoy unique factorization into primes for a simpler reason than the former, namely because the algebraic integers in the algebraic number field $\mathbb{Q}[\sqrt{-11}]$ are the numbers of the form $a + b(\frac{1+\sqrt{-11}}{2})$. See an earlier footnote. *Trans.*

might say that the key to factorizing an integer of the form a^2+b^2 into factors that are also sums of two squares is furnished by the factorization of numbers of the form $a+bi$ as products of Gaussian primes.[70] And there are other such examples.

Kummer's attempt to prove that there are no nontrivial integer solutions to Fermat's equation $x^n+y^n=z^n$ for $n>2$, led him inevitably to the investigation of the ring of all numbers of the form $a+b\rho+c\rho^2+\cdots+k\rho^{p-1}$, where p is an odd prime (dividing n), ρ is a root of the so-called cyclotomic (i.e., circle-division) equation $x^{p-1}+x^{p-2}+\cdots+x+1=1$, and $a,b,c,\ldots,k$ are rational integers. It turned out that for certain p the elements of this ring, like the Gaussian integers, do enjoy unique factorization into primes, while for others this is not the case. It was at first not clear what approach to take in the case of "bad" primes p. After many years spent investigating this problem Kummer found the following resolution of the difficulty: He introduced certain supplementary factors not actually existing in the ring, which he called "ideal". If in addition to the elements of the ring these ideal factors are admitted, then some formerly irreducible elements become reducible as products of ideal factors, amongst which there will be *prime* ideal factors which do not further decompose, even into ideal factors. Kummer's crowning result was then that each ordinary or ideal element of the ring decomposes uniquely, up to insertion of unit factors, as a product of prime ideal factors.

Thus Kummer's ring of complex integers depending on a root of the cyclotomic equation, once equipped with ideal numbers, has restored to it unique factorization of its elements, together with the ideal numbers, as products of prime ideal numbers. The former nonuniqueness of decomposition of an element as a product of irreducible elements is now seen to result from different groupings of prime ideal factors yielding different products of irreducibles of the ring.

Suppose for example that in a ring we have $A=B\cdot C=B'\cdot C'$, where A,B,C,B',C' are irreducible elements of the ring. This may derive from A having a decomposition $A=\alpha\cdot\beta\cdot\gamma\cdot\delta$ as a product of prime ideal numbers $\alpha,\beta,\gamma,\delta$, with $B=\alpha\cdot\beta, C=\gamma\cdot\delta, B'=\alpha\cdot\gamma$ and $C'=\beta\cdot\delta$. While the ideal factors are not actually elements of the ring, certain of their products can be taken to represent actual elements. This is roughly analogous to the situation in chemistry where radicals cannot be separated out individually, but nonetheless in combination form specific compounds.[71]

Kummer's discovery made a considerable impression on contemporary number-theorists. It seemed that this was the path one had to follow to do arithmetic in more recondite rings. The next step was to try to extend Kummer's results for the

---

[70]Historically, this order was reversed, since Fermat's theorem that an odd prime p is expressible as a sum of two squares if and only if $p\equiv 1(\text{mod } 4)$, came well before the introduction of the Gaussian integers. *Trans.*

[71]The seeming mystery suggested here was dispelled by Dedekind, to whom, in essence, the modern formulation is due, originating however in Kummer's ideal factors. An *ideal* I of a (commutative) ring R is an additive subgroup such that for all $a\in I$ and all $r\in R$, we have $ar\in I$. The product IJ of two ideals is the smallest ideal containing all products ab with $a\in A$ and $b\in B$. If an ideal is generated by a single element, so that it has the form aR, it is called a *principal* ideal, and as in the description in the main text such an ideal can be taken to represent the actual element a of the ring. All the ideals of the ring $\mathbb{Z}$ of ordinary integers are principal, as is also the case for the Gaussian integers, and certain other rings of integers of algebraic number fields. The subtlety enters in when there are "prime" ideals that are not principal, a *prime* ideal P being one other than R and $\{0\}$ such that whenever a product ab is in P, then at least one of a and b is in P. See any modern text on abstract algebra. *Trans.*

ring of complex integers depending on a root of the cyclotomic equation[72] to rings depending on roots of more complicated polynomial equations. It was convenient that for general such rings the theory of their units already existed, created by Dirichlet in 1846 in the course of generalizing Lagrange's solution of Pell's equation. (Recall that the *units* of a unital ring are the elements a such that $ab = 1$ for some element b of the ring, i.e., the multiplicatively invertible elements. In the ring of ordinary integers they are just ± 1, and in the Gaussian integers, $\pm 1, \pm i$.) Dirichlet showed that in the ring of algebraic integers depending on a root ξ of an arbitrary irreducible monic polynomial

$$F(x) = x^n + c_1 x^{n-1} + c_2 x^{n-2} + \cdots + c_{n-1}x + c_n \tag{1}$$

over the rational integers,[73] there are in general, as indeed in the rings considered by Kummer, infinitely many units, i.e., numbers dividing every number in the ring. Multiplication by a unit transforms the ring into itself, and as far as questions of divisibility are concerned multiplying an element of a ring by a unit has essentially no effect, like, for example, multiplication by -1 in the ring of integers. Thus with respect to divisibility the units play a special role.[74] The precise result of Dirichlet is as follows:

Let R denote the ring of algebraic integers (as they are now called) of an algebraic number field $\mathbb{Q}[\xi]$, where ξ is a root of a monic irreducible polynomial $F(x)$ of degree n over the rational integers (as in (1)). Denote by σ the number of real roots of F and by τ the number of complex conjugate pairs of nonreal roots, and set $r = \sigma + \tau - 1$. Then there exist r ("basic") units $e_1, e_2, \ldots, e_r$ in R such that every unit in R can be expressed uniquely in the form

$$\epsilon \cdot e_1^{m_1} e_2^{m_2} \cdots e_r^{m_r},$$

where ϵ is a particular root of unity in R and $m_1, m_2, \ldots, m_r$ are rational integers.

Thus, to repeat, it was not the units of a given ring R of algebraic integers that presented a problem for the arithmetic of that ring, but rather the need to introduce ideal factors, as Kummer had done in special cases, so that all the nonzero numbers of R might be expressible uniquely as products of ideal prime factors. This problem was solved in its most general form for rings R of algebraic integers independently by Dedekind and Zolotarev in the 1870s. In 1926 Emmy Noether investigated in an abstract setting the class of commutative unital rings whose nonzero elements enjoy unique factorization as products of primes,[75] i.e., abstracted the unique-factorization property of the rational integers in order to discover exactly which further kinds of rings have this desirable feature.

At the time of Dedekind and Zolotarev it was necessary first of all to establish precisely what to take as the most appropriate definition of the ring of algebraic integers depending on a root ξ of an equation of the form (1). As we have seen, initially, in his doctoral dissertation, Zolotarev proposed defining them to be all

[72] That is, $\mathbb{Z}[\rho]$, where ρ is as earlier. *Trans.*

[73] In more modern terminology, the ring of algebraic integers of the algebraic number field $\mathbb{Q}[\xi]$. *Trans.*

[74] A minor role, perhaps. Two elements a, b of a unital ring are called *associates* if $a = ub$, where u is a unit. Associates may be considered essentially the same as far as divisibility questions are concerned. *Trans.*

[75] Called "unique-factorization domains". *Trans.*

complex numbers of the form

$$\phi(\xi) = a_0 + a_1\xi + a_2\xi^2 + \cdots + a_{n-1}\xi^{n-1}, \tag{2}$$

where the coefficients $a_0, a_1, \ldots, a_{n-1}$ are rational integers. It is easy to verify that indeed this set of numbers forms a ring. Thus they add and subtract coefficientwise:

$$\begin{aligned}\phi(\xi) + \phi'(\xi) &= (a_0 + a_1\xi + \cdots + a_{n-1}\xi^{n-1}) + (a_0' + a_1'\xi + \cdots + a_{n-1}'\xi^{n-1}) \\ &= (a_0 + a_0') + (a_1 + a_1')\xi + \cdots + (a_{n-1} + a_{n-1}')\xi^{n-1},\end{aligned}$$

and under multiplication we obtain the integer polynomial expression $\phi(\xi) \cdot \phi'(\xi)$ of degree $2(n-1)$ in ξ, which reduces to one of the form (2) by using the fact that $F(\xi) = 0$, i.e., by dividing $\phi(x) \cdot \phi'(x)$ by $F(x)$, to obtain

$$\phi(x) \cdot \phi'(x) = F(x) \cdot \gamma(x) + \phi''(x),$$

where both $\phi''(x)$ and $\gamma(x)$ are integer polynomials (since $F(x)$ is monic), and $\phi''(x)$ is either 0 or has degree less than n. Then since $F(\xi) = 0$, it follows that $\phi(\xi) \cdot \phi'(\xi) = \phi''(\xi)$, which has the form (2).

In the situation where the irreducible monic polynomial $F(x)$ is such that among the primes p dividing its discriminant there are none that are "exceptional" for that polynomial, in the sense defined in Section 45 of his doctoral dissertation, Zolotarev constructs in that dissertation a general theory of ideal factors yielding unique factorization in the ring of numbers of the form (2).

Then in his articles of 1878 and 1885 Zolotarev changes, or rather generalizes, the concept of integrality, proposing that a complex number depending on a root ξ of the monic irreducible polynomial[76] $F(x)$ be considered *integral* not just if it has the form (2), but if it satisfies an integer monic polynomial, i.e., is of the form (2) but now with rational, not necessarily integer, coefficients allowed, and is the root of some monic integer polynomial. For this ring, the ring of all algebraic integers depending on a root of a given irreducible monic integer polynomial in this revised sense, Zolotarev constructs in his 1885 memoir a complete theory of ideal factors, now admitting no exceptions.

We shall now give a geometric interpretation, allowing a more visual understanding of this situation. Let R be a ring of the sort just defined. Thus each of the numbers in R can be expressed as in (2), where however now the coefficients a_i, although all rational, may sometimes happen not to be integers, the defining condition for admittance to R being that the number be the root of a monic integer polynomial. For a given such number $\phi(\xi)$ from R, such a polynomial can be obtained from $F(x)$ and the form (2) of $\phi(\xi)$ by means of a "Tschirnhausen transformation". If $\xi_1 (= \xi), \xi_2, \ldots, \xi_n$ are the n (necessarily distinct) roots of $F(x)$, then the roots of the resulting monic polynomial satisfied by $\phi(\xi)$ are $\phi(\xi_1), \phi(\xi_2), \ldots, \phi(\xi_n)$. We now associate with the given algebraic integer from R the point

$$(\phi(\xi_1), \phi(\xi_2), \ldots, \phi(\xi_n))$$

of the n-dimensional complex vector space $\mathbb{C}^n$.

It is not difficult to show that the totality of points of $\mathbb{C}^n$ corresponding in this way to numbers in R forms an n-dimensional point lattice in $\mathbb{C}^n$, i.e., a collection of

[76]That is, a number of the form $q_0 + q_1\xi + \cdots + q_{n-1}\xi^{n-1}$, $q_0, q_1, \ldots, q_{n-1} \in \mathbb{Q}$, which numbers comprise the "algebraic number field" $\mathbb{Q}[\xi]$. *Trans.*

the points of $\mathbb{C}^n$ having rational integer coordinates relative to some frame $\mathcal{E}_n$ for $\mathbb{C}^n$ attached to the origin O. This lattice can, as usual, be produced by adding and subtracting coordinatewise (relative to the standard frame E_n) its points (considered as vectors attached to O), starting with the vectors of the frame $\mathcal{E}_n$—which is, of course, always the case by the very definition of point lattice. However, it is obvious that this lattice has the less familiar property that the coordinatewise *product* (coordinates always with respect to the standard frame E_n) of any two points of the lattice is again a point of the lattice. Thus if $(y_1, y_2, \dots, y_n)$ and $(z_1, z_2, \dots, z_n)$ are two points of the lattice, then the points $(y_1 \pm z_1, y_2 \pm z_2, \dots, y_n \pm z_n)$ and $(y_1 \cdot z_1, y_2 \cdot z_2, \dots, y_n \cdot z_n)$ likewise belong to the lattice. Moreover it can be shown that this lattice is *irreducible*, i.e., that it is not the vector sum of two other lattices lying in subspaces of $\mathbb{C}^n$ of complementary dimensions. It is also *maximal* in the sense that it is not a sublattice of some denser lattice in $\mathbb{C}^n$ closed under multiplication.

It can be proved that these properties characterize lattices defined in terms of rings of algebraic integers:

Every lattice in $\mathbb{C}^n$ with the above-described properties of closure under multiplication, irreducibility, and maximality, arises in the above-described manner from some ring R of algebraic integers.

A lattice in $\mathbb{C}^n$ obtained using Zolotarev's earlier definition of integrality, requiring the coefficients of $\phi(x)$ to be integers, will also be closed under multiplication relative to a basic frame, and irreducible (as always, provided that $F(x)$ is irreducible), but may not be maximal, i.e., may be a proper sublattice of some maximal lattice.[77] This makes graphically clear the superiority of Zolotarev's revised 1885 definition of integrality (formulated also by Dedekind in articles dating from 1879).

The ring of integers R' say, of an algebraic number field according to Zolotarev's earlier definition, i.e., of those numbers $\phi(\xi)$ as in (2), with the coefficients restricted to being rational integers, depends on the accidental, as it were, choice of irreducible polynomial $F(x)$, i.e., of "irreducible" point in the corresponding irreducible maximal lattice, closed under multiplication. It can be shown that the lattice in $\mathbb{C}^n$ corresponding to R' is generated by the vectors from O to the points corresponding to $\mathbf{1}, \mathbf{v}, \mathbf{v}^2, \dots, \mathbf{v}^{n-1}$, where $\mathbf{1} = (1, 1, \dots, 1)$ and $\mathbf{v} = (\xi_1, \xi_2, \dots, \xi_n)$, the vector with components the roots of $F(x)$. There are infinitely many lattices of this sort, i.e., with generating frames made up of powers, contained as sublattices in the full, maximal lattice corresponding to R. On the other hand, the possibly larger ring R of algebraic integers in the revised (and modern) sense, depending on a root of a given monic irreducible integer polynomial, corresponds to a completely natural and unambiguous object in $\mathbb{C}^n$, namely an irreducible, maximal lattice, closed under coordinatewise multiplication.

Let Γ denote such a lattice. The product $y_1 \cdot y_2 \cdots y_n$ of the coordinates of a point y is called the *norm* of the point, denoted by $N(y)$. The norm is always an integer; in fact it is up to sign just the constant term of the integer monic polynomial of degree n having $y_1, y_2, \dots, y_n$ as its roots. The norm of a product $(y_1 \cdot z_1, y_2 \cdot z_2, \dots, y_n \cdot z_n)$ of two points is the product of their norms since

$$y_1 z_1 \cdot y_2 z_2 \cdots y_n z_n = (y_1 y_2 \cdots y_n)(z_1 z_2 \cdots z_n).$$

[77]Take, for instance, $F(x) = x^2 + 3$, $x^2 + 7$, or $x^2 + 11$. (See earlier footnotes.) *Trans.*

Points $\mathbf{e}$ of the lattice Γ with norms ± 1, are the units of the lattice, since $\mathbf{e} \cdot \Gamma = \Gamma$. For in the first place all points of $\mathbf{e} \cdot \Gamma$ (i.e., the points $\mathbf{e} \cdot \gamma$ where γ ranges over the points of Γ) are contained in Γ, since Γ is closed under coordinatewise multiplication; on the other hand multiplication of the lattice Γ by $\mathbf{e}$ yields as image a lattice whose fundamental parallelepiped has the same complex volume as that of Γ, since this multiplication simply "stretches" $\mathbb{C}^n$ along its standard axes by the factors $e_1, e_2, \ldots, e_n$ respectively, where $\mathbf{e} = (e_1, e_2, \ldots, e_n)$, and since $e_1 e_2 \cdots e_n = \pm 1$, the lattice Γ will be transformed not onto a proper sublattice but onto itself. It follows that the units of Γ are just those points with coordinates units of R, so that Dirichlet's theorem on the units of R applies to those of Γ.

Since the norm of a divisor of a point of Γ is a divisor of the norm of the point, each (nonzero, nonunit) point is either *irreducible*, i.e., not expressible as a product of two points neither of which is a unit, or else is the product of a finite number of such irreducible points. Nonuniqueness of the decomposition as a product of irreducible points means of course that there are finitely many irreducible points whose product is a point which is also expressible as a product of a different set of irreducible points.[78] Thus given two points A and B of Γ, can we decide by means of a single decomposition of each of A and B as a product of irreducible points, whether A divides B? Clearly not.

So how can we resuscitate uniqueness of decomposition of the points of Γ into products of "primes"? With this we come to the crux of this commentary, where all will be made clear.

It turns out that the following situation always occurs: In $\mathbb{C}^n$ one can supplement the lattice Γ with certain additional points in such a way that this larger system of points, called a *principal configuration*, while not a lattice, is still discrete and closed under multiplication, and has the further property that each nonzero point of the configuration decomposes uniquely (to within multiplication by units of Γ) as a product of prime points. These additional points themselves form finitely many lattices $\Gamma_1, \Gamma_2, \ldots, \Gamma_{h-1}$, having only the origin O in common with Γ and with each other. In addition the fundamental parallelepipeds of these lattices all have the same complex volume as Γ. The number h of these lattices (including Γ) is called the *number of classes of ideal numbers* or the *ideal class number* of the ring R in question, and the supplementary points, i.e., the points of $\Gamma_1, \Gamma_2, \ldots, \Gamma_{h-1}$, are called *auxiliary* or *nonprincipal.* These points correspond to the ideal factors introduced by Dedekind and Zolotarev.

This vivid geometric picture of the situation was first described by Klein in 1896 in the case of a 2-dimensional lattice; however, it can be realized in any dimension. Although this geometrical theory of ideal numbers is incomparably more graphic than all others, it has not so far become widely known among specialists.

How does the work of Zolotarev and Dedekind on ideal numbers fit into this picture? For every "nonexceptional" ordinary prime p, Zolotarev considered each ideal factor of p as "belonging to" some prime factor $V(x)$ of the given integer monic irreducible polynomial $F(x)$ modulo p, i.e., in his version an ideal prime factor is determined, for nonexceptional p, by a pair $p, V(x)$. For exceptional p on the other hand, an ideal prime factor of p is determined by p together with an algebraic integer ν from the list (6) of the above summary of Zolotarev's articles of 1878 and 1885 "having exactly one ideal prime factor in common with p," as this phrase is

[78] Each counted according to its multiplicity. *Trans.*

defined towards the end of that summary. The interpretation of this in terms of the above-described geometric picture is as follows: It can be shown that each point of the principal configuration $\Gamma \cup \Gamma_1 \cup \Gamma_2 \cup \cdots \cup \Gamma_{h-1}$ is the greatest common divisor of two points of Γ (in fact of infinitely many different pairs of points of Γ). The pair $p, V(\xi)$ (where as usual ξ is a root of $F(x)$) in the case where p is nonexceptional, or p, ν when p is exceptional, corresponds in each case to such a pair of points, with the ideal prime factor of p that the pair determines corresponding to a point of the configuration outside Γ (i.e. in one of the auxiliary lattices) which is the greatest common divisor of the original pair of points of Γ. Dedekind, on the other hand, uses the concept of an "ideal of a ring". It can be shown that corresponding to each of the lattices $\Gamma, \Gamma_1, \Gamma_2, \dots, \Gamma_{h-1}$ of the principal configuration there is among those lattices an "inverse" Γ_i^{-1}, in the sense that if all points of Γ_i^{-1} are multiplied by any point η of Γ_i, one obtains a sublattice of Γ, geometrically congruent, of course, to the lattice Γ_i^{-1}. This sublattice of Γ turns out to correspond to the ideal of the ring R (as defined by Dedekind) corresponding to the point η of Γ_i. It turns out that a sublattice of the principal lattice Γ corresponds to such an ideal precisely if multiplication of any point of the sublattice by any point of Γ yields again a point of the sublattice. Dedekind takes this as the defining property of an ideal (here of the lattice Γ). The concept of an ideal of a ring, as defined by Dedekind for these particular rings, has come to play a very important role in various branches of algebra.

If A and B are two points of the lattice Γ, then the totality of points of the form $AX+BY$, where X and Y range over all points of Γ, is an ideal, and corresponds to that point of the principal configuration representing the greatest common divisor of points A and B of Γ. This shows the equivalence of the two theories.

The theories of both Dedekind and Zolotarev may be regarded as in essence attempts to define ideal divisors without straying beyond the bounds of the lattice Γ. From the computational point of view Zolotarev's theory is superior since starting from a given p and corresponding V or ν, together defining an ideal prime divisor, it is clear how to perform calculations involving that ideal prime. On the other hand Dedekind's theory is better on the theoretical side, since it operates with more general concepts, and in particular includes a perfectly clear idea of what a nonprime ideal divisor should be—which doubtless Zolotarev's approach might be extended to encompass also. The incorporation of the geometric picture expounded above into either of these theories produces a very complete picture of the situation, although for actual calculations Zolotarev's theory is the more convenient.

Andreĭ Andreevich Markov (1856–1922)

Andreĭ Andreevich MARKOV

Andreĭ Andreevich Markov was born in 1856. His father, son of a country deacon, completed studies at a church seminary, and then served first as clerk, and then, when living in St. Petersburg, worked privately managing households and estates. He married twice, the first marriage producing, apart from daughters, a son Andreĭ, and the second another son Vladimir, who, although he died at 25 from tuberculosis, also became an outstanding mathematician well known at home and abroad. The son of Andreĭ Andreevich Markov, also Andreĭ Andreevich, is a prominent contemporary[79] Soviet mathematician.

Markov received his highschool education at Petersburg Gymnasium No. 5, and his higher education in the mathematics department of the faculty of physics and mathematics of St. Petersburg University.

On completing university in 1878 with the degree of "candidate", he was awarded the gold medal for his essay "On the integration of differential equations by means of continued fractions", the topic having been set by the faculty.

Throughout the 1860s and 70s the department of mathematics of St. Petersburg University could boast of such outstanding scholars as Chebyshev, Korkin and Zolotarev, as well as the prominent professors Posse and Bobylev. Through their lectures Chebyshev, Korkin and Zolotarev exerted an especially strong influence on their students; Chebyshev in particular often proposed to his students new questions and problems for investigation. Markov was fortunate in having continual access to such gifted mathematicians.

In this favourable atmosphere Markov's mathematical talent developed quickly, and two years after completing his courses he submitted his splendid dissertation for the degree of *Magister*, entitled "On binary quadratic forms of positive determinant". This work, very highly esteemed by Chebyshev, represents one of the finest achievements of the St. Petersburg school of number theory, and perhaps even of all Russian mathematics. It is enough to recall the sorts of questions in the field of rational approximation which at that time preoccupied the most prominent number-theorists of France and Germany, to appreciate how much deeper into the field Markov had penetrated. It is therefore perhaps not surprising that, although the dissertation was published immediately (in French in *Mathematische Annalen*), it did not become generally absorbed by west-European mathematicians, until from 1910 to the 1920s the Berlin mathematicians Frobenius and Remak attempted to master the set of ideas contained in Markov's work. In the last 10 to 15 years[80] several outstanding number-theorists from England and America have given Markov's achievement its due, and are attempting to extend his work further, albeit with only partial success.

[79]1947. *Trans.*

[80]Prior to 1947. *Trans.*

In 1884 Markov defended his doctoral dissertation "On certain applications of continued fractions". From 1880 he was *Privatdozent*, and from 1886 professor of St. Petersburg University. In 1886 also he was proposed by Chebyshev as adjunct of the Academy of Sciences, becoming in 1890 an extraordinary member, and in 1896 ordinary academician.

In 1905 Markov was awarded the title of distinguished professor of the university, but almost at once retired, in order, so he said, to make way for other, younger, mathematicians of ability. However, thanks to his perquisites as academician he was able to continue lecturing at the university, almost up to his death, on such topics as the theory of probability and continued fractions, among others. Andreĭ Andreevich died in 1922.

Apart from number theory his main work was carried out in the theory of algebraic continued fractions and its applications, in differential equations, in finite-difference and interpolation theory, in the theory of "Chebyshev functions", i.e., functions deviating minimally from zero, on problems of Chebyshev concerning limiting values of definite integrals, on the question of approximate computation of definite integrals, and especially in the theory of probability.

Markov was the most eloquent spokesman for Chebyshev's ideas and directions of research in probability theory. Especially remarkable is his research relating to the theorem of Jakob Bernoulli known as the Law of Large Numbers, to two fundamental theorems of probability theory due to Chebyshev, and to the method of least squares. Markov founded a completely new branch of probability theory, namely the study of the probabilities associated with sequences or chains of events, now of great importance in statistics and physics.

S. N. Bernshtein writes: "A. A. Markov's classic course on the computation of probabilities, and his original memoirs, models of accuracy and clarity of exposition, contributed to a very large extent to the transformation of the theory of probability into one of the most perfected areas of mathematics, and to the wide dissemination of Chebyshev's methods and directions of research. His profound analysis in the spirit of Chebyshev of the dependencies among observed random phenomena allowed Markov to extend probability theory in an essential way through the introduction and investigation of dependent random quantities."

In each field in which he worked, Markov chose for research unsolved problems of the first importance, and produced complete and rigorous solutions of them. Noteworthy is the special "Markovian style", marked by keenness of vision and rigour, characteristic of all of his work.

On Binary Quadratic Forms of Positive Determinant

Master's Dissertation

The remark of Korkin and Zolotarev concerning indefinite binary quadratic forms at the beginning of the second of their memoirs on positive quadratic forms examined above, provided the starting point for one of the finest number-theoretic works of the St. Petersburg school, namely Markov's master's dissertation "On binary quadratic forms of positive determinant".

Denoting the general binary quadratic form by

$$ax^2 + 2bxy + cy^2,$$

Markov defines the *discriminant*[81] D of the form by

$$D = -\begin{vmatrix} a & b \\ b & c \end{vmatrix} = b^2 - ac.$$

If $D < 0$ the form will be definite, if $D > 0$ indefinite. Thus by "binary quadratic forms of positive determinant" Markov means simply indefinite binary quadratic forms. He confines himself to the case where a, b, c are arbitrary real numbers.

Markov makes the following prefatory remarks. It was known at the time that the exact upper bound for the minima[82] of all definite binary quadratic forms of prescribed (negative) discriminant D is $\sqrt{\frac{4}{3}|D|}$, which is the minimum of the form

$$\sqrt{\frac{4}{3}|D|}(x^2 + xy + y^2)$$

(which indeed has discriminant D), and that the exact upper bound for the minima of all indefinite binary quadratic forms of prescribed (positive) discriminant D is $\sqrt{\frac{4}{5}D}$, which is the minimum of the form

$$\sqrt{\frac{4}{5}D}(x^2 - xy - y^2),$$

which is of discriminant D. However, as Korkin and Zolotarev observed, there is a large difference in behaviour between binary quadratic forms of positive and negative discriminant, consisting in the following: For any given positive number $l \leq \frac{4}{3}$ one can always find among the definite binary quadratic forms (i.e., with negative

[81] "Determinant" in the Russian original. *Trans.*

[82] Here the *minimum* of a quadratic form, definite or indefinite, is understood to be the minimum of the absolute value of the form under substitution of integer values, not all zero, for the variables. *Trans.*

discriminant D), those having minimum $\sqrt{l|D|}$, while in the case of positive discriminant this is not the case. If the form $\sqrt{\frac{4}{5}D}(x^2 - xy - y^2)$, which achieves the largest minimum $\sqrt{\frac{4}{5}D}$, is excluded, together with all forms equivalent to it, then the exact upper bound for the minima of the remaining indefinite binary forms of discriminant D is strictly less than $\sqrt{\frac{4}{5}D}$, namely $\sqrt{\frac{1}{2}D}$, achieved by the form

$$\sqrt{\frac{1}{2}D}(x^2 - 2xy - y^2)$$

(which indeed has discriminant D).

Markov continued this line of investigation further, discovering that if one excludes all forms equivalent to either $\sqrt{\frac{4}{5}D}(x^2 - xy - y^2)$ or $\sqrt{\frac{1}{2}D}(x^2 - 2xy - y^2)$, then the exact upper bound for the minima of all remaining binary forms of positive discriminant D, is different again from $\sqrt{\frac{1}{2}D}$, being in fact equal to $\sqrt{\frac{100}{221}D}$, achieved by the form

$$\sqrt{\frac{4}{221}D}(5x^2 - 11xy - 5y^2),$$

and so on. The numbers $\frac{4}{5}$, $\frac{1}{2}$, and $\frac{100}{221}$ turn out to be just the first three terms of an infinite (strictly) decreasing sequence

$$N_1, N_2, N_3, \ldots, N_k, N_{k+1}, \ldots,$$

of positive rational numbers, with the following properties:

1) *Corresponding to each of these numbers N_k there is an equivalence class of indefinite binary quadratic forms of discriminant D with minimum $\sqrt{N_k D}$.*

2) *If the (absolute) values taken by an indefinite binary quadratic form (or equivalence class of such forms) of determinant D as x and y range over all pairs of integers other than $0, 0$, are all greater than or equal to $\sqrt{N_k D}$, then the minimum of this form is one of the numbers*

$$\sqrt{N_1 D}, \sqrt{N_2 D}, \sqrt{N_3 D}, \ldots, \sqrt{N_k D}.$$

3) *As $k \to \infty$, the numbers $N_k \to \frac{4}{9}$. There exist infinitely many equivalence classes of binary quadratic forms of positive discriminant D with minimum $\frac{2}{3}\sqrt{D}$.*

The numbers N_k depend on certain periodic continued fractions, all with partial quotients (or terms) 1 or 2, which are closely related to particular sequences of positive numbers encountered by Johann Bernoulli in connection with his solution of a different problem.

This completes the preliminary summary of Markov's discoveries concerning indefinite binary quadratic forms. However, further investigation of the above-mentioned continued fractions led him to the following application.

It turns out that each number N_k has the form

$$\frac{1}{\frac{9}{4} - \frac{1}{Q_k^2}},$$

for some integer Q_k. Markov proves that if the equation

$$x^2 + y^2 + z^2 = 3xyz$$

has

$$x = l, \; y = m, \; z = n$$

as a solution in positive integers, then the numbers

$$\frac{1}{\frac{9}{4} - \frac{1}{l^2}}, \; \frac{1}{\frac{9}{4} - \frac{1}{m^2}}, \; \frac{1}{\frac{9}{4} - \frac{1}{n^2}}$$

must figure among the N_k, and, conversely, for each $N_k = \frac{1}{\frac{9}{4} - \frac{1}{l^2}}$, say, where l is a positive integer, there are infinitely many positive integers m, n such that

$$l^2 + m^2 + n^2 = 3lmn.$$

Thus Markov's investigations of continued fractions enabled him to find without especial difficulty all solutions of the Diophantine equation

$$x^2 + y^2 + z^2 = 3xyz$$

(i.e., all integer solutions) not exceeding in absolute value an arbitrary prescribed bound. We see, therefore, that the solution of this Diophantine equation is closely connected with certain periodic continued fractions with partial quotients all 1 or 2.

We now examine Markov's dissertation, published as a memoir, in detail. It is divided into two chapters.

In §1 of Chapter I Markov recalls how one goes from a given indefinite binary quadratic form to an equivalent "reduced" form using Lagrange's method for achieving this (Lagrange, "Oeuvres", Tome III, "Recherches d'arithmétique", §24).

For a given binary quadratic form

$$(a_0, b_0, c_0) = a_0 x_0^2 + 2b_0 x_0 y_0 + c_0 y_0^2, \tag{1}$$

with real coefficients a_0, b_0, c_0 and positive discriminant $D = b_0^2 - a_0 c_0$, the roots of the quadratic equation

$$a_0 \xi^2 + 2b_0 \xi + c_0 = 0 \tag{2}$$

will be real and distinct. If we develop one of these roots ξ_0 as a continued fraction

$$\pm\xi_0 = \alpha_0 + \cfrac{1}{\alpha_1 + \cfrac{1}{\alpha_2 + \cfrac{1}{\ddots + \cfrac{1}{\alpha_{n-1} + \frac{1}{\xi_n}}}}}, \tag{3}$$

where $\alpha_0, \alpha_1, \alpha_2, \ldots, \alpha_{n-1}$ are integers satisfying

$$\alpha_0 \geq 0, \; \alpha_1 > 0, \; \alpha_2 > 0, \ldots, \alpha_{n-1} > 0; \text{ and } \xi_n \geq 1,$$

and apply the successive changes of variables[83]

$$\begin{array}{lll} x_0 \to \pm(\alpha_0 x_1 + y_1), & & y_0 \to x_1, \\ x_1 \to & \alpha_1 x_2 + y_2, & y_1 \to x_2, \\ x_2 \to & \alpha_2 x_3 + y_3, & y_2 \to x_3, \\ \vdots & \vdots & \vdots \\ x_{n-1} \to & \alpha_{n-1} x_n + y_n, & y_{n-1} \to x_n, \end{array} \tag{4}$$

[83]Note that these linear transformations all have integer coefficients and determinant ± 1, so that the forms (a_i, b_i, c_i) are all equivalent. *Trans.*

then the form (1) is transformed into a form

$$(a_n, b_n, c_n) = a_n x_n^2 + 2b_n x_n y_n + c_n y_n^2, \tag{5}$$

with the property that one of the roots of the equation

$$a_n \xi^2 + 2b_n \xi + c_n = 0 \tag{6}$$

is ξ_n.

In the case that ξ_n is rational, say $\xi_n = A/B$ where A and B are integers, the form (a_n, b_n, c_n) takes the value zero for $x_n = A, y_n = B$, so that its minimum as (x, y) ranges over all pairs of integers other than $(0, 0)$ is zero, and so also for the equivalent form (a_0, b_0, c_0). Since we seek the largest of the minima of the absolute value of (indefinite) binary quadratic forms, this case is not of interest to us.

Hence we may assume that ξ_n is irrational, in which case it is known (see Lagrange, "Traité de la résolution des équations numériques", 1826, Chap. III, §19) that for sufficiently large n one of the roots of the equation (6) will lie strictly between -1 and 0. In view of this we may assume henceforth that our form (a_0, b_0, c_0) in (2) is such that both roots of the equation $a_0\xi^2 + 2b_0\xi + c_0 = 0$ are irrational, with one larger than 1 and the other negative and less than 1 in absolute value. Such a form is called *reduced.*

In §2 Markov considers the doubly infinite chain F

$$\dots (a_{-2}, b_{-2}, c_{-2}),\ (a_{-1}, b_{-1}, c_{-1}),\ (a_0, b_0, c_0),\ (a_1, b_1, c_1),\ (a_2, b_2, c_2), \dots$$

of reduced forms obtained from the form (a_0, b_0, c_0) in (2) (now assumed reduced) via the changes of variables

$$\begin{aligned} x_0 &\to \alpha_0 x_1 + y_1, & y_0 &\to x_1, \\ x_1 &\to \alpha_1 x_2 + y_2, & y_1 &\to x_2, \\ x_2 &\to \alpha_2 x_3 + y_3, & y_2 &\to x_3, \\ &\vdots & &\vdots \end{aligned}$$

where $\alpha_0, \alpha_1, \alpha_2, \dots$ are the terms of the continued fraction expansion of the positive root ρ_1 of equation (2):

$$\rho_1 = \alpha_0 + \cfrac{1}{\alpha_1 + \cfrac{1}{\alpha_2 + \cfrac{1}{\ddots}}},$$

and the changes of variables

$$\begin{aligned} x_0 &\to y_{-1}, & y_0 &\to -\alpha_{-1} y_{-1} + x_{-1}, \\ x_{-1} &\to y_{-2}, & y_{-1} &\to -\alpha_{-2} y_{-2} + x_{-2}, \\ x_{-2} &\to y_{-3}, & y_{-2} &\to -\alpha_{-3} y_{-3} + x_{-3}, \\ &\vdots & &\vdots \end{aligned}$$

where $\alpha_{-1}, \alpha_{-2}, \alpha_{-3}, \dots$ are the successive terms of the continued fraction expansion of the negative root ρ_2 of equation (2):

$$\rho_2 = -\cfrac{1}{\alpha_{-1} + \cfrac{1}{\alpha_{-2} + \cfrac{1}{\alpha_{-3} + \cfrac{1}{\ddots}}}}.$$

Furthermore it turns out that the same chain F is obtained starting from any other one of its members.

In §3 it is shown that the chain F contains all reduced forms equivalent to (a_0, b_0, c_0), as does the chain obtained by replacing each form (a, b, c) in F by $(c, -b, a)$.

In §4 Markov studies the chain

$$\ldots, \alpha_{-3}, \alpha_{-2}, \alpha_{-1}, \alpha_0, \alpha_1, \alpha_2, \alpha_3, \ldots,$$

which he denotes by I, whose members are (as noted above) the terms of the continued fraction expansions of the positive and negative roots of the equation $a_0\xi^2 + 2b_0\xi + c_0 = 0$, where (a_0, b_0, c_0) may be taken to be any reduced form equivalent to the original one. Denoting as before the discriminant of this form by D, Markov shows that the terms of this chain satisfy the equations[84]

$$d_\mu = \frac{2\sqrt{D}}{a_\mu} = \alpha_\mu + \cfrac{1}{\alpha_{\mu+1} + \cfrac{1}{\alpha_{\mu+2} + \cfrac{1}{\ddots}}} + \cfrac{1}{\alpha_{\mu-1} + \cfrac{1}{\alpha_{\mu-2} + \cfrac{1}{\ddots}}} = \frac{2}{L_\mu}.^{85}$$

In this connection he also shows that if the least in absolute value of the values taken by the form (a_0, b_0, c_0) at integer pairs (x_0, y_0) other than $(0,0)$, is $l\sqrt{D}$, then the quantity $\frac{2}{l}$ is the least upper bound of the d_μ.

Next the converse of this statement is proved, namely that an arbitrary doubly infinite chain I of positive integers arises as above from a chain F of reduced forms, with the μth form given by

$$L_\mu\sqrt{D}\left\{x - y\left(\alpha_\mu + \cfrac{1}{\alpha_{\mu+1} + \cfrac{1}{\alpha_{\mu+2} + \cfrac{1}{\ddots}}}\right)\right\} \times \left\{x + y\left(\cfrac{1}{\alpha_{\mu-1} + \cfrac{1}{\alpha_{\mu-2} + \cfrac{1}{\ddots}}}\right)\right\}.$$

Having established this equivalence, Markov writes: "On these grounds in our further discussion we will almost exclusively refer only to chains I and the related quantities L_μ."

Thus the first four sections are essentially preparatory. The remaining sections §§5, 6, and 7 of Chapter I are devoted to a complete investigation of those chains I for which for all μ

$$L_\mu \geq \frac{2}{3}, \quad \text{i.e.,} \quad d_\mu \leq 3.$$

Section 5 begins with the important observation that all such chains I must consist only of 1s and 2s, since if for some μ we had $\alpha_\mu \geq 3$, then from the above formula defining d_μ, we should obtain

$$d_\mu \geq 3 + \cfrac{1}{\ddots} > 3.$$

Furthermore chains for which $d_\mu \leq 3$ cannot contain isolated 1s or 2s, since if

$$\alpha_{\mu-1} < 3, \;\; \alpha_\mu = 2, \;\; \alpha_{\mu+1} = 1, \;\; \alpha_{\mu+2} = 2,$$

---

[84]Markov does not in fact use the symbol d_μ for this sum of continued fractions, employing throughout the notation $\frac{2}{L_\mu}$ instead.

[85]Here a_μ is the first coefficient of the form (a_μ, b_μ, c_μ). *Trans.*

then

$$d_\mu = 2 + \cfrac{1}{1 + \cfrac{1}{2 + \cfrac{1}{\ddots}}} + \cfrac{1}{\alpha_{\mu-1} + \cfrac{1}{\ddots}} > 2 + \frac{1}{1+\frac{1}{2}} + \frac{1}{3} = 3,$$

and similarly, if

$$\alpha_{\mu-1} = 1, \ \alpha_\mu = 2, \ \alpha_{\mu+1} = 1,$$

then

$$d_\mu = 2 + \cfrac{1}{1 + \cfrac{1}{\ddots}} + \cfrac{1}{1 + \cfrac{1}{\ddots}} > 2 + \frac{1}{1+1} + \frac{1}{1+1} = 3.$$

By means of similar, but more complex, arguments, which we omit, involving in particular the identity

$$\frac{1}{2+\frac{1}{g}} + \cfrac{1}{1 + \cfrac{1}{1+\frac{1}{g}}} = 1,$$

which appears in Lagrange's work, Markov ultimately arrives at the result that the only chains for which $d_\mu \leq 3$ for all μ, are

$$(I_0) \qquad \ldots, 1, 1, 1, 1, 1, 1, 1, 1, 1, \ldots,$$

consisting only of 1s,

$$(I_1) \qquad \ldots, 2, 2, 2, 2, 2, 2, 2, 2, 2, \ldots,$$

consisting only of 2s,

$$(I_\infty) \qquad \ldots, 1, 1, 1, 1, 2, 2, 1, 1, 1, 1, \ldots,$$

containing two consecutive 2s, with all the rest 1s,

$$(I^\infty) \qquad \ldots, 2, 2, 2, 2, 1, 1, 2, 2, 2, 2, \ldots,$$

containing two consecutive 1s, with all the rest 2s, and finally certain chains among those of the form

$$(I') \qquad \ldots, \underbrace{1, \ldots, 1}_{2r_{-2}}, 2, 2, \underbrace{1, \ldots, 1}_{2r_{-1}}, 2, 2, \underbrace{1, \ldots, 1}_{2r_0}, 2, 2, \underbrace{1, \ldots, 1}_{2r_1}, 2, 2, \underbrace{1, \ldots, 1}_{2r_2}, \ldots,$$

where $2r_k$ indicates the number of 1s appearing between pairs $2, 2$ of consecutive 2s, and at least one of these even numbers is nonzero.[86]

The chains I_0 and I_1 both satisfy the condition $d_\mu \leq 3$, since for I_0 we have

$$d_\mu = 1 + \cfrac{1}{1 + \cfrac{1}{1 + \cfrac{1}{\ddots}}} + \cfrac{1}{1 + \cfrac{1}{1 + \cfrac{1}{\ddots}}} = \sqrt{5},$$

and for I_1

$$d_\mu = 2 + \cfrac{1}{2 + \cfrac{1}{2 + \cfrac{1}{\ddots}}} + \cfrac{1}{2 + \cfrac{1}{2 + \cfrac{1}{\ddots}}} = 2\sqrt{2}.$$

[86]If just one such number is nonzero, and equal to 2, then of course we obtain the earlier type of chain denoted by I^∞. *Trans.*

The chains I_∞ and I^∞ also satisfy the condition $d_\mu \le 3$ for all μ, with the largest of the d_μ actually equal to 3, since

$$2+\cfrac{1}{1+\cfrac{1}{1+\cfrac{1}{\ddots}}}+\cfrac{1}{2+\cfrac{1}{1+\cfrac{1}{1+\cfrac{1}{\ddots}}}}=3,$$

and

$$2+\cfrac{1}{2+\cfrac{1}{2+\cfrac{1}{\ddots}}}+\cfrac{1}{1+\cfrac{1}{1+\cfrac{1}{2+\cfrac{1}{2+\cfrac{1}{\ddots}}}}}=3.$$

For chains of the form I' it can be shown that the condition $d_\mu \le 3$ is satisfied if and only if the chain

$$(R) \qquad \dots, r_{-3},\ r_{-2},\ r_{-1},\ r_0,\ r_1,\ r_2,\ r_3, \dots$$

satisfies the following conditions:

1) the differences

$$\dots, r_{-3}-r_{-2},\ r_{-2}-r_{-1},\ r_{-1}-r_0,\ r_0-r_1,\ r_1-r_2, \dots$$

should have only the values $-1, 0$ or 1;

2) for each k such that $r_k - r_{k+1} = 1$, the first of the numbers

$$r_{k+2}-r_{k-1},\ r_{k+3}-r_{k-2},\ r_{k+4}-r_{k-3}, \dots$$

that is nonzero should be positive, or else they should all be zero (to infinity);

3) for each k such that $r_k - r_{k+1} = -1$, the first nonzero number of the sequence

$$r_{k+2}-r_{k-1},\ r_{k+3}-r_{k-2},\ r_{k+4}-r_{k-3}, \dots$$

should be negative, or else they should all be zero (to infinity).

It can be inferred from these conditions that the absolute value of the difference between any two members of the chain R is at most 1, and then that the chain R must have one of the following forms:

$$(R_1) \qquad \dots, r,\ r,\ r,\ r,\ r,\ r,\ r,\ r,\ r, \dots,$$

$$(R_\infty) \qquad \dots, r-1,\ r-1,\ r-1,\ r,\ r-1,\ r-1,\ r-1, \dots,$$

$$(R^\infty) \qquad \dots, r,\ r,\ r,\ r-1,\ r,\ r,\ r, \dots,$$

$$(R') \qquad \dots, \underbrace{r-1,\dots}_{s_{-2}}, r, \underbrace{r-1,\dots}_{s_{-1}}, r, \underbrace{r-1,\dots}_{s_0}, r, \underbrace{r-1,\dots}_{s_1}, r, \underbrace{r-1,\dots}_{s_2}, \dots .$$

It is easy to verify that chains of types R_1, R_∞ and R^∞ satisfy conditions 1), 2) and 3), so that it only remains to determine for which chains

$$(S) \qquad \dots, s_{-4},\ s_{-3},\ s_{-2},\ s_{-1},\ s_0,\ s_1,\ s_2,\ s_3,\ s_4, \dots$$

chains of type R' satisfy them. A short argument shows that for this it is necessary and sufficient that the chain S itself satisfy those conditions.

Iterating the above, now with the chain S, we obtain a chain T say, and continuing in this way we obtain an infinite sequence

$$(\Phi) \qquad R,\ S,\ T, \dots$$

of sequences required to satisfy the conditions 1), 2), 3). From the above it is clear that if any one of these sequences satisfies those conditions, then those preceding it will also satisfy them, and so, finally, the original chain of type I' will satisfy the condition $d_\mu \leq 3$ for all μ.

In §6 the case where some chain of the sequence (Φ) is constant, i.e., has all its members equal, is considered. In this case the chain preceding this one will be periodic, and then, clearly, so also will the one in turn preceding it, and so on, whence finally we shall have that the original chain I' must be periodic. Hence in this case the number of distinct d_μ, and therefore L_μ, is finite, whence one may infer[87] that there are infinitely many chains of type I' satisfying $d_\mu < 3$, i.e. such that the least of the L_μ is greater than $\frac{2}{3}$.

We conclude therefore that:

There exist infinitely many equivalence classes of indeterminate binary quadratic forms (a, b, c) all of the same discriminant D, the least of whose (absolute) values at integer pairs (x, y) other than $(0, 0)$, is greater than $\frac{2}{3}\sqrt{D}$.

In the second half of §6 Markov considers the situation where one of the chains of the sequence (Φ) has the form

$$\ldots, \delta,\ \delta,\ \delta,\ \delta,\ \delta,\ \delta \pm 1,\ \delta,\ \delta,\ \delta,\ \delta,\ \delta, \ldots,$$

where δ is any positive integer. Under this assumption all earlier terms of the sequence (Φ) turn out to have the form

$$\ldots, \omega_4,\ \omega_3,\ \omega_2,\ \omega_1,\ \omega,\ \omega \pm 1,\ \omega_1,\ \omega_2,\ \omega_3,\ \omega_4, \ldots,$$

i.e., to be symmetric with respect to a segment of the form $\omega,\ \omega \pm 1$, and the chain I' to contain a segment 2, 2, 1, 1 (or its reverse), about which the other 1s and 2s are arranged symmetrically. It follows that at one of these places (at the first 2) we shall have $d_\mu = 3$, from which we conclude that:

There exist infinitely many equivalence classes of indefinite binary quadratic forms (a, b, c), all of the same discriminant D, with least absolute value at integer pairs $(x, y) \neq (0, 0)$ exactly $\frac{2}{3}\sqrt{D}$.

In §7 Markov considers those chains of type I' which for all μ are such that

$$L_\mu \geq l,$$

where l is a given number greater than $\frac{2}{3}$, i.e., such that $d_\mu \leq d$ where d is a given positive real number less than 3. By means of a very clever argument Markov shows that such a condition will be satisfied by only finitely many chains of type I'. In other words:

Given any number $l > \frac{2}{3}$, there are only finitely many equivalence classes of indefinite binary quadratic forms (a, b, c) of prescribed discriminant D, whose least absolute value at integer pairs $(x, y) \neq (0, 0)$ is greater than or equal to $l\sqrt{D}$. Moreover the chains of type I' corresponding to these forms are periodic.

It is not difficult to infer from the periodicity of these chains of type I' that the triples (a, b, c) of coefficients of the forms in question are proportional to triples of integers.

In view of the high degree of complexity of the arguments of Chapter II, in which Markov establishes a connection between forms whose minimum $l\sqrt{D}$ among the

[87] In the original this is claimed to be "obvious". *Trans.*

numbers they represent is larger than $\frac{2}{3}\sqrt{D}$, and gives a complete solution of the Diophantine equation

$$x^2 + y^2 + z^2 = 3xyz$$

(as noted earlier), we will forego examining his proofs. Instead we shall indicate how his final result can be given the following reformulation (different from Markov's).

Let $M_1, M_2, M_3, \ldots$ be indefinite binary quadratic forms of discriminants D_1, D_2, $D_3, \ldots$ respectively, such that for all i, M_i has minimum $m_i > \frac{2}{3}\sqrt{D_i}$, arranged in order of decreasing $m_i/\sqrt{D_i}$. By the above we may also suppose them normalized, i.e., multiplied by a suitable factor, so that each has integer coefficients of greatest common divisor 1. In this situation it turns out that each form M_i has an integer multiple of the form

$$Qx^2 + (3Q - 2q)xy + (r - 3q)y^2, \tag{7}$$

where Q, q, r, are positive integers with r defined by $Qr - q^2 = 1$, and with the property that there exist two integers Q', Q'' such that

$$Q^2 + Q'^2 + Q''^2 = 3QQ'Q'', \tag{8}$$

and $\pm q$ is equal to the least positive residue $\frac{Q'}{Q''} \pmod{Q}$. Furthermore the ratio of the minimum of this form divided by the square root of its discriminant turns out to be equal to

$$\sqrt{\frac{1}{\frac{9}{4} - \left(\frac{1}{Q}\right)^2}}.$$

The positive integers Q are called "Markov numbers". Here are the smallest of these numbers, yielding the smallest solutions (in absolute value) of the Diophantine equation (8):

$$Q = 1,\ 2,\ 5,\ 13,\ 29,\ 34,\ 89,\ 169,\ 194, \ldots.$$

The corresponding "Markov forms" (7) are respectively

$$\begin{aligned}
M_1 &= x^2 - xy - y^2,\\
M_2 &= x^2 - 2xy - y^2,\\
M_3 &= 5x^2 - 11xy - 5y^2,\\
M_4 &= 13x^2 - 29xy - 13y^2,\\
M_5 &= 29x^2 - 63xy - 31y^2,\\
M_6 &= 17x^2 - 38xy - 17y^2,\\
M_7 &= 89x^2 - 199xy - 89y^2,\\
M_8 &= 169x^2 - 367xy - 181y^2,\\
M_9 &= 97x^2 - 216xy - 97y^2,\\
&\vdots \qquad\qquad \vdots
\end{aligned}$$

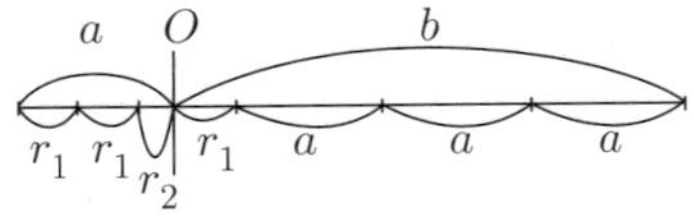

FIGURE 6

and the ratios of the minima of these quadratic forms to the square root of their discriminants are, by the above formula:

$$l = \sqrt{\frac{4}{5}}, \sqrt{\frac{1}{2}}, \sqrt{\frac{100}{221}}, \sqrt{\frac{676}{1517}}, \sqrt{\frac{3364}{7565}}, \sqrt{\frac{289}{650}}, \sqrt{\frac{31684}{71285}}, \sqrt{\frac{114244}{257045}}, \sqrt{\frac{9409}{21170}}, \dots .$$

Commentary

In order to obtain a more visual understanding of what is involved in Markov's dissertation, we consider his topic from a geometric viewpoint. We begin with the geometry of continued fractions. Recall how one looks for a common unit of measure for two given straight line segments a and b. Supposing the segment a is the shorter, one lays off a as many times, say α_0, as it will go into the larger segment b, obtaining a segment r_1 left over, the *first remainder*, shorter than a. Then one lays off the segment r_1 on the segment a as many times as possible, say α_1 times, obtaining a left-over *second remainder* r_2. Next one lays off the segment r_2 on the first remainder r_1 as many times, say α_2, as possible, obtaining a left-over *third remainder* r_3, and so on. If at any stage, say the kth, the process of laying off of r_k on the preceding remainder r_{k-1} yields zero remainder, i.e., r_k goes into r_{k-1} an exact whole number of times, then the procedure stops, and it is easy to verify by examining the procedure (or algorithm) in reverse that r_k is the largest possible common unit of measure for a and b, i.e., the longest segment which can be laid off on both a and b a whole number of times. If the algorithm does not terminate, i.e., if one continues obtaining nonzero remainders, the segments a and b are *incommensurable*. This is the (more general) geometric version of the well-known Euclidean algorithm involving successive division of integers.

This construction may be carried out in an especially convenient manner by positioning the segments a and b appropriately. As shown in Figure 6, one lays the segments a and b end to end in a straight line with their ends abutting at O, and lays off the length a on the segment b from the free end of the latter; the first remainder r_1 then forms a subsegment of b with one of its endpoints at O. Next the length r_1 is laid off on a similarly, starting from the free end of a, leaving the second remainder r_2 as a subsegment of a with right-hand endpoint O, and so on.

This method of construction of successive remainders can be further modified into one taking place not just on the line, but in the plane. As shown in Figure 7, take any straight line v in the plane, lay off on opposite sides from any two distinct points of v line segments a and b perpendicular to v, and consider the frame $(\mathfrak{A}_{-1}, \mathfrak{A}_0)$ attached to any point O of v as origin and with the tips of the vectors at the respective free ends of the segments b and a, labelled by A_{-1} and A_0 in Figure 7.

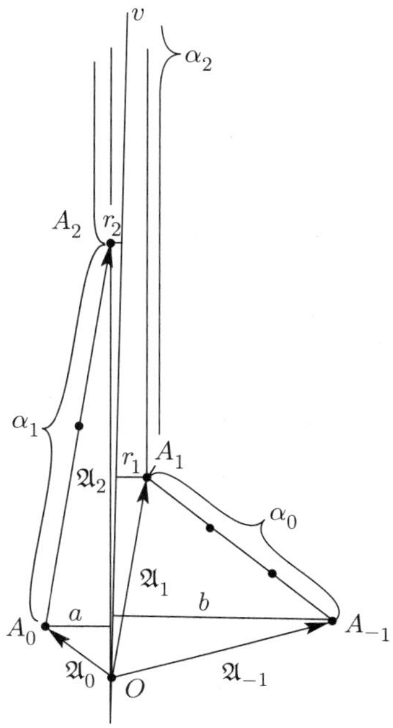

FIGURE 7

Now lay off from the point A_{-1} the vector $\mathfrak{A}_0$ as many times as possible (in fact it will be α_0 times) without crossing the line v, and label the point reached this way A_1, i.e., draw from A_{-1} a line segment parallel to $\mathfrak{A}_0$ and α_0 times its length. Thus A_1 is on the same side of the line v as A_{-1}. Write $\mathfrak{A}_1$ for the vector $\overrightarrow{OA_1}$. Next lay off the vector $\mathfrak{A}_1$ from the tip A_0 of $\mathfrak{A}_0$ as many times as possible (namely α_1) without crossing the line v, i.e., so that the point A_2 reached in this way lies on the same side of v as the vector $\mathfrak{A}_0$, and denote the vector $\overrightarrow{OA_2}$ by $\mathfrak{A}_2$. Next lay off the vector $\mathfrak{A}_2$ from the tip A_1 of the vector $\mathfrak{A}_1$ as many time as possible (namely α_2 times) without crossing v; label the point reached in this way A_3, and write $\mathfrak{A}_3$ for the vector $\overrightarrow{OA_3}$, and so on. We shall call this algorithm "construction of beaks". It is clear that the perpendicular distances $r_1, r_2, r_3, \ldots$ from the tips of the beaks, i.e., from the points $A_1, A_2, A_3, \ldots$, to the line v, will be, in order, the remainders obtained by means of successive subdivision starting with the segments a and b as in the earlier algorithm. In fact if in Figure 7 one projects parallel to v the segments labelled $r_1, r_2, r_3, \ldots$ together with those labelled a and b, onto any line perpendicular to v, one obtains exactly Figure 6.

Let A_1' be the point at which the line segment $A_{-1}A_1$ extended, meets v, as shown in Figure 8, and denote by ξ_0 the length of the vector $\overrightarrow{A_{-1}A_1'}$ as a multiple of the parallel vector $\mathfrak{A}_0$ taken as having unit length; thus $\overrightarrow{A_{-1}A_1'} = \xi_0\mathfrak{A}_0$. Similarly define the number λ_0 by $\overrightarrow{A_1A_1'} = \lambda_0\mathfrak{A}_0$, as in Figure 8. Analogously, denote by ξ_1 and λ_1 the lengths of the segments A_0A_2' and A_2A_2', which are parallel to the vector $\mathfrak{A}_1$, in terms of multiples of the length of $\mathfrak{A}_1$. (Here A_2' is the point where the

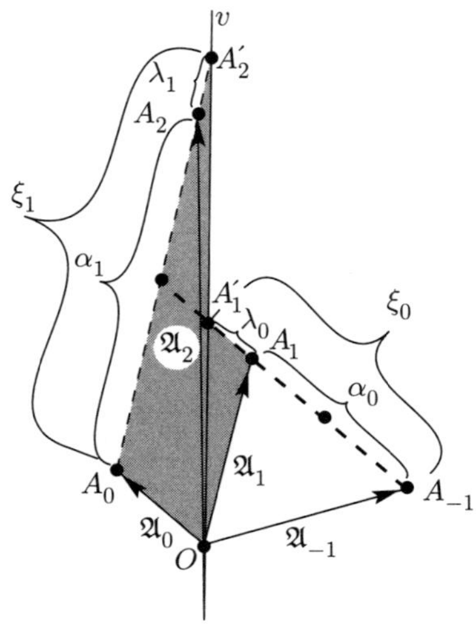

FIGURE 8

segment A_0A_2 extended, meets v, as in Figure 8.) Continue in this way, defining $\xi_2, \lambda_2, \xi_3, \lambda_3, \ldots$, and so on.

From the similar triangles OA_1A_1' and $A_2'A_0O$ shaded in in Figure 8, we have that λ_0 is less than OA_0, the length of $\mathfrak{A}_0$, by the same factor by which ξ_1 is larger than OA_1, the length of $\mathfrak{A}_1$, and then, analogously, λ_1 is less than the length of $\mathfrak{A}_1$ by the same factor by which ξ_2 is larger than the length of $\mathfrak{A}_2$, and so on.[88] It follows that

$$\begin{aligned}\xi_0 = \alpha_0 + \lambda_0 = \alpha_0 + \frac{1}{\xi_1} &= \alpha_0 + \frac{1}{\alpha_1 + \lambda_1}\\ = \alpha_0 + \frac{1}{\alpha_1 + \frac{1}{\xi_2}} &= \alpha_0 + \frac{1}{\alpha_1 + \frac{1}{\alpha_2+\lambda_2}}\\ = \cdots = \alpha_0 &+ \cfrac{1}{\alpha_1 + \cfrac{1}{\alpha_2 + \cfrac{1}{\ddots + \cfrac{1}{\alpha_{n-1} + \cfrac{1}{\underbrace{\alpha_n + \lambda_n}_{\xi_n}}}}}},\end{aligned}$$

where α_0 is the integer part of ξ_0, i.e., of $\frac{b}{a}$; α_1 is the integer part of $\xi_1 = \frac{1}{\lambda_0}$, i.e., of $\frac{a}{r_1}$; and so on. Thus we have obtained the expansion of ξ_0 as a continued fraction, i.e., of $\frac{b}{a}$, since it is easy to see from the appropriate similar triangles that $\xi_0 = \frac{b}{a}$. Hence the "algorithm for constructing beaks" yields, starting from the given line segments a and b, the successive terms of the continued fraction expansion of their ratio $\frac{b}{a}$.

The number ξ_0 has the following further geometric meaning. From Figure 8 we see that relative to the frame $(\mathfrak{A}_{-1}, \mathfrak{A}_0)$, the point A_1' of the line v through

[88] That is, since, in Figure 8, $\frac{OA_0}{A_1A_1'} = \frac{A_0A_2'}{OA_1}$, which is the same as $\frac{|\mathfrak{A}_0|}{\lambda_0|\mathfrak{A}_0|} = \frac{\xi_1|\mathfrak{A}_1|}{|\mathfrak{A}_1|}$, we infer that $\frac{1}{\lambda_0} = \xi_1$, and then, analogously, $\frac{1}{\lambda_1} = \xi_2$, and so on. *Trans.*

the origin O of that frame, has ordinate (i.e., $\mathfrak{A}_0$-coordinate) ξ_0 and abscissa 1, so that the number ξ_0 may be regarded as the *angular coefficient*[89] of the line v with respect to the frame $(\mathfrak{A}_{-1}, \mathfrak{A}_0)$. Thus the number ξ_0 expanded via the beak-construction algorithm as a continued fraction, is the angular coefficient measuring the inclination of the initially chosen line v relative to the frame $(\mathfrak{A}_{-1}, \mathfrak{A}_0)$.

We shall now prove geometrically that an arbitrary infinite continued fraction converges. To this end we choose any frame $(\mathfrak{A}_{-1}, \mathfrak{A}_0)$ in the plane, with origin O, and a completely arbitrary sequence

$$\alpha_0,\ \alpha_1,\ \alpha_2,\ \alpha_3, \ldots, \alpha_n, \ldots$$

of positive integers, and proceed to execute the algorithm for constructing beaks of successive "lengths" $\alpha_0, \alpha_1, \alpha_2, \alpha_3, \ldots, \alpha_n, \ldots$, as in Figure 7, but now without reference to any segments a or b or initial line v. We shall prove that the resulting sequence of vectors

$$\mathfrak{A}_{-1}, \mathfrak{A}_0, \mathfrak{A}_1, \mathfrak{A}_2, \mathfrak{A}_3, \ldots, \mathfrak{A}_n, \ldots$$

all emanating from O, will converge in direction to some limiting direction. For on the one hand the lengths of these vectors increase to infinity since their coordinates relative to the frame $(\mathfrak{A}_{-1}, \mathfrak{A}_0)$ are obtained from the coordinates $(1,0), (0,1)$ of the vectors of the frame $(\mathfrak{A}_{-1}, \mathfrak{A}_0)$ by means of successive addition and multiplication by positive integers,[90] while on the other hand the parallelograms determined by the pairs of vectors $(\mathfrak{A}_{-1}, \mathfrak{A}_0), (\mathfrak{A}_0, \mathfrak{A}_1), (\mathfrak{A}_1, \mathfrak{A}_2), \ldots$, all have the same area since the parallelograms determined by successive pairs $(\mathfrak{A}_{i-1}, \mathfrak{A}_i), (\mathfrak{A}_i, \mathfrak{A}_{i+1})$ share the base $\mathfrak{A}_i$ and have the same height. Hence the angles between the vectors $\mathfrak{A}_i$ and $\mathfrak{A}_{i+1}$ must tend to zero as i tends to infinity (since the area of a parallelogram is equal to the product of the lengths of two nonparallel sides and the sine of the angle between them). Furthermore each such angle after the first is contained in its predecessor. Hence by a well-known lemma of analysis there is exactly one direction contained in all these angles, which therefore the directions of the vectors $\mathfrak{A}_{-1}, \mathfrak{A}_0, \mathfrak{A}_1, \mathfrak{A}_2, \ldots$, approach as a limit.

Consider the straight line v through the origin O of the frame $(\mathfrak{A}_{-1}, \mathfrak{A}_0)$ having this direction. It is now clear that our arbitrarily chosen continued fraction

$$\alpha_0 + \cfrac{1}{\alpha_1 + \cfrac{1}{\alpha_2 + \cfrac{1}{\ddots + \frac{1}{\alpha_n}}}}$$

has as limit as $n \to \infty$ the above-mentioned "angular coefficient" of the line v with respect to the frame $(\mathfrak{A}_{-1}, \mathfrak{A}_0)$.

Note that the tips of the vectors $\mathfrak{A}_{-1}, \mathfrak{A}_0, \mathfrak{A}_1, \mathfrak{A}_2, \mathfrak{A}_3, \ldots$, are points of the planar lattice determined by the frame $(\mathfrak{A}_{-1}, \mathfrak{A}_0)$, since by construction they have integer coordinates relative to that frame.

Having considered the geometry of continued fractions, we now turn to the investigation of the distribution of the points of an arbitrary planar lattice Γ with respect to a cross formed by straight lines u and v passing through the origin O of the lattice. We shall consider only the most interesting case, namely when these lines are incident with no points of the lattice other than O, their point of

[89] The analogue of the usual slope of a line relative to rectangular axes. *Trans.*

[90] Thus in Figure 7, the tips $A_{-1}, A_0, A_1, A_2, \ldots$ of the vectors have coordinates $(1,0)$, $(0,1), (1, \alpha_0), (\alpha_1, \alpha_1\alpha_0), \ldots$. *Trans.*

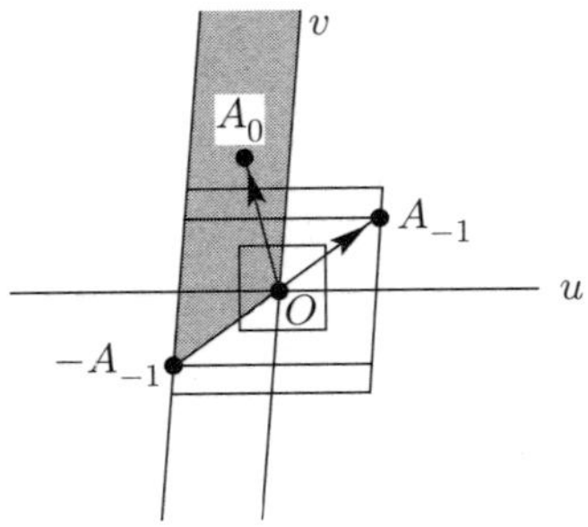

FIGURE 9

intersection. Let $(\mathfrak{A}, \mathfrak{B})$ be a generating frame for the lattice, and consider the adjoining frame $(-\mathfrak{A}, \mathfrak{B})$,[91] both attached to O. The lines u and v either both pass through the first "quadrant" of one of these frames[92] or else one passes through the first quadrant of one frame, the other through the first quadrant of the other. It can be shown that for a straight line through O to be incident with no points of the lattice other than O, it is necessary and sufficient that the angular coefficient of that line relative to a generating frame of the lattice with respect to which the line passes through the first quadrant, be irrational.

Consider now a parallelogram with centre O and sides parallel to u and v, sufficiently small to contain no points of the lattice Γ except O. Expand the parallelogram via a homothety[93] with centre O until it first encounters a point A_{-1} say, of the lattice, and then truncate it, if need be, so as to obtain finally a parallelogram with centre O, sides parallel to u and v, and with the point A_{-1} at a vertex, as in Figure 9. This parallelogram will then have the point $-A_{-1}$ as the vertex diagonally opposite A_{-1}, but no points of Γ other than these and O either in its interior or on its boundary. For if there were some other point of Γ on the boundary of the parallelogram, then there would be a lattice vector, i.e., a vector with tail and tip points of the lattice, parallel to one of u and v, and then by laying this vector off from O along whichever of u and v it is parallel to, its tip would define a point of the lattice on one of the two lines, contrary to our assumption. A line containing a row of points of the lattice Γ parallel to OA_{-1} and closest to it, will meet the shaded region shown in Figure 9, whence, taking into account its width, that region will contain a point A_0 of the row, lying of course above the parallelogram since the latter contains only the points O, A_{-1} and $-A_{-1}$ of Γ.

Denoting the vectors $\overrightarrow{OA_{-1}}$ and $\overrightarrow{OA_0}$ by $\mathfrak{A}_{-1}$ and $\mathfrak{A}_0$ respectively, we now apply the beak-constructing algorithm as initially described, starting with these two vectors and the line v, assumed, as in Figure 9, to pass between them. Since A_0 lies closer to the line v than A_{-1}, we shall have $\alpha_0 > 0$, and then, as is easily seen, all subsequent α_i will likewise be positive. We now carry out a somewhat modified version of the algorithm in reverse, starting with the frame $(\mathfrak{A}_0, -\mathfrak{A}_{-1})$ and the line u. As the first step we lay off the vector $-\mathfrak{A}_{-1}$ $(= \overrightarrow{OA_{-1}})$ from A_0

[91]The original has $(\mathfrak{B}, \mathfrak{A})$ here. *Trans.*

[92]In the sense of containing points with both coordinates relative to the frame in question positive. *Trans.*

[93]That is, dila(ta)tion. *Trans.*

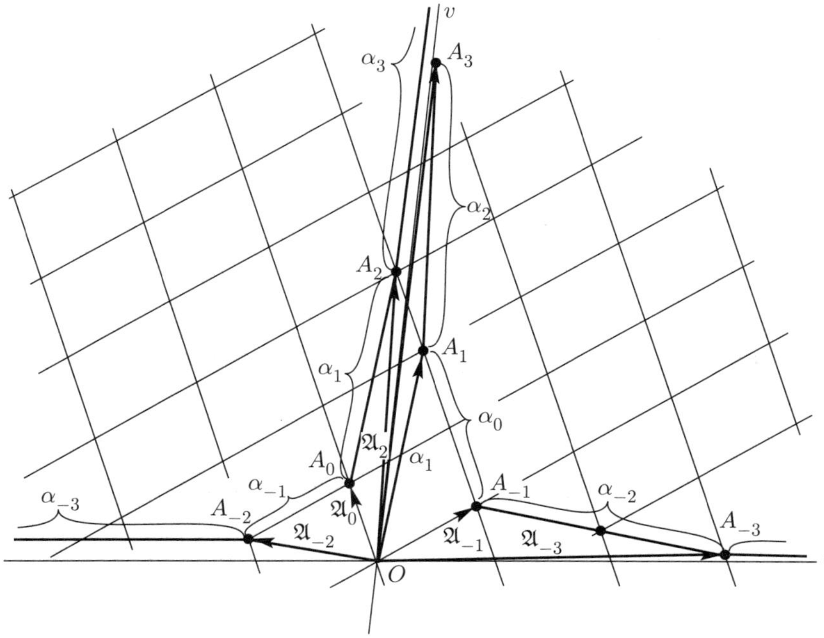

FIGURE 10

as many times as possible, say α_{-1} times, without crossing the line u (in Figures 9 and 10, remaining above u), reaching a point A_{-2}, and write $\mathfrak{A}_{-2} = \overrightarrow{OA_{-2}}$, as in Figure 10. Then, as the second step of the algorithm, from the point A_{-1} we lay off the vector $-\mathfrak{A}_{-2}$ as many times as possible, say α_{-2} times, again while still remaining above u, reaching a point A_{-3} (see Figure 10), and denoting $\overrightarrow{OA_{-3}}$ by $\mathfrak{A}_{-3}$, and so on. Since A_0 lies further from the axis u than does A_{-1}, the integer α_{-1} will be positive, and then all subsequent α_{-i} will also be positive.

The two infinite polygons

$$(\Pi_I) \qquad \ldots A_{-3}A_{-1}A_1A_3\ldots$$

and

$$(\Pi_{II}) \qquad \ldots A_{-4}A_{-2}A_0A_2A_4\ldots$$

are called the *Kleinian polygons* of the lattice Γ relative to the asymptotes u and v. We shall now show that these polygons are the boundaries of the convex hulls of the sets Γ_I and Γ_{II} consisting of the points of Γ lying in the "quadrants" I and II respectively, determined by the axes u and v.

We begin with the observation that the strip bounded by the segment OA_0 (extended) on one side and by $A_{-1}A_1$ (extended) on the other, contains no points of the lattice Γ in its interior, since $(\mathfrak{A}_{-1}, \mathfrak{A}_0)$ is a generating frame for the lattice. Hence the triangle cut off from the corner of quadrant I by the line segment $A_{-1}A_1$ extended, has no points of the lattice in its interior, so that all the points of Γ_I are contained in the intersection of quadrant I with the half-plane with boundary $A_{-1}A_1$ extended, not containing O. The analogous assertion holds via similar reasoning for the lines containing the segments A_1A_3, $A_3A_5, \ldots$, and also for the lines containing $A_{-3}A_{-1}$, $A_{-5}A_{-3}, \dots$. Thus all the points of Γ_I lie in the infinite convex polygonal region with boundary the polygon Π_I. However, since the vertices $\dots, A_{-3}, A_{-1}, A_1, A_3, \dots$ of this polygon are all points of Γ_I, it is immediate that this polygonal region is indeed the smallest convex set containing all points of Γ_I, i.e., the convex hull of Γ_I. One shows completely analogously that (the region bounded by) the polygon Π_{II} is the convex hull of the set Γ_{II}.

We shall call the vectors $\mathfrak{A}_i$ with tails at the origin O and tips the vertices of the polygons Π, the *vectors of the polygons* Π_I and Π_{II}. We have the following remarkable theorem:

Between the polygonal boundaries Π_I and Π_{II} of the convex hulls of the subsets Γ_I and Γ_{II} of the lattice Γ lying in quadrants I and II, one has the following correspondence: Each side of one of these polygons corresponds to a unique vector of the other, parallel to it, and furthermore the length of the side is an integer multiple of the length of the corresponding vector, and the same holds for the other polygon mutatis mutandis.

Ordering the positive integers α_i according to their indices, we obtain a doubly infinite chain

$$(\alpha) \qquad \dots, \alpha_{-3}, \ \alpha_{-2}, \ \alpha_{-1}, \ \alpha_o, \ \alpha_1, \ \alpha_2, \ \alpha_3, \dots,$$

of positive integers, which are the lengths of successive sides of the polygons Π_I and Π_{II} alternately, measured as multiples of the lengths of the corresponding vectors. From the above it follows that to each planar lattice Γ and cross uv with its point of intersection a point of Γ but no other point of Γ on it, there corresponds in the manner described above a doubly infinite chain of positive integers.

We shall now prove the converse:

To each doubly infinite chain of positive integers (α) there corresponds a cross uv situated as above in a planar lattice Γ, giving rise to that chain in the manner just described.

To see this take an arbitrary frame $(\mathfrak{A}_{-1}, \mathfrak{A}_0)$ in the plane, and the "adjacent" frame $(-\mathfrak{A}_0, \mathfrak{A}_{-1})$. Now implement the beak-construction algorithm, starting with the frame $(\mathfrak{A}_{-1}, \mathfrak{A}_0)$ and the sequence of positive integers $\alpha_0, \alpha_1, \alpha_2, \dots,$ as described earlier, to obtain ultimately a straight line v (determined by the limiting direction and the origin O of the frame). Now repeat the beak-construction algorithm starting with the frame $(-\mathfrak{A}_0, \mathfrak{A}_{-1})$ and the sequence $\alpha_{-1}, \alpha_{-2}, \alpha_{-3}, \dots,$ as in Figure 10. Ultimately this yields a limiting direction and hence a line u through the origin O. The lattice Γ generated by the frame $(\mathfrak{A}_{-1}, \mathfrak{A}_0)$, and the cross uv, will then give rise to the given chain (α) in the manner described earlier.

Note that the above constructions are affine, i.e., the same conclusions hold if the figures are subjected to any affine transformation,[94] so that by applying a suitable affine transformation if necessary, we may always arrange for the lines u and v to be perpendicular. Hence there is no loss in generality in all of the above in assuming that u and v are perpendicular.

We next investigate the extent to which the frame $(\mathfrak{A}_{-1}, \mathfrak{A}_0)$ is determined by a given chain (α).[95] From earlier we know that the chain (α) determines the angular coefficients of the lines u and v relative to the frames $(\mathfrak{A}_{-1}, \mathfrak{A}_0)$ and $(-\mathfrak{A}_0, \mathfrak{A}_{-1})$

[94]That is, a bijection of the plane preserving straight lines and parallelism. *Trans.*

[95]And cross uv? —which as just noted may be assumed right-angled. *Trans.*

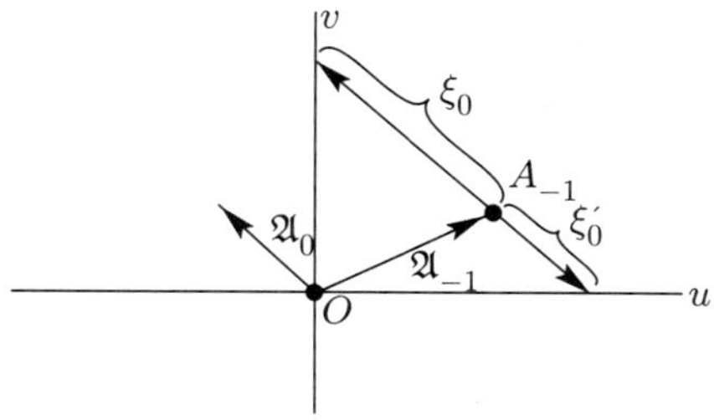

FIGURE 11

respectively. (The angular coefficient of v with respect to the frame $(\mathfrak{A}_{-1}, \mathfrak{A}_0)$ is

$$\xi_0 = \alpha_0 + \cfrac{1}{\alpha_1 + \cfrac{1}{\alpha_2 + \cfrac{1}{\ddots}}},$$

and that of u with respect to $(-\mathfrak{A}_0, \mathfrak{A}_{-1})$ is

$$\xi_0' = \cfrac{1}{\alpha_{-1} + \cfrac{1}{\alpha_{-2} + \cfrac{1}{\alpha_{-3} + \cfrac{1}{\ddots}}}},$$

as was shown earlier.) Hence if we choose the vector $\mathfrak{A}_{-1}$ arbitrarily (emanating from the point O where u and v cross), then, as shown in Figure 11, through its tip A_{-1} we must choose that (clearly unique) line segment with endpoints on u and v, with the property that the point A_{-1} divides it in the ratio ξ_0/ξ_0', and then the vector $\mathfrak{A}_0$ is essentially determined by having to be parallel to that line segment and to be of such a length that the two parts of the line segment through A_{-1} have lengths ξ_0 and ξ_0' when measured in terms of multiples of $\mathfrak{A}_0$. Thus if $\mathfrak{A}_{-1}$ is chosen arbitrarily, then the vector $\mathfrak{A}_0$ is determined.[96] We conclude that a doubly infinite chain (α) of positive integers determines the lattice Γ only to within an arbitrary "stretching" of the axes u and v, corresponding to a different choice of the vector $\mathfrak{A}_{-1}$. In other words, a given chain (α) determines the lattice to within a hyperbolic rotation relative to u and v as asymptotes,[97] and a homothety centred at the origin O.

We shall say that a frame attached to O is *reduced along the axis* v if it satisfies the following four conditions:

1° Its vectors lie on opposite sides of the axis v;

2° Its vectors lie to one side of the axis u (perpendicular to v and also passing through O);

3° The tip of the second vector of the frame lies closer to the axis v than the tip of the first;

4° The tip of the second vector lies further from the axis u than the tip of the first.

---

[96] Up to sign? *Trans.*

[97] This is defined to be a transformation of the form $u \to a^{-1}u, v \to av$ $(a > 0, a \neq 1)$, for all points (u, v) of the plane. *Trans.*

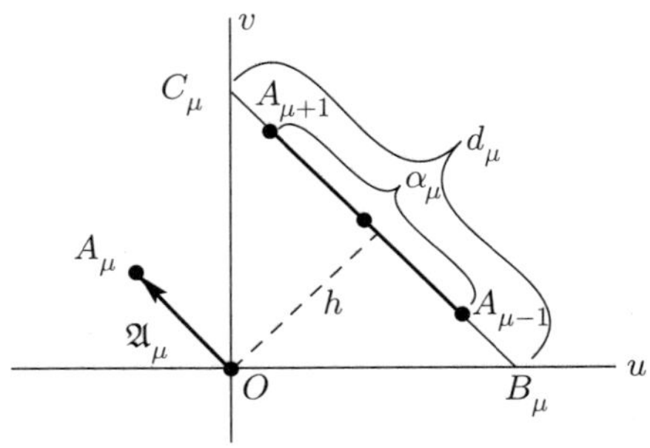

FIGURE 12

The frame $(\overrightarrow{OA_{-1}}, \overrightarrow{OA_0})$ constructed in Figure 9 is reduced along the axis v, and in the ensuing argument based on that frame, only the properties $1°, 2°, 3°, 4°$ were actually used. It turned out as a consequence of those properties that the tips of this frame form an adjacent pair among the converging pairs of vertices of the polygons Π_I and Π_{II}, in the sense of having consecutive indices. It is easy to see from the beak-construction algorithm that the converse holds: vertices comprising such an adjacent convergent pair are the tips of a reduced frame. Thus:

Reduced frames are exactly those frames with the property that the tips of their vectors form an adjacent convergent pair of vertices of the polygons Π_I and Π_{II}.

We now bring to light another relation important for us in the sequel. Assuming the axes u and v perpendicular, we apply an appropriate hyperbolic rotation to bring the side $A_{\mu-1}A_{\mu+1}$ of one of the polygons Π into a position where it is parallel to the bisector of the adjacent right angle, as in Figure 12. In the notation of the diagram, we then have $h = \frac{B_\mu C_\mu}{2}$, where $B_\mu C_\mu$ denotes the usual Euclidean length of the chord formed by the side $A_{\mu-1}A_{\mu+1}$ extended to meet the axes u and v. Hence the area of the fundamental parallelogram of the lattice is

$$S = |\mathfrak{A}_\mu| \cdot \frac{B_\mu C_\mu}{2}, \quad \text{whence } |\mathfrak{A}_\mu| = \frac{2S}{B_\mu C_\mu}, \text{ or } |\mathfrak{A}_\mu|^2 = \frac{2S}{d_\mu},$$

where d_μ is the length of the chord $B_\mu C_\mu$ measured in multiples of the vector $\mathfrak{A}_\mu$, i.e., $B_\mu C_\mu = d_\mu |\mathfrak{A}_\mu|$. Denoting by u_μ, v_μ the coordinates of A_μ, we have $|\mathfrak{A}_\mu|^2 = u_\mu^2 + v_\mu^2$, whence, in view of the fact that OA_μ bisects the right angle in which it lies, $|\mathfrak{A}_\mu|^2 = 2|u_\mu v_\mu|$. Now the parameter $N(A_\mu) = u_\mu v_\mu$ is the constant term of the hyperbola through A_μ having the u- and v-axes as asymptotes,[98] and therefore represents the hyperbolic distance[99] of the point A_μ from the origin O. Thus we infer the relation

$$(*) \qquad |N(A_\mu)| = \frac{S}{d_\mu}.$$

Finally we express the "length" d_μ of the chord $B_\mu C_\mu$ as measured in units of $|\mathfrak{A}_\mu|$, in terms of the members α_i of the chain (α). Since ξ_μ is the angular coefficient

[98] Namely $xy = N(A_\mu)$. *Trans.*

[99] The *hyperbolic distance* of a point of the lattice from the origin O is defined as the product of the uv-coordinates of the point—see below. *Trans.*

of the line v relative to the frame $(\mathfrak{A}_{\mu-1}, \mathfrak{A}_\mu)$, and $|\xi'_\mu|$ is the angular coefficient of the line u relative to the frame $(\mathfrak{A}_{\mu-1}, -\mathfrak{A}_\mu)$, and $d_\mu = \xi_\mu + |\xi'_\mu|$, we have

$$d_\mu = \alpha_\mu + \cfrac{1}{\alpha_{\mu+1} + \cfrac{1}{\alpha_{\mu+2} + \cfrac{1}{\ddots}}} + \cfrac{1}{\alpha_{\mu-1} + \cfrac{1}{\alpha_{\mu-2} + \cfrac{1}{\alpha_{\mu-3} + \cfrac{1}{\ddots}}}}.$$

This is the geometric version of the important formula, constituting the basis of Markov's investigations.

We are now ready to explicate the geometric import of the results of Markov's dissertation. In connection with the above theorem that each doubly infinite chain (α) of positive integers corresponds in a particular way to a (perpendicular) cross uv and a planar lattice Γ, and conversely, we observe first that if the chain (α) has arbitrarily large members, then the lattice Γ contains points hyperbolically arbitrarily close to the origin O, since we always have $d_\mu > \alpha_\mu$, so that by the formula (*) the larger α_μ, the smaller $|N(A_\mu)|$, i.e., in the case of unbounded (α), the greatest lower bound of the absolute hyperbolic distances of the points of Γ from the origin O is zero. On the other hand when the chain (α) is bounded above, the greatest lower bound of the absolute hyperbolic distances of the points of Γ from O will be positive, since in view of the inequalities $d_\mu < \alpha_\mu + 2$, the set of all d_μ is then also bounded above. In particular the latter situation will occur when the chain (α) is periodic.

From any lattice for which the greatest lower bound of the absolute hyperbolic distances of its points from the origin is positive, we can by applying a suitable homothety obtain a lattice for which this greatest lower bound is 1. Hence there exist lattices for which the greatest lower bound of the numbers $|N(A_\mu)|$ is 1.

Consider the region γ of the plane consisting of those points the product of whose u- and v-coordinates is less than 1 in absolute value, i.e., briefly, for which $|u \cdot v| < 1$. We shall call a lattice *admissible* if it has only the point O in the region γ. An *extremal* lattice will then be one that is admissible and in addition admits no continuous deformation[100] throughout which the area of its fundamental parallelogram decreases while admissibility is preserved.

We now consider how one may continuously change an admissible lattice Γ into a lattice Γ' in such a way that throughout the change all intermediate lattices are likewise admissible. In order for admissibility to be preserved it is of course necessary in the first place that the point of the lattice at the centre O of the hyperbolic cross $|u \cdot v| < 1$ remains there, and secondly that no other of its points enter that hyperbolic cross during the transformation.

Any such change of the lattice may be achieved by means of a continuous family of affine transformations[101] of the plane of the lattice. Of course such affine transformations must fix the origin O. We may also assume without loss of generality that they fix the lines u and v as wholes, since any continuous "rotational" motion of the axes u or v about O relative to the points of the lattice would involve some lattice points entering the "forbidden" region γ, in view of the existence of points of the lattice arbitrarily close to these lines sufficiently far out along them. (Note that

---

[100]Or rather a continuous family of transformations parametrized by an interval, i.e., a homotopy, where the functions are of a restricted kind. *Trans.*

[101]That is, transformations of the plane onto itself preserving collinearity. *Trans.*

we are throughout assuming all lattices arise as earlier from a doubly infinite chain of positive integers.) Since the only affine transformations of the plane fixing u and v as wholes are hyperbolic rotations relative to the axes u and v, and homotheties centred at O, it follows that these are the only types of transformations preserving admissibility.

Hence if we agree not to distinguish lattices obtained from one another by means of a hyperbolic rotation relative to the axes u and v, then we have the following theorem:

Any given admissible lattice can be transformed in the prescribed way into one and only one extremal lattice, namely that with greatest lower bound of the hyperbolic distances of its points from the origin O equal to 1, obtained by applying to the original lattice a suitable homothety centred at O.

Thus corresponding to each bounded chain (α) there is essentially just one extremal lattice. Since distinct such chains correspond to distinct extremal lattices, it follows that the cardinality of the set of extremal lattices is equal to that of the continuum.

The problem Markov set himself in his dissertation—and which he solved completely—is equivalent to that of finding all extremal lattices having fundamental parallelograms of area S less than or equal to 3. Couched in geometric terms, what Markov showed was that the extremal lattices satisfying $S < 3$ form an infinite sequence $M_1, M_2, M_3, \ldots$ (provided we distinguish them only up to a hyperbolic rotation relative to the axes u and v) with fundamental parallelograms $S_1, S_2, S_3, \ldots$ forming an increasing sequence with first term $S_1 = \sqrt{5}$, tending to 3 in the limit, while for $S = 3$ there are continuously many extremal lattices. (He did not investigate what happens for $S > 3$.)

We shall now show that Markov's solution is equivalent to the determination of certain special chains (α).

We begin by noting that the greatest lower bound of the absolute values of the hyperbolic distances of the points of a lattice from the origin coincides with the greatest lower bound for the subset made up of the vertices of the polygons Π. This can be shown to follow essentially from the fact that a hyperbola in a given quadrant (I say) "embracing" all points of the polygon Π_I will embrace the polygon itself, and hence its convex hull, which contains all points of the lattice lying in that quadrant.

The following important consequence of the foregoing furnishes the geometric basis for Markov's results:

Let Γ be the (essentially) unique extremal lattice corresponding as in the preceding italicized result to a chain (α), and let S denote the area of its fundamental parallelogram. Then for any given positive number t, one has $S \leq t$ if and only if $d_\mu \leq t$ for all μ.

From the fact that $|N(A_\mu)| \geq 1$ and the formula $d_\mu = \frac{S}{|N(A_\mu)|}$ established earlier, it is immediate that if $S \leq t$, then $d_\mu \leq t$. Conversely, since $S = d_\mu |N(A_\mu)|$, and the greatest lower bound of the quantities $|N(A_\mu)|$ is 1, we have that $d_\mu \leq t$ implies $S \leq t$.

In view of this corollary and the earlier formula

$$d_\mu = \alpha_\mu + \cfrac{1}{\alpha_{\mu+1} + \cfrac{1}{\alpha_{\mu+2} + \cfrac{1}{\ddots}}} + \cfrac{1}{\alpha_{\mu-1} + \cfrac{1}{\alpha_{\mu-2} + \cfrac{1}{\alpha_{\mu-3} + \cfrac{1}{\ddots}}}},$$

the problem of finding all extremal lattices satisfying $S \le t$ becomes equivalent to that of finding all doubly infinite chains

$$\dots, \alpha_{-3},\ \alpha_{-2},\ \alpha_{-1},\ \alpha_0,\ \alpha_1,\ \alpha_2,\ \alpha_3, \dots$$

of positive integers such that $d_\mu \le t$ for all μ.

Having established these remarkable preliminary geometric results, we now apply them to the elucidation of Markov's method. (Note that his algebraic approach to the question, via the theory of quadratic forms, was more usual for his time.)

Suppose that the discriminant $D = b_0^2 - a_0c_0$ of a given binary quadratic form $(a_0, b_0, c_0) = a_0x^2 + 2b_0xy + c_0y^2$, is positive, i.e., that the form is indefinite. In this case the roots of the equation[102] $a_0\xi^2 + 2b_0\xi + c_0 = 0$, namely

$$\xi_0 = \frac{-b_0 + \sqrt{D}}{a_0},\ \xi_0' = \frac{-b_0 - \sqrt{D}}{a_0},$$

are real and distinct, and therefore the form may be decomposed as a product of real linear factors:

$$(a_0, b_0, c_0) = a_0\left(x - \frac{-b_0 + \sqrt{D}}{a_0}y\right)\left(x - \frac{-b_0 - \sqrt{D}}{a_0}\right),$$

i.e., the form (a_0, b_0, c_0) is equal to a product of the form $(u_0x + u_{-1}y)(v_0x + v_{-1}y)$, where u_0, u_{-1}, v_0, v_{-1} are real; for example one may take

$$u_0 = a_0,\ u_{-1} = b_0 - \sqrt{D},\ v_0 = 1,\ v_{-1} = \frac{b_0 + \sqrt{D}}{a_0}.$$

As well as this particular factorization, there are those obtained from it by multiplying one of these factors by any real number τ and the other by τ^{-1}:

$$\tau(u_0x + u_{-1}y) \cdot \frac{1}{\tau}(v_0x + v_{-1}y),$$

and clearly this accounts for all factorizations of the form into linear factors.

Instead of the xy-coordinate system for the plane, we shall use coordinates u, v given in terms of x, y by

$$u = u_0x + u_{-1}y, \quad v = v_0x + v_{-1}y.$$

Thus the points $(x, y) = (0, 0), (1, 0)$ and $(0, 1)$, become in the uv-system of coordinates respectively $(0, 0), (u_0, v_0), (u_{-1}, v_{-1})$. Write $\mathfrak{A}_0$ for the vector from $O = (0, 0)$ to (u_0, v_0), and $\mathfrak{A}_{-1}$ for the vector from O to (u_{-1}, v_{-1}) (in the uv-system), and consider the planar lattice generated by the frame $(\mathfrak{A}_0, \mathfrak{A}_{-1})$. The points of this lattice are exactly those with coordinates (x, y) (x and y integers) relative to this frame; with respect to the uv-system of coordinates, they will of course have coordinates $u = u_0x + u_{-1}y$, $v = v_0x + v_{-1}y$. Since the quadratic form $a_0x^2 + 2b_0xy + c_0y^2 = u \cdot v$, the variable number σ represented by the form at integer values of x and y, may be

[102] Markov denotes the second root ξ_0' by $-\frac{1}{\eta_0}$ from considerations which will become clear in the sequel.

considered as parametrizing the hyperbolae $u \cdot v = \sigma$ on which the points (x, y) of the lattice lie. We shall call the number σ the *hyperbolic distance* from the origin O to the lattice point (x, y). Thus with the given indefinite form we have associated the canonical frame $(\mathfrak{A}_0, \mathfrak{A}_{-1})$.

If we apply an integral unimodular change of variables

$$x = \alpha\bar{x} + \beta\bar{y},$$
$$y = \gamma\bar{x} + \delta\bar{y},$$

i.e., where $\alpha, \beta, \gamma, \delta$ are integers and the determinant $\alpha\delta - \beta\gamma = \pm 1$, then we obtain the *equivalent* form

$$\begin{aligned} &a_0(\alpha\bar{x} + \beta\bar{y})^2 + 2b_0(\alpha\bar{x} + \beta\bar{y})(\gamma\bar{x} + \delta\bar{y}) + c_0(\gamma\bar{x} + \delta\bar{y})^2 \\ &\quad = [u_0(\alpha\bar{x} + \beta\bar{y}) + u_{-1}(\gamma\bar{x} + \delta\bar{y})][v_0(\alpha\bar{x} + \beta\bar{y}) + v_{-1}(\gamma\bar{x} + \delta\bar{y})] \\ &\quad = [(\alpha u_0 + \gamma u_{-1})\bar{x} + (\beta u_0 + \delta u_{-1})\bar{y}][(\alpha v_0 + \gamma v_{-1})\bar{x} + (\beta v_0 + \delta v_{-1})\bar{y}], \end{aligned}$$

from which we see that the canonical frame $(\bar{\mathfrak{A}}_0, \bar{\mathfrak{A}}_{-1})$ corresponding to the new variables $\bar{x}, \bar{y}$ is related to the former frame by

$$\bar{\mathfrak{A}}_0 = \alpha\mathfrak{A}_0 + \gamma\mathfrak{A}_{-1}, \quad \bar{\mathfrak{A}}_{-1} = \beta\mathfrak{A}_0 + \delta\mathfrak{A}_{-1}.$$

The set of all forms equivalent to a particular one is called a *class* of forms.[103] The canonical frames of the forms in a single class comprise all generating frames of the lattice generated by any one of them; in fact this association of a frame with each form of the class yields a one-to-one correspondence between the set of generating frames of the lattice (which thus belongs to the class) and the forms of the class.

For the discriminant of the form we have

$$\begin{aligned} D = b_0^2 - a_0c_0 &= \frac{1}{4}(u_0v_{-1} + u_{-1}v_0)^2 - u_0v_0u_{-1}v_{-1} \\ &= \frac{1}{4}(u_0v_{-1} - u_{-1}v_0)^2 = \frac{1}{4}S^2, \end{aligned}$$

where

$$S = \begin{vmatrix} u_0 & v_0 \\ u_{-1} & v_{-1} \end{vmatrix}$$

is the area of the fundamental parallelogram of the lattice generated by the frame $(\mathfrak{A}_0, \mathfrak{A}_{-1})$.

The fact that our indefinite form decomposes as a product of linear factors where the decomposition is unique only to within multiplication of one factor by any nonzero real number τ and the other by $\frac{1}{\tau}$, signifies from a geometric point of view that the frame corresponding to the form, and therefore also the lattice generated by the frame, is determined by the form only up to "stretchings" of the u- and v-axes respectively by factors τ and $\frac{1}{\tau}$ with product 1 (assuming τ positive). Such an affine transformation of the plane, expanding it in the u-direction by a factor τ and compressing it in the v-direction by the same coefficient τ (i.e., expanding it by the factor $\frac{1}{\tau}$), is called a *hyperbolic rotation* of the plane relative to the u- and v-asymptotes. (Clearly if $0 < \tau < 1$, the transformation is more aptly described as compressing the plane in the u-direction, and expanding it in the v-direction.) If $\tau < 0$ then the transformation involves an additional reflection of the plane in the

[103] The point here is that two equivalent forms have the same range for integer values of their respective variables. *Trans.*

origin O.[104] Under a hyperbolic rotation every hyperbola having the u- and v-axes as asymptotes is sent onto itself, since such a hyperbola has equation $u \cdot v = k$ for some constant k, and this equation is clearly invariant under transformations of the form $x \to \tau x$, $y \to \frac{1}{\tau} y$.

We conclude that the planar lattice obtained as above from a given indefinite binary quadratic form is actually determined by the form only to within a hyperbolic rotation. However, in accordance with the usual mathematical principle, if we are investigating a particular lattice obtained from the form, and we wish to observe the geometric effect of transformations of the form, i.e., of changes of variables, then we should consider the effect on that particular lattice, and not on some other obtained from it by means of some hyperbolic rotation.

We now turn to the transformations of the quadratic form (a_0, b_0, c_0) considered by Markov in §§ 1–3 of his dissertation. As above we have

$$a_0\xi^2 + 2b_0\xi + c_0 = (u_0\xi + u_{-1})(v_0\xi + v_{-1}).$$

The factors $u_0\xi + u_{-1}$ and $v_0\xi + v_{-1}$ are the u- and v-coordinates respectively, of the vector $\xi\mathfrak{A}_0 + \mathfrak{A}_{-1}$. The roots ξ_0 and ξ_0' of the above polynomial in ξ are the respective values of ξ for which the u-coordinate and v-coordinate of this vector vanish. In other words:

The quantity ξ_0 represents the "length" as measured in multiples of $\mathfrak{A}_0$, of the line segment parallel to $\mathfrak{A}_0$ from the tip of the vector $\mathfrak{A}_{-1}$ to the v-axis, and similarly, ξ_0' is the "length", again as measured in multiples of $\mathfrak{A}_0$, of the line segment from the tip of $\mathfrak{A}_{-1}$ to the u-axis.

It is easy to see that ξ_0 is positive precisely when the vectors $\mathfrak{A}_{-1}$ and $\mathfrak{A}_0$ lie on different sides of the v-axis.

One of the two equivalent forms (a_0, b_0, c_0) and $(a_0, -b_0, c_0)$, where the second is obtained from the first for example by replacing the vector $\mathfrak{A}_{-1}$ by $-\mathfrak{A}_{-1}$, must therefore have ξ_0 positive. Since it is immaterial which form we choose to represent an equivalence class of forms, we may assume ξ_0 positive. The ambiguity in sign in the first change of variables in (4) above (which we have omitted in reproducing (4) below) in effect indicates Markov's desire to have ξ_0 positive.

Thus we shall from now on assume that $\xi_0 > 0$, so that the vectors of the frame $(\mathfrak{A}_0, \mathfrak{A}_{-1})$ lie on different sides of the v-axis. We now carry out the beak-construction algorithm. As in Figure 7, for instance, we lay off the vector $\mathfrak{A}_0$ from the tip of $\mathfrak{A}_{-1}$ as many (nonnegative) times, say α_0, as possible without crossing the v-axis, i.e., so that the vector $\mathfrak{A}_1 = \alpha_0\mathfrak{A}_0 + \mathfrak{A}_{-1}$ lies on the same side of the v-axis as $\mathfrak{A}_0$. Next we lay off the vector $\mathfrak{A}_1$ from the tip of $\mathfrak{A}_0$ as many times as possible, say α_1 times, without crossing the v-axis, i.e., so that the vector $\mathfrak{A}_2 = \alpha_1\mathfrak{A}_1 + \mathfrak{A}_0$ lies on the same side of the v-axis as $\mathfrak{A}_0$, and so on. As described earlier this procedure yields the continued fraction expansion of the number ξ_0.[105]

Note that the forms

$$(a_0, b_0, c_0),\ (a_1, b_1, c_1),\ (a_2, b_2, c_2), \ldots$$

corresponding to the frames $(\mathfrak{A}_{-1}, \mathfrak{A}_0), (\mathfrak{A}_0, \mathfrak{A}_1), (\mathfrak{A}_1, \mathfrak{A}_2), \ldots$, are equivalent to one another since their frames are all generating for the same lattice. The coordinate

[104] These are included here under the name "hyperbolic rotation", although they are also called "cross hyperbolic rotations". *Trans.*

[105] If ξ_0 is rational, the procedure will terminate after finitely many steps. *Trans.*

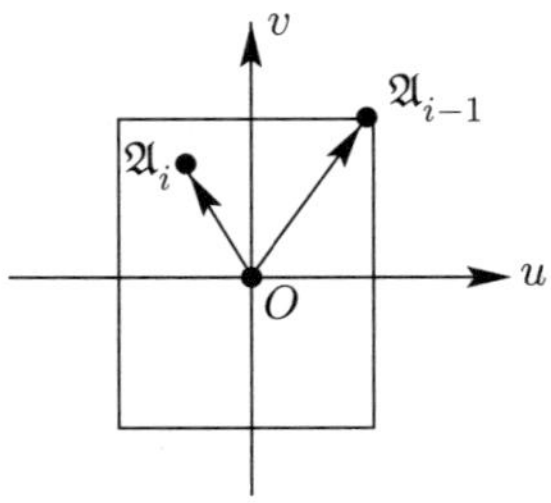

FIGURE 13

changes

$$
\begin{array}{ll}
x_0 \to \alpha_0 x_1 + y_1, & y_0 \to x_1, \\
x_1 \to \alpha_1 x_2 + y_2, & y_1 \to x_2, \\
x_2 \to \alpha_2 x_3 + y_3, & y_2 \to x_3, \\
\vdots & \vdots \\
x_{n-1} \to \alpha_{n-1} x_n + y_n, & y_{n-1} \to x_n \\
\vdots & \vdots
\end{array} \tag{4'}
$$

are those that transform the form (a_0, b_0, c_0) into the form (a_1, b_1, c_1), and then (a_1, b_1, c_1) into the form (a_2, b_2, c_2), and so on. We write $\xi_0, \xi_1, \xi_2, \ldots, \xi_n, \ldots$ for the "first" roots of the equations

$$a_0\xi^2 + 2b_0\xi + c_0 = 0,\ a_1\xi^2 + 2b_1\xi + c_1 = 0,$$
$$a_2\xi^2 + 2b_2\xi + c_2 = 0, \ldots,\ a_n\xi^2 + 2b_n\xi + c_n = 0, \ldots,$$

respectively.

If we assume that both roots ξ_0, ξ_0' of the equation $a_0\xi^2 + 2b_0\xi + c_0 = 0$ are irrational, then the lattice corresponding to the form (a_0, b_0, c_0) will have no points other than the origin O on the u- and v-axes, since the points of the lattice are by definition those of the form $(u_0x + u_{-1}y, v_0x + v_{-1}y)$ where x and y are integers, and $\xi_0 = -\frac{u_{-1}}{u_0}$, $\xi_0' = -\frac{v_{-1}}{v_0}$. It then follows that no two points of the lattice have the same u- or v-coordinate, since if for instance the lattice contained two points A and B with the same u coordinate, then the vector $\overrightarrow{AB}$ would be parallel to the v-axis, and the points on that axis obtained by laying it off from O would all be points of the lattice.[106]

As is clear from Figure 7, under the successive transformations in (4′), starting from $i = 1$, the absolute values of the u-components of the vectors $\mathfrak{A}_i$ decrease strictly with increasing i. If the absolute values of the v-coordinates also decreased, then the tips of all of them would lie in the "coordinate rectangle" of $\mathfrak{A}_1$, i.e., in the rectangle with the tip of $\mathfrak{A}_1$ as one vertex, with sides parallel to the coordinate axes, and with centre O. (In Figure 13 it is shown how in this case each $\mathfrak{A}_i$ would lie in the coordinate rectangle of $\mathfrak{A}_{i-1}$.) However, since the tips of the $\mathfrak{A}_i$ are all points of a lattice, namely that generated by the frame $(\mathfrak{A}_{-1}, \mathfrak{A}_0)$, this would mean

[106] From here on it is assumed that the roots are both irrational. *Trans.*

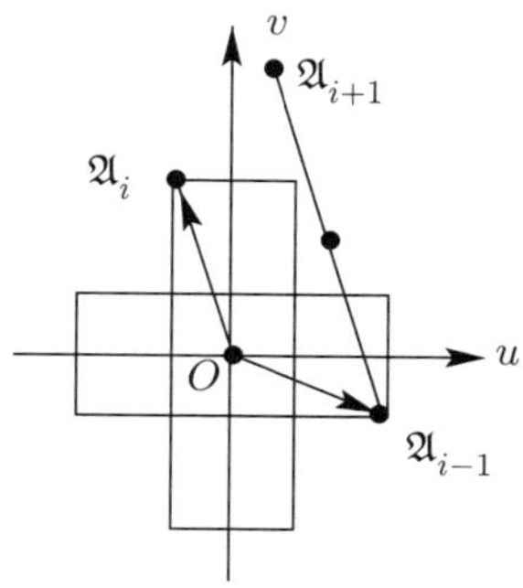

FIGURE 14

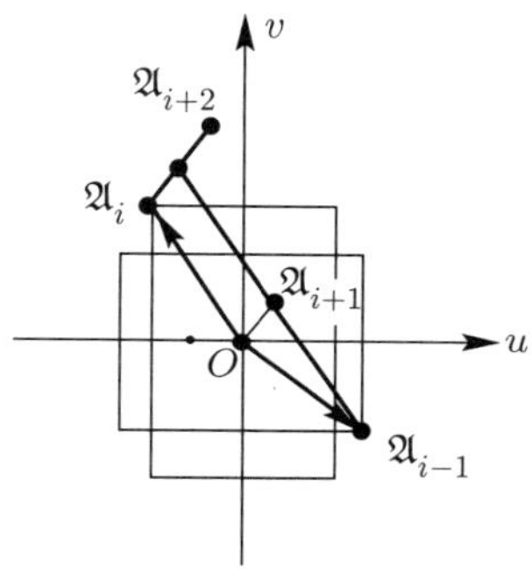

FIGURE 15

that a finite rectangle contained infinitely many points of the lattice, which is of course not possible.

It follows that while the tips of the vectors $\mathfrak{A}_i$ move closer to the v-axis as i increases, on the other hand for infinitely many i the tip of the vector $\mathfrak{A}_i$ must lie further from the u-axis than that of its predecessor $\mathfrak{A}_{i-1}$. If the vectors $\mathfrak{A}_{i-1}$, $\mathfrak{A}_i$ lie on the same side of the u-axis, we call the frame $(\mathfrak{A}_{i-1}, \mathfrak{A}_i)$ *reduced.*[107] If they lie on opposite sides of the u-axis, then after one or two further steps in the algorithm we shall obtain a reduced frame. Thus if in constructing $\mathfrak{A}_{i+1}$ we have $\alpha_i > 1$, then, as shown in Figure 14, we obtain a reduced frame $(\mathfrak{A}_i, \mathfrak{A}_{i+1})$, while if $\alpha_i = 1$, then only after yet one further step do we obtain a reduced frame $(\mathfrak{A}_{i+1}, \mathfrak{A}_{i+2})$, as shown in Figure 15.

Hence after a finitely many steps of the algorithm we shall always arrive at a reduced frame, i.e., one having its vectors on the same side of the u-axis, but on different sides of the v-axis, and with the tip of the second vector of the frame closer to the v-axis than that of the first.[108]

Once we have reached a reduced frame $(\mathfrak{A}_{n-1}, \mathfrak{A}_n)$, it then follows from the form of the beak-construction algorithm that all subsequent frames

$$(\mathfrak{A}_n, \mathfrak{A}_{n+1}),\ (\mathfrak{A}_{n+1}, \mathfrak{A}_{n+2}), \dots$$

will also be reduced.

[107] Compare with the earlier definition. *Trans.*

[108] This result had, of course, already been proved arithmetically by Lagrange.

For a reduced frame $(\mathfrak{A}_{n-1}, \mathfrak{A}_n)$, it is clear that $\xi_n > 1$ and $-1 < \xi_n' < 0$.[109] In fact these conditions are also sufficient for a frame to be reduced:

The frame $(\mathfrak{A}_{n-1}, \mathfrak{A}_n)$ *is reduced if and only if*

$$\xi_n > 1 \text{ and } -1 < \xi_n' < 0.$$

For the sufficiency, observe that the condition $\xi_n > 1$ implies that the vectors $\mathfrak{A}_{n-1}$ and $\mathfrak{A}_n$ lie on opposite sides of the u-axis, the condition $\xi_n' < 0$ implies that they lie on the same side of the v-axis, and then the condition $\xi_n > 1$ implies that the tip of $\mathfrak{A}_n$ is closer to the v-axis than that of $\mathfrak{A}_{n-1}$, and finally from $|\xi_n'| < 1$ it follows that the tip of $\mathfrak{A}_n$ is further from the u-axis than that of $\mathfrak{A}_{n-1}$.

In view of this fact, we also call[110] an indefinite quadratic form (a_0, b_0, c_0) *reduced* if its roots ξ_0, ξ_0' are irrational and satisfy the inequalities $\xi_0 > 1$, $-1 < \xi_0' < 0$. Henceforth we shall assume that the original form (a_0, b_0, c_0) is reduced.

As we have seen earlier, the α_i are the terms of the continued fraction expansion of the positive root ξ_0:

$$\xi_0 = \alpha_0 + \cfrac{1}{\alpha_1 + \cfrac{1}{\alpha_2 + \cfrac{1}{\ddots}}}.$$

We now carry out the "reverse" variant of the beak-construction algorithm (as described earlier—see Figure 10) starting with the frame $(\mathfrak{A}_{-1}, \mathfrak{A}_0)$, so that we first lay off $-\mathfrak{A}_{-1}$ from the tip of $\mathfrak{A}_0$ as (positively) many times as possible, say α_{-1} times, without crossing the u-axis, reaching a point A_{-2}, say, and defining $\mathfrak{A}_{-2} = \overrightarrow{OA_{-2}}$, and then lay off the negative of the latter vector from the tip of $-\mathfrak{A}_{-1}$ as many times α_{-2} as possible without crossing the u-axis, to obtain the next vector $\mathfrak{A}_{-3}$, and so on. This algorithm yields a sequence

$$(\alpha_{-i}),\ i = 1, 2, 3, \ldots$$

of positive integers, and a sequence

$$(\mathfrak{A}_{-2}, \mathfrak{A}_{-1}),\ (\mathfrak{A}_{-3}, \mathfrak{A}_{-2}),\ (\mathfrak{A}_{-4}, \mathfrak{A}_{-3}), \ldots$$

of reduced frames corresponding respectively to reduced forms (a_{-i}, b_{-i}, c_{-i}), $i = 1, 2, 3, \ldots$, obtained from (a_0, b_0, c_0) by means of the successive transformations

$$\begin{aligned}
x_0 &\to y_{-1}, & y_0 &\to -\alpha_{-1} y_{-1} + x_{-1}, \\
x_{-1} &\to y_{-2}, & y_{-1} &\to -\alpha_{-2} y_{-2} + x_{-2}, \\
&\vdots & &\vdots \\
x_{-m} &\to y_{-m-1}, & y_{-m} &\to -\alpha_{-m-1} y_{-m-1} + x_{-m-1} \\
&\vdots & &\vdots
\end{aligned}$$

The positive integers $\alpha_{-1}, \alpha_{-2}, \alpha_{-3}, \ldots$ are then the terms of the continued fraction expansion of $-\frac{1}{\xi_0'}$:

$$-\frac{1}{\xi_0'} = \alpha_{-1} + \cfrac{1}{\alpha_{-2} + \cfrac{1}{\alpha_{-3} + \cfrac{1}{\ddots}}}.$$

---

[109] See the above geometric interpretation of the roots ξ_0, ξ_0' of the equation $a_0\xi^2 + 2b_0\xi + c_0 = 0$, and also Figure 11. *Trans.*

[110] As earlier, at the beginning of the summary of Markov's dissertation. *Trans.*

Thus we have a doubly infinite chain of reduced frames

$$\ldots, (\mathfrak{A}_{-3}, \mathfrak{A}_{-2}),\ (\mathfrak{A}_{-2}, \mathfrak{A}_{-1}),\ (\mathfrak{A}_{-1}, \mathfrak{A}_0),\ (\mathfrak{A}_0, \mathfrak{A}_1),\ (\mathfrak{A}_1, \mathfrak{A}_2),\ (\mathfrak{A}_2, \mathfrak{A}_3), \ldots$$

corresponding in order to the reduced forms of the doubly infinite chain

$$\ldots, (a_{-2}, b_{-2}, c_{-2}),\ (a_{-1}, b_{-1}, c_{-1}),\ (a_0, b_0, c_0),\ (a_1, b_1, c_1),\ (a_2, b_2, c_2), \ldots,$$

with the property that if we start the above-described two-way algorithm at any frame of the chain (or form), we shall obtain the same doubly infinite chain of frames (or forms).

As noted in the summary of Markov's dissertation, he uses the notation L_μ for $\frac{2}{d_\mu}$, so that his investigation of the forms for which $L_\mu \geq \frac{2}{3}$ for all μ, is equivalent to the investigation of the lattices arising from indefinite binary quadratic forms for which $d_\mu \leq 3$ for all μ, or, as we have shown above, to the investigation of those chains (α) (which Markov denotes by I) for which all $d_\mu \leq 3$.[111]

In terms of the geometric picture delineated above, the results of Markov's investigations of such chains (α) in Chapter I of his dissertation, amount to the following: First, there exists a particular such chain (α_1) whose corresponding lattice M_1 (unique to within a hyperbolic rotation) is extremal, with fundamental cell of area $S = \sqrt{5}$. Then, if this lattice, together with all those obtained from it by means of a hyperbolic rotation, is excluded from consideration, i.e., if the chain (α_1) is set apart, there exists a chain (α_2) which corresponds to an extremal lattice M_2 (as always, unique up to a hyperbolic rotation), second in density to M_1, with fundamental cell of area $S = \sqrt{8}$. Continuing in this way we obtain further extremal lattices $M_3, M_4, M_5, \ldots$[112] with respective fundamental cells of areas $S_3 = \sqrt{\frac{221}{25}}, S_4 = \sqrt{\frac{1517}{169}}, S_5 = \sqrt{\frac{7566}{841}}, \ldots$, which approach 3 in the limit.

As noted in the summary of his dissertation above, Markov shows that the chains (α) for which $d_\mu < 3$ must be periodic. We shall now show geometrically that this implies that the forms (a, b, c) corresponding to them are rational. Thus suppose (α) is a periodic chain with period m, so that for all i we have $\alpha_{i+m} = \alpha_m$. Let $\mathfrak{A}_\mu$ be any vector of the corresponding lattice. If we apply the affine transformation defined by appropriate expansions (stretchings) of the coordinate axes so that the vector $\mathfrak{A}_\mu$ coincides with, i.e., is transformed into, $\mathfrak{A}_{\mu+m}$, then the lattice must transform onto itself, since the chain (α) is the same whether considered from the location indexed by $\mu+m$ or μ. It follows that such affine transformations certainly do not change the area of a fundamental cell of the lattice, so that they must be hyperbolic rotations, whence we infer that the lattice admits hyperbolic rotations. The forms (a_μ, b_μ, c_μ) and $(a_{\mu+m}, b_{\mu+m}, c_{\mu+m})$, corresponding to the generating frames $(\mathfrak{A}_{\mu-1}, \mathfrak{A}_\mu)$ and $(\mathfrak{A}_{\mu+m-1}, \mathfrak{A}_{\mu+m})$ respectively, must therefore be equal. However, if the coordinates of the tips, i.e., the components, of the vectors $\mathfrak{A}_{\mu+m-1}$ and $\mathfrak{A}_{\mu+m}$ as multiples of the basis vectors $\mathfrak{A}_{\mu-1}$ and $\mathfrak{A}_\mu$, are (α, γ) and (β, δ) respectively, then the second form is obtained from the first by means of the transformation $x \to \alpha x + \beta y,\ y \to \gamma x + \delta y$, so that this transformation preserves the form (a_μ, b_μ, c_μ).

---

[111] See Figure 12 for the geometric import of the numbers d_μ. *Trans.*

[112] In Markov's dissertation, the M_i denote the corresponding forms; see the summary of his dissertation above. *Trans.*

By comparing coefficients we obtain the following homogeneous system of linear equations with integer coefficients in the unknowns a_μ, b_μ, c_μ:

$$\begin{aligned} a_\mu &= a_\mu\alpha^2 + 2b_\mu\alpha\gamma + c_\mu\gamma^2, \\ b_\mu &= a_\mu\alpha\beta + b_\mu(\alpha\delta + \beta\gamma) + c_\mu\gamma\delta, \\ c_\mu &= a_\mu\beta^2 + 2b_\mu\beta\delta + c_\mu\delta^2. \end{aligned}$$

The determinant of this homogeneous system must be zero since it has a nonzero solution for a_μ, b_μ, c_μ. It is immediate that such solutions are proportional to rational ones, and therefore to integral ones.

It follows from the fact that the Markovian forms (i.e., those with lattices $M_1, M_2, M_3, \ldots$) are proportional to forms with integer coefficients, that each of them attains the minimum of its absolute values under substitution of integer pairs $(x, y) \neq (0, 0)$. In other words each of the corresponding "Markovian lattices" has points, in fact infinitely many points, on the boundary of the hyperbolic cross $|uv| \leq 1$, i.e., on the hyperbolae $u \cdot v = \pm 1$. Incidentally this can be seen to follow directly from the periodicity of these lattices.

We shall not expound the geometric ideas underlying Markov's further analysis of the the chains (α), in particular because this has been done by Gorshkov in work still to be published.[113] We shall instead devote the rest of this commentary to describing research into other aspects of this fascinating problem. The Markov numbers[114] Q were the subject of memoirs by Frobenius and then Remak. After Markov's results became known, Remak, building on Frobenius' work, managed to find a new approach via induction, avoiding the use of continued fractions, i.e., of the chains (α).[115] Although Remak's proof of Markov's formula[116]

$$\phi = Qx^2 + (3Q - 2q)xy + (r - 3q)y^2$$

for the forms $M_1, M_2, M_3, \ldots$, is of great interest, it fails to reveal the essential reasons behind it.

In a subsequent work published in 1926 in the *Jahresbericht der Deutscher Math. Ver.*, Remak gives the following elegant geometric explanation—falling short of a proof—of the results from Chapter I of Markov's dissertation.

By the above we may assume that the Markovian lattices have points on the boundary of the hyperbolic cross $|uv| \leq 1$, but only the origin O inside it. Since it makes no theoretical difference whether we consider a given lattice or one obtained from it by means of a hyperbolic rotation, we may normalize all of these lattices by hyperbolically rotating each of them so that among those of its points on the boundary hyperbolae it has one, O_1 say, at a vertex of one of those hyperbolae.

Once this normalization is carried out, the Markovian lattices will have in common a line of points $\ldots, O_{-3}$, O_{-2}, O_{-1}, O_0, O_1, O_2, $O_3, \ldots$ as shown in Figure 16. Since the hyperbolic cross $|uv| \leq 1$ centred at O has no points of any of these lattices in its interior (except for O), in view of the (appropriate) translation invariance of lattices, for every i the image of this hyperbolic cross under translation along the vector $\overrightarrow{OO_i}$ will likewise have in its interior no points

[113] As of 1947. *Trans.*

[114] See the conclusion of the summary of Markov's dissertation above. *Trans.*

[115] I in Markov's notation. *Trans.*

[116] In this formula q, r are defined in terms of Q—see the conclusion of the summary of Markov's dissertation above. *Trans.*

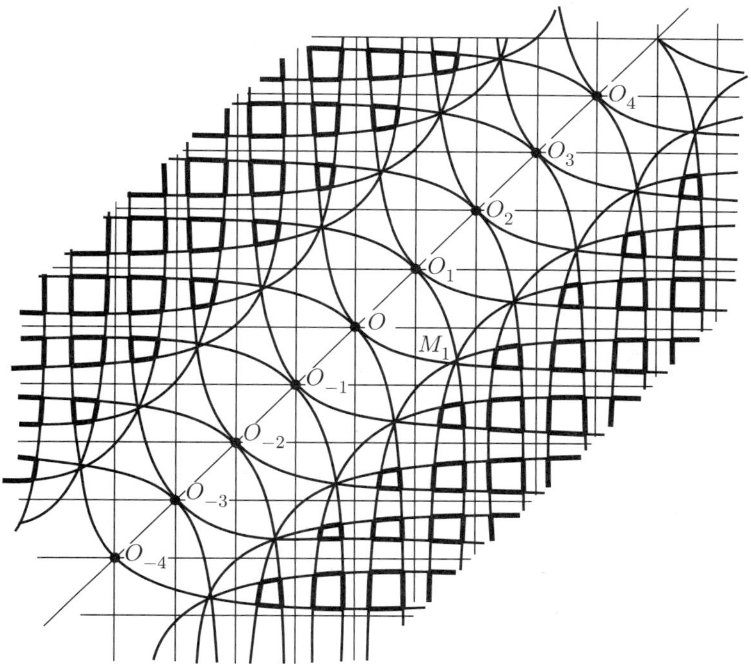

FIGURE 16

(except O) of any of the lattices in question. Each normalized Markovian lattice then determines[117] a row of points adjacent to the row $\dots O_{-2}O_{-1}O_0O_1O_2\dots$, and the lattice is determined by any one point of an adjacent row. A necessary, but not sufficient, restriction is of course that no points of the lattice can be in the interior of the chain of hyperbolic crosses.

If a point A of the line of such an adjacent row of the lattice (see Figure 17) does not lie in the interior of any hyperbolic cross, then all of its translates by integer multiples of the vector $\overrightarrow{OO_1}$ will of course also not lie in the interior of any hyperbolic cross. However, the points then determined latticewise in the next row (the row of points B_i in Figure 17) might well all turn out to lie in the interiors of some of the hyperbolic crosses. If the point B of this row lay in the interior of the hyperbolic cross with centre O_i for some even $i = 2j$, then the point A_j would lie inside a similar and parallel hyperbolic cross centred at O_j and half the size of the former, and conversely. On the other hand if the point B lay in the interior of the hyperbolic cross centred at O_i for some odd $i = 2j + 1$, then the point A_j would lie inside a similar and parallel hyperbolic cross centred at the midpoint of the segment $O_{2j}O_{2j+1}$ of half the size, and conversely.[118]

Hence in order to avoid having any of the points $\dots, B_{-2}, B_{-1}, B_0, B_1, B_2, \dots$ of the second row parallel to the row $\dots, O_{-2}, O_{-1}, O_0, O_1, O_2, \dots,$ in the prohibited interiors of crosses, it is necessary and sufficient not just that the points of the first parallel row $\dots, A_{-2}, A_{-1}, A_0, A_1, A_2, \dots$ lie outside the open crosses

[117] And is determined by. *Trans.*

[118] Presumably one needs to draw a diagram to see this. *Trans.*

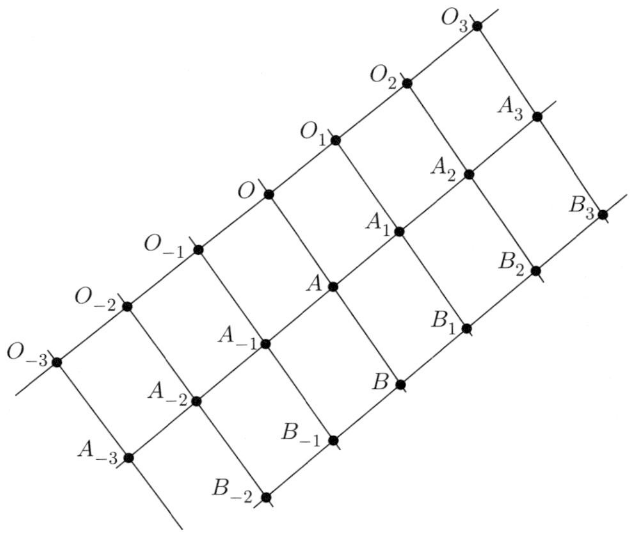

FIGURE 17

centred at the points $\dots, O_{-2}, O_{-1}, O_0, O_1, O_2, \dots$, but also outside the parallel and similar open crosses of half the size centred at the midpoints of the segments $\dots, O_{-2}O_{-1}$, $O_{-1}O$, OO_1, $O_1O_2, \dots$. By iterating this argument one ultimately sees that for the lattice in question not to have any points in the interior of any hyperbolic cross, it is necessary and sufficient that the points A_i of a row adjacent to the row $\dots, O_{-2}, O_{-1}, O, O_1, O_2, \dots$ do not lie in the interior of any hyperbolic cross of a net comprised of the crosses of the following rows: the row of basic crosses centred at the points O_i; the row of similar and parallel crosses of half the dimensions of these, centred at the midpoints of the segments $O_{i-1}O_i$; the row of similar and parallel crosses of dimensions one-third those centred at the points O_i, centred at the points trisecting the segments $O_{i-1}O_i$ into equal thirds, and so on. Since this net of hyperbolic crosses is invariant as a whole under translations of the plane by integer multiples of the vectors $\overrightarrow{OO_1}$ and $\overrightarrow{OA}$, this necessary and sufficient condition is equivalent to the condition that just the point A lie outside the interiors of the crosses of this net. Furthermore since we are concerned only with lattices having points on the boundary of the hyperbolic cross $|uv| \leq 1$, we also require that A lie on the boundary of some cross of this net. By studying this net of crosses and using Markov's results, Remak discovered that if the fundamental parallelogram OO_1A_1A of Figure 17, determined by the generating frame $(\overrightarrow{OO_1}, \overrightarrow{OA})$ of the lattice, has area 3, then it will be covered by the net of hyperbolic crosses except for isolated points of its boundary, lying also on the boundaries of several crosses of the net, like for example the point M_1 in Figure 16. It is these isolated points that determine the Markovian lattices $M_1, M_2, M_3, \dots$, corresponding to Markov's extreme forms.

Very recently,[119] Gorshkov has shown using Lobachevskian geometry how to

[119] Around 1947. *Trans.*

obtain all of Markov's results, from both Chapters I and II, by direct geometric means.

In his article "On indefinite ternary quadratic forms" (*Izvestiya Akad. Nauk*, 1901), Markov found the first two extremal indefinite ternary forms, and then in an article in *Mathematische Annalen* in 1903, the third. This showed that the indefinite ternary quadratic forms of given discriminant achieving the largest of the absolute values of the minima of such forms, are distributed discretely as in the binary case, at least initially. Later Dickson and Oppenheim found the fourth extremal indefinite ternary form (see the book: Dickson, "Studies in the theory of numbers", 1930).

Recently B. A. Venkov has found the next seven extremal indefinite ternary quadratic forms, so that at the time of writing (1947) the first 11 are known. Venkov's method is based on his concept of an "exterior vertex". We shall now indicate what this involves.

It can be shown by means analogous to those used above for indefinite binary quadratic forms, that the problem of finding the largest among the minima of indefinite ternary quadratic forms has the following equivalent geometric formulation. Consider in Euclidean 3-space coordinatized by u, v, w, the region consisting of those points (u, v, w) satisfying the inequality

$$|u^2 + v^2 - w^2| \leq 1,$$

which we shall call a *hyperboloidal cross*. Here, analogously to the binary case, the 3-dimensional lattices of interest are those with the origin O as one of their points, but without any other points in the interior of the hyperboloidal cross, and either having points on the boundary of the cross or else arbitrarily close to it, i.e., those lattices for which the greatest lower bound of the *hyperboloidal* distances from O of its points other than O is 1. For present purposes two such lattices are considered equivalent if one can be obtained from the other by means of a *hyperboloidal rotation*, i.e., an affine transformation fixing the hyperboloids $u^2 + v^2 - w^2 = \pm 1$ as wholes. The problem is then first to find among all such lattices that with fundamental cell of least volume (if there is one), and then to find among the rest that with least volume (again if such a one exists), and so on.

Consider in each of the two regions of space enclosed by the nappes of the asymptotic cone

$$u^2 + v^2 - w^2 = 0,$$

the convex hull of the set of all points of the lattice in question contained within or on the boundary of that region. The boundary of each of these convex hulls will be an infinite polyhedron analogous to the Kleinian polygons of the planar case.

The vertices of this polyhedron will be, analogously to the planar case, the points of closest approach to the cone; but what might the appropriate such points be for the region outside the cone? Venkov proposes that a point P of the lattice lying outside the cone be considered a point of closest approach, or an *exterior vertex* as he calls it, if it is a vertex of the Kleinian polygon arising from any section by a plane through OP containing a planar sublattice of the spatial lattice in question. By means of the concept of exterior vertices Venkov is able to bring to bear the known results in the binary case on the problem of finding the ternary extremal forms.

In conclusion, we give the most general formulation of the question. In its most general form, the "Markovian problem" is the appropriate name for the following problem: Given some region γ of n-dimensional (Euclidean) space, consider those n-dimensional lattices having no points in γ, except under a certain "supplementary condition". These are then the *admissible* lattices, and an admissible lattice with the property that there is no continuous family of transformations throughout which the volume of the lattice's fundamental cell decreases while admissibility is preserved, is an *extremal* lattice. Thus Markov's problem in this most general sense consists in finding all extremal lattices.

From this point of view the problem of finding an upper bound for the minima of positive n-ary quadratic forms, investigated so fruitfully by Korkin and Zolotarev, and effectively solved by Voronoĭ in his memoir on "perfect forms", represents that particular Markovian problem in the above wide sense, where the region γ is the interior of the n-ball, and the supplementary condition is just that the centre O of the n-ball should be a point of the lattice. Thus here the admissible lattices are those with no points other than O in the interior of the ball, and the extremal lattices are those corresponding to classes of extremal forms. In this case the spectrum of extremal forms is discrete, in fact finite. For $n = 1, 2, 3, 4, 5$, the numbers of inequivalent such extremal n-dimensional lattices are $1, 1, 1, 2, 3$ respectively, found by Korkin and Zolotarev. Later Voronoĭ (see below) created for every n an algorithm for finding in a finite number of steps all such n-dimensional extremal lattices. Blichfeldt found the densest extremal lattices for $n = 6$ and 7, and Korkin–Zolotarev and Hofreiter found a few other extremal lattices for $n = 6$ and 7, as well as for some other values of n.[120]

In Markov's dissertation, $n = 2$, γ is the interior of the hyperbolic cross $|u \cdot v| < 1$, and the supplementary condition is that the centre O of the hyperbolic cross be a point of the lattice. As was shown by Markov, in this case among all admissible lattices with fundamental cell of area $S < 3$, the extremal ones form an infinite but discrete spectrum.

In the work of Markov in 1901 and 1903, of Dickson and Oppenheim, and then of Venkov in 1944, γ is the open region bounded by the two hyperboloids $u^2 + v^2 - w^2 = \pm 1$, and the supplementary condition is again that the origin O at the centre of these hyperboloids be the only point of the lattice in γ. Markov showed that in this case, at least in the early stages, the spectrum of extremal lattices is discrete. The above-mentioned authors found altogether the first 11 extremal lattices. In his work of 1902, Markov considered the analogous problem for $n = 4$, finding the first two extremal lattices, whence once again it appears that the spectrum of all such lattices is discrete.

In the case where as above $n = 2$, and the region γ is the hyperbolic cross $|u \cdot v| < 1$, but now the supplementary condition is that one of the asymptotes contains a linear row of points of the lattice, including the centre O of the cross, and in addition that apart from these points the region should contain only finitely many points of the lattice, one has the situation corresponding to the celebrated problem of the accuracy to which irrational numbers are approximable by rational numbers.[121]

[120] At the time of writing (1947) the densest extremal lattices have been found also for $n = 8$ and 9.

[121] This presumably refers to the line of investigation starting from Liouville's result of this type. *Trans.*

It was shown by Hurwitz that the latter problem reduces to a direct application of the results of Markov's dissertation. If here the supplementary condition is slightly weakened by dropping just the requirement that the centre of the cross $|u \cdot v| < 1$ be a point of the lattice, the extremal lattice problem becomes equivalent to a problem of Chebyshev. Recently[122] Khinchin found a direct link between this extremal lattice problem and the results in Markov's dissertation.

In the Markovian problem with $n = 3$, the region γ defined by $|u \cdot v \cdot w| < 1$, and the supplementary condition that the only point of the lattice in γ is the point O at the centre of γ, the first two extremal lattices were found by Mordell and Davenport; the first of these was also found by the present author using different methods. At the time of writing the analogues for $n \geq 4$ remain a mystery; even for $n = 4$ the first extremal lattice has not yet been found.

In the case where $n = 2$, the region γ is defined by $|uv(u + v)| < 1$, and the supplementary condition is that the only point of the lattice in γ is its centre, the Markovian problem is equivalent to that of bounding the minima of cubic binary forms with real roots. Here the first extremal lattice was found by Davenport and Mordell, and by other means by the present author. The analogous problem for binary forms of degree higher than 3 remains completely unsolved at the time of writing.

If for general n we take γ to be defined by $|u_1 u_2 \cdots u_n| < 1$, and demand that *no* points of the lattice lie in γ, we have the well-known Minkowskian problem concerning the product of n inhomogeneous linear expressions in n variables. Here the first extremal lattice was found by Minkowski for $n = 2$, and for $n = 3$ by Remak in a very difficult article of 1923, subsequently partially simplified by Davenport. For $n > 3$ even the first extremal lattice of this problem has not been found. In work now in press,[123] the present author shows that in the case $n = 2$ of this problem, the first extremal lattice as a function of the area of its fundamental parallelogram is not discretely distributed.

These problems form a distinctive chapter in the geometry of numbers, attracting at the present time the attention of such researchers as Mordell, Mahler, Siegel, and others. All of these problems may in fact be labelled "Markovian", in honour of our great scholar, who proposed and so brilliantly and completely solved the first of them, whereas till now only the first few steps have been made towards solutions of the remainder.

[122]That is, in relation to 1947. *Trans.*

[123]In 1947. *Trans.*

Georgiĭ Fedoseevich Voronoĭ (1868–1908)

Georgiĭ Fedoseevich VORONOĬ

Georgiĭ Fedoseevich Voronoĭ was born in 1868 on his father's property Zhuravka in Poltava province. His father held the position of professor at the Nezhinsk *lycée*, and was later director of the gymnasia in Kishinev, Berdyansk, and Priluki. It was in Priluki that Voronoĭ completed his schooling, preserving thenceforth the very best of memories of his mathematics teacher Bogoslovskiĭ. As a gymnasium pupil he had already demonstrated his remarkable mathematical ability. He especially liked algebra. At that time Professor Vasiliĭ Petrovich Ermakov of Kiev University proposed in his *Journal of Elementary Mathematics* the topic of factorization of a polynomial using a property of the roots of a quadratic equation. Though still a gymnasist, Voronoĭ set to work on this topic, as a result of which there appeared in Yermakov's journal an article of moderate length furnished with a significant number of examples. After this Voronoĭ tested his prowess against a more difficult problem: to find the positive integer solutions x, y, z of the Diophantine equation

$$x^2 + y^2 + z^2 = 2mxyz,$$

for a given positive integer m. Here he did not succeed.

Voronoĭ continued to ponder this problem, even after entering Petersburg University in 1885. Concerning it he wrote in his diary: "I myself have lost hope of ever solving this problem."

During his first two years at the university Voronoĭ attended lectures faithfully, and studied Serret's algebra text, Faa de Bruno's book on binary forms, and Chebyshev's articles. The means provided by his father were insufficient, especially after the latter's retirement, and Voronoĭ had to resort to giving lessons, sometimes for very little payment. He found these lessons exhausting. In his diary he constantly complains about life in the student residence, which was not in the least conducive to devotion to the study of mathematics. Disturbances were constant.

At that time he wrote in his diary: "The pure mathematics lectures captivate me more and more. I prefer Professor Sokhotskiĭ's lectures in the special course on higher algebra to all the others... The main thing that concerns me is whether I have enough talent." To test himself he solved many quite difficult problems involving calculations with complicated symmetric functions, hard definite integrals, and integration of differential equations.

During his senior years Voronoĭ wrote his "candidate's" essay on the Bernoulli numbers, in which in particular he proved the following result, generalizing Adams' theorem:

If the index m of the mth Bernoulli number is divisible by $k = p_1^{\alpha} p_2^{\beta} \cdots p_l^{\lambda}$, where the p_i are primes not dividing the denominator of that Bernoulli number, then its numerator is divisible by k.

Markov expressed his liking for this work, which was published in the *Communications of the Khar'kov Mathematical Society* for 1900.

In 1889 Voronoĭ was kept at the university in order to prepare for a professorial vocation. He soon passed the exams for the Master's degree, and began to work on his dissertation "On integral algebraic numbers depending on a root of an irreducible equation of the third degree", which he defended in 1894. He was then appointed professor at Warsaw University in the department of pure mathematics.

In Warsaw Voronoĭ worked at finding a new algorithm that would allow him to solve for cubic fields the same problems that the continued fraction algorithm is used to solve for quadratic fields. In May of 1897 his doctoral dissertation, entitled "On a generalization of the continued fraction algorithm", was brilliantly defended at Petersburg University. The Academy of Sciences awarded both of his dissertations the Bunyakovskiĭ prize.

Voronoĭ's creative life now becomes more intensive. However, he does not rush to publish his work; he waits for it to mature, revising repeatedly in an effort to find the simplest and most elegant methods through which to express his deep ideas.

One after another his marvellous articles appear: "Sur un problème du calcul des fonctions asymptotiques" (*Crelle*, Bd. 126); "Sur une fonction transcendente et ses applications à la sommation de quelques séries" (*Ann. de l'Éc. Norm. Sup.*, t. 20, 1904); "Sur quelques propriétés des formes quadratiques positives parfaites" (*Crelle*, Bd. 133, 1907); and "Recherches sur les paralleloèdres primitifs" (*Crelle*, Bde. 134, 136, 1908, 1909).

The last two of these articles, the fruit of many years of sustained thought, gave Voronoĭ much pleasure and pride.

In 1904 Voronoĭ attended the International Congress of Mathematicians in Heidelberg, where he met Minkowski, who was of course very close to him in mathematical interests, and who showed him great interest and respect.

His research on positive quadratic forms completed, Voronoĭ turned his attention to indefinite quadratic forms. However, by this time his illness (gallstones) was doing its work. In his diary Voronoĭ, then working in Novocherkassk, writes: "I am making great progress with the question under study; however, at the same time my health is becoming worse and worse. Yesterday I had for the first time a clear idea of the algorithm in the theory of forms I am investigating, but also suffered a strong attack of bilious colic, which prevented me from working in the evening and from sleeping the whole night. I am so afraid that the results of my enduring efforts, obtained with such difficulty, will perish along with me."

Unfortunately his "Note on indefinite quadratic forms", occupying 28 large pages, which Voronoĭ's friends remembered seeing, and which contained an outline of the results of this ambitious work, has hitherto not been found. Results of B. A. Venkov contained in two memoirs in *Izvestiya Akademii Nauk* to some extent fill the gap.

Voronoĭ died in 1908.

Voronoĭ worked exclusively in number theory. In his brief life he published only a few articles: the six large memoirs mentioned above, and six shorter notes. However, the depth and importance of these spacious works is such that they have had a profound influence on modern number theory. Voronoĭ was in fact the co-founder, along with Minkowski, of the geometry of numbers.

Voronoĭ's work of 1903 on the number of points under a hyperbola represents one of the landmarks from which the modern analytic theory of numbers took its bearings. In particular the research of Vinogradov in analytic number theory began with an elaboration of the ideas of this work. In Voronoĭ's work on an algorithm for computing cubic units he proposes a series of problems on the distribution of relative minima, one of which—a very difficult one—he solved himself, but the majority of which remain unsolved to this day (1947). Voronoĭ's work on quadratic forms and the partition of space far from exhaust the important questions raised there, and in connection with which he produced such fundamental results.

Of subsequent investigations, those most closely related to Voronoĭ's are the earliest of Vinogradov's articles, and certain work of Zhitomirskiĭ, Venkov, and the present author. Outside Russia almost no one has continued his line of research, so that his work remains essentially in the same state as that of Markov in number theory 30 years ago.[124]

[124]That is, around 1917. *Trans.*

Voronoĭ's Dissertations on Algebraic Numbers of the Third Degree

On Integral Algebraic Numbers Depending on a Root of an Equation of the Third Degree

Master's dissertation

Already in the 1860s mathematicians of the Chebyshev school began to take an interest in the theory of algebraic integers. In 1869 there appeared Zolotarev's Master's dissertation "On an indeterminate [i.e., Diophantine] equation of the third degree", where he describes a subtle algorithm for computing a basic unit in a purely cubic field; in 1874 his doctoral dissertation "The theory of integral complex numbers with an application to the computation of integrals"; in 1878 his note "Sur les nombres complexes"; and then in 1885, 8 years after his death, his long article "Sur la théorie des nombres complexes". In the latter three works Zolotarev created a complete theory of ideal numbers for an arbitrary algebraic number field.

In 1891 the Master's dissertation of Ivanov was published, showing the equivalence of Dedekind's and Zolotarev's theories of algebraic integers; in 1892 Markov's article "Sur les nombres entiers dépendants d'une racine cubique d'un nombre ordinaire", containing the complete theory of purely cubic fields; and in 1893 Sokhotskiĭ's dissertation "The principle of the greatest common divisor as applied to the divisibility theory of algebraic numbers", giving a new, modified treatment of the theory of ideal factors.

Thus just at the time of Voronoĭ's graduation from St. Petersburg University interest in questions in algebraic number theory was very lively there. Hence it is not surprising that, following on his first short but subtle article "On the Bernoulli numbers", published in 1890 in the *Communications of the Khar'kov Mathematical Society*, he should turn his attention to algebraic number theory, in fact to the theory of number fields of degree 3, which theory he brought to perfection, from the algorithmic point of view, in the fundamental investigations recorded in his two dissertations.

In his Master's dissertation of 1894 "On integral algebraic numbers depending on a root of an equation of the third degree", Voronoĭ sets himself the goal of analysing the general cubic field in detail in the sense of finding a basis for all of its integers, and their decompositions as products of prime ideals. In his doctoral dissertation "On a generalization of the continued fraction algorithm", he gives the best algorithm known up to the present time (1947) for computing basic units in a general cubic field of negative or positive discriminant. In both of these works

Voronoĭ goes to the heart of the questions he considers and effects the most complete solution possible in the computational sense, and in addition, in the second dissertation, obtains splendid results on lattices.

In the preface to his Master's dissertation, Voronoĭ writes: "In the essay I am now presenting, results from the general theory of algebraic integers are applied to the particular case of numbers depending on the root of an irreducible equation $\rho^3 = r\rho + s$. The results obtained turn out to be very graphic. All integers of the field in question have a certain form. Using the form taken by the integers, it is not difficult to find a form embracing all the [integral] numbers divisible by a given ideal number, or, in other words, to find the ideal corresponding to that given ideal number. In our exposition the resolution of these questions is based on a detailed study of the solutions of third-degree equations relative to a prime and a composite modulus."

In Chapter I a method for solving an equation

$$X^2 - rX - s \equiv 0 \pmod{p}$$

of degree 3 is given, based on the introduction of *complex numbers modulo* p, i.e., numbers of the form $A + Bi$ where A and B are rational integers and i denotes a "nonexistent in fact" solution of an equation of the form $i^2 \equiv N \pmod{p}$, where N is a quadratic nonresidue mod p. Thus complex numbers mod p are "Galois imaginaries" for the polynomial $X^2 - N$ modulo p.[125] The *norm* of a mod p complex number $A + iB$ is defined to be $(A + iB)(A - iB) = A^2 - NB^2$. If the norm of a complex number mod p is congruent to 1 modulo p it is called a *complex unit*.[126] A mod p complex number E is a unit if and only if it satisfies the congruence $E^{p+1} \equiv 1 \pmod{p}$. For each choice of p and N there are precisely $p + 1$ complex units. The least positive integer δ for which $E^\delta \equiv 1 \pmod{p}$ is called the *exponent to which the unit E belongs*; it is a divisor of $p+1$. It is proved in the usual way that there are complex units belonging to $p + 1$, i.e., that there exist so-called *primitive* units. The powers up to the $(p + 1)$st of any primitive unit yield all the complex units. If for given p and N we fix on a primitive unit E, then the exponents $0, 1, \dots, p$ are called the *indices* (relative to E) of the units $E^0, E^1, \dots, E^p$ respectively. For such an E (and hence for all units) there always exists a mod p complex number ω with norm congruent to N modulo p, satisfying the congruence

$$\omega^2 \equiv NE \pmod{p}.$$

As an appendix to his essay Voronoĭ gives a "Table of indices of complex units for [all] prime moduli less than 200", compiled according to the following system: For each prime $p < 200$ he specifies a quadratic nonresidue N, a primitive unit E, and a complex number ω (as above), indicating also ω^{-1},[127] and lists the indices of all the units relative to E, and conversely the units according to their indices.

Voronoĭ shows that each mod p complex number $L + L'i$ can be expressed in the form $a\omega^\mu(l + l'i)$, where a is a residue mod p, i.e., is one of $0, 1, 2, \dots, p-1$, μ is 0 or 1, and $l + l'i$ is a complex unit. On the basis of this representation he gives a method for solving two-term congruences of degrees 2 and 3 in terms of complex

[125] In modern terminology they are the elements of the quotient ring $\mathbb{Z}_p[x]/(x^2 - N)$ where $(x^2 - N)$ denotes the ideal of $\mathbb{Z}_p[x]$ generated by $x^2 - N$. *Trans.*

[126] It is easy to check that such elements are precisely the units of the ring in question in the usual sense, i.e., multiplicative invertibles. *Trans.*

[127] Not necessarily representing an element of the ring in question. *Trans.*

numbers mod p using the above-mentioned table of indices. The complete solution of the original congruence $X^3 - rX - s \equiv 0 \pmod{p}$ then reduces to solving a finite sequence of two-term congruences mod p of degrees 2 or 3, involving the examination of four separate cases.

In particular, Voronoĭ obtains the following theorem:

If the discriminant $-\Delta = 4r^3 - 27s^2$ *of a given congruence* $X^3 - rX - s \equiv 0 \pmod{p}$ *is a quadratic nonresidue modulo p, then the congruence has exactly one (rational)*[128] *solution modulo p. On the other hand if the discriminant is a quadratic residue, then the congruence has either three solutions or none.*

Chapter II is devoted to finding an integral basis for the algebraic integers of a cubic number field.[129]

For the sake of clarity we begin with some definitions (not given by Voronoĭ). We denote by O any subring of the ring of integers of the given cubic number field, containing 1 and some primitive element ρ of the field.[130] We shall call an integral basis for such a subring a *unit* basis if it has the form $(1, \omega_1, \omega_2)$, *normal* if in addition $\omega_1 \cdot \omega_2$ is rational, and *graded* with respect to ρ if it has the form

$$\left(1,\ \frac{u\rho + v}{\Delta},\ \frac{\lambda\rho^2 + \mu\rho + \nu}{\Delta}\right),$$

where $u, v, \lambda, \mu, \nu, \Delta$ are rational integers. It can be shown that every such ring O has a normal graded basis, i.e., there always exist rational integers $u, v, \lambda, \mu, \nu, \Delta$, such that each element of O can be uniquely expressed in the form

$$X + X'\frac{u\rho + v}{\Delta} + X''\frac{\lambda\rho^2 + \mu\rho + \nu}{\Delta},$$

with X, X', X'' rational integers.

Voronoĭ proves the following:

If D_ρ *is the discriminant of the cubic equation*

$$F(\rho) = \rho^3 - s\rho^2 - q\rho - n = 0$$

satisfied by ρ, *and* δ, a, D_0, t *are rational integers satisfying the equation* $D_\rho = \delta^6 a^2 D_0$, *and the congruences*

$$\frac{1}{2}F''(t) \equiv 0 \pmod{\delta},\ \ F'(t) \equiv 0 \pmod{\delta^2 a},\ \ F''(t) \equiv 0 \pmod{\delta^3 a^2},$$

then the numbers

$$1,\quad \frac{\rho - t}{\delta},\quad \frac{\rho^2 + (t-s)\rho + (t^2 - st - q)}{\delta^2 a}$$

form a normal graded (with respect to ρ*) integral basis of a subring* O *of the ring of all algebraic integers of the field generated by* ρ *over the rationals, containing* 1 *and* ρ *and of discriminant* D_0. *Moreover every normal graded (in* ρ*) integral basis of every such subring of algebraic integers has this form. Finally, one may further assume that*

$$-\frac{a\delta}{2} < t \leq \frac{a\delta}{2}.$$

---

[128] That is, among the residues $0, 1, \ldots, p-1$. *Trans.*

[129] That is, a field of degree 3 over the rationals. *Trans.*

[130] The "primitivity" of ρ here signifies that the given field is $\mathbb{Q}(\rho)$, i.e., is generated over $\mathbb{Q}$ by ρ. *Trans.*

In particular, if δ and a are chosen so that $\delta^6 a^2$ is as large as possible, then the above is a basis for the ring of all algebraic integers of the field generated by ρ over the rationals, i.e., an integral basis for the field.

Consider the special case where the minimal polynomial of ρ has zero quadratic term, so that

$$F(\rho) = \rho^3 - q\rho - n = 0.$$

We may assume here that the rational integer q is square-free, i.e., is not divisible by the square of any rational integer exceeding 1, and that n is cube-free, since this can always be ensured by taking $\tau\rho$ in place of ρ for appropriate rational integral τ. In this case the formula for the integral basis of the cubic field depends on whether or not the system of congruences

$$(*)\qquad \begin{aligned} 3 - q &\equiv 0 \pmod{9}, \\ n \pm (q-1) &\equiv 0 \pmod{27}, \end{aligned}$$

is satisfied:

1° *If the system of congruences (*) is not satisfied, then one finds the largest perfect square a dividing the discriminant D_ρ of the equation $F(\rho) = 0$, for which the system of congruences*

$$(1)\qquad \begin{aligned} F'(t) &\equiv 0 \pmod{a}, \\ F(t) &\equiv 0 \pmod{a^2}, \end{aligned}$$

has a solution satisfying $-\frac{a}{2} < t \leq \frac{a}{2}$, and then the desired basis is

$$(\mathrm{I})\qquad 1,\ \rho,\ \frac{\rho^2 - t\rho + (t^2 - q)}{a}.$$

2° *If the system of congruences (*) is satisfied, then one must find the largest perfect square dividing $\frac{D_\rho}{729}$ (which turns out in this case to be an integer) for which the system of congruences*

$$(2)\qquad \begin{aligned} F'(t) &\equiv 0 \pmod{9a}, \\ F(t) &\equiv 0 \pmod{27a^2}, \end{aligned}$$

has a solution satisfying $-\frac{3a}{2} < t \leq \frac{3a}{2}$, and then the desired basis is

$$(\mathrm{II})\qquad 1,\ \frac{\rho - t}{3},\ \frac{\rho^2 + t\rho + (t^2 - q)}{9a}.$$

The systems of congruences (1) and (2) are to be solved, Voronoĭ proposes, by means of the above-described method.

These two integral bases are called *Voronoĭ bases.* For their derivation we refer the reader to Voronoĭ's dissertation itself, or to the book "The theory of irrationalities of the third degree" by B. Delone and D. Faddeev, from which the above notation for the bases is taken.

In Chapter III Voronoĭ gives convenient computational methods, based on his method (summarized above) for finding solutions of congruences of degree 3 using the table of indices appended to his dissertation, for solving the following sorts of problems:

1) Given a rational prime p, find its factorization as a product of prime ideal factors;

2) Decompose a given algebraic integer of a given cubic number field as a product of prime ideal factors;

3) Find all algebraic integers of a given cubic number field divisible by a given ideal number, i.e., find an integral basis for the corresponding ideal.

In summary, one may say of Voronoĭ's Master's dissertation that, although it contains no results of general theoretical importance, it nonetheless constitutes a very detailed investigation of basic algorithmic questions in the theory of general cubic number fields, excluding those concerning computation of its automorphisms. In particular he reduces the solution of cubic congruences, needed for these algorithms, to the solution of chains of two-term congruences some of which are solved using readily available tables, while others are solved by means of the extensive tables (appended to the dissertation) of the indices of "complex units" $l + l'i$ modulo p, where i is an "imaginary" solution of the congruence $X^2 \equiv N \pmod{p}$ for a prescribed quadratic nonresidue N modulo p.

On a Generalization of the Continued Fraction Algorithm

Doctoral dissertation

In 1896, two years after the defence of his Master's dissertation, Voronoĭ defended his doctoral dissertation "On a generalization of the continued fraction algorithm", a work of the first rank. Here Voronoĭ gives an algorithm for computing basic units for a general cubic number field, in the case of both negative and positive discriminant, penetrating deeply into the geometry of the situation. Only just now (1947), 50 years after this work was written, are some of the results contained in it being reproved by German mathematicians as important new discoveries, being unaware of the details of its contents. (It was published only in Russian.)

A convenient algorithm for finding the nontrivial basic unit of a quadratic number field (equivalently, the basic solution of Pell's equation $x^2 - \Delta y^2 = 1$) using the decomposition of Δ as a continued fraction, was published by Wallis as early as 1657 in the *Commercium aepistolicum*. A proof of the fact that for any rational integral Δ which is not a perfect square this algorithm yields a periodic continued fraction, and hence the desired solution, which evaded Wallis and even Euler, who devoted much effort to it, was at last obtained by Lagrange in 1768. In a short but remarkable article Dirichlet generalized the theory of Pell's equation to decomposable n-ary forms in n variables, i.e., forms

$$\begin{aligned} &F(x_1, x_2, \ldots, x_n) \\ &\qquad = (\alpha_{11}x_1 + \alpha_{12}x_2 + \cdots + \alpha_{1n}x_n)\cdots(\alpha_{n1}x_1 + \alpha_{n2}x_2 + \cdots + \alpha_{nn}x_n), \end{aligned}$$

in n variables $x_1, x_2, \ldots, x_n$, which decompose as indicated into products of n linear forms, and furthermore where the coefficients of F are rational integers. (Compare $x^2 - \Delta y^2 = (x + \sqrt{\Delta}y)(x - \sqrt{\Delta}y)$.) Dirichlet's results were subsequently elaborated upon by Kronecker in an extensive memoir. In view of Dirichlet's theorem on the basic units of the ring of algebraic integers of an algebraic number field, which we stated earlier,[131] from a theoretical point of view the question of the structure of those units was settled. Moreover it follows from Dirichlet's proof of his structure theorem that if one knows the coefficients of the minimal polynomial of a number

[131] In the chapter on Zolotarev. *Trans.*

ρ whose adjunction to the rational field gives the number field in question, then one can in a finite number of steps, or rather by means of a finite number of trials, calculate a full set of independent basic units for the ring of algebraic integers of that field. However, even for cubic fields this is far from constituting a convenient algorithm for computing such a set of units comparable to the continued fraction algorithm in the case of quadratic fields.

Several researchers attempted to find generalizations of the continued fraction algorithm that would play a role for cubic fields analogous to the one played by that algorithm in the quadratic case. In the preface to his work, Voronoĭ writes: "In his article 'De relatione inter ternas pluresve quantitates instituenda' Euler gave the first generalization of the continued fraction algorithm.... Jacobi changed Euler's algorithm somewhat in order to unify the calculations. The altered algorithm amounts to the following.... If one uses Jacobi's algorithm to transform [linear] forms with coefficients algebraic numbers depending on a root of an irreducible equation of the third degree, and to represent the forms obtained in this way in the normal form

$$X + X'\phi + X''\psi,$$

then in certain cases it turns out that the forms of this kind begin to repeat periodically.

"The following question is of great interest and importance: Do applications of the Jacobi algorithm always lead to periodically repeating forms if the coefficients depend on a root of an equation of the third degree[?]—this remains unanswered up to the present time." He continues: " In those cases when application of Jacobi's algorithm leads to a sequence of periodically repeating forms, the product of the corresponding substitutions yields a substitution which leaves a certain form

$$X + X'\phi + X''\psi$$

unchanged, i.e., this form transforms to the form $XE + X'E\phi + X''E\psi$. If we now denote this substitution [as indicated by]

$$\begin{vmatrix} \alpha & \beta & \gamma \\ \alpha' & \beta' & \gamma' \\ \alpha'' & \beta'' & \gamma'' \end{vmatrix} = \pm 1,$$

then the number E defined by $E = \alpha + \alpha'\phi + \alpha''\psi$, is an algebraic unit, depending on a root of the same equation as ϕ and ψ.

"A substitution which does not change the form but which is not the identity often becomes obvious in the course of calculating a few forms using Jacobi's algorithm; however, [it may happen that] calculation using Jacobi's algorithm has to be continued for a long time, and often it even remains unclear whether the forms will eventually become periodically repeating or not."

Voronoĭ goes on to note the unsuccessful attempts by Poincaré and Hurwitz at generalizing the continued fraction algorithm along the same lines as Jacobi, and draws the conclusion that the generalizations obtained by Euler, Jacobi, and Poincaré are merely formal and not useful for answering those fundamental questions of the theory of algebraic integers of a cubic field which in the case of a quadratic field are settled by the continued fraction algorithm.

Turning to Dirichlet's approach, as set out in the above-mentioned remarkable article of 1846, which was based on "Dirichlet's principle" or the "pigeon-hole principle", as it is called, i.e., the fact that if $n + 1$ objects are placed in n boxes, then at least one box will contain at least two of the objects, Voronoĭ shows that as the coefficients grow, the number of applications of this principle needed increases very rapidly, and that the principle does not furnish an algorithm, but merely a method using trial and error. Voronoĭ pays more detailed attention to the method of Hermite involving the introduction of an auxiliary continuously varying parameter. This approach in combination with the determination of the minimum of a certain positive quadratic form for various values of the parameter (on which the coefficients of the quadratic form depend) does afford algorithms for computing the units.

One such algorithm based on Hermite's approach is described in Zolotarev's Master's dissertation "On an indeterminate equation of the third degree", and another was given by Hermite's student Charve in 1880 for a general cubic field both in the case of negative discriminant (when there is just one real root) and positive discriminant (when all three roots are real).

"An essential shortcoming of Hermite's approach from a practical point of view," writes Voronoĭ, "consists in the fact that as the parameter varies, some of the form's coefficients increase rapidly while others decrease...," and further on: "In order to determine the algebraic units depending on the root of an equation of the third degree with positive discriminant, Charve considers [quadratic] forms with two variable parameters, but the question of finding a basic system of units is only touched on, and in our opinion [his approach] does not afford a method for answering it."

Turning to his own investigations, Voronoĭ writes: "Taking as the basis of our investigations a special way of viewing the continued fraction algorithm, we propose a new generalization of it."

We have dwelt in some detail on the preface to Voronoĭ's dissertation in order to communicate to the reader the importance which mathematicians of an algorithmic bent attach to the quality of an algorithm. Dirichlet's proof in his 1846 article does after all contain a method for computing for any algebraic number field a system of basic units of its ring of algebraic integers by means of a finite number of operations, not, however, furnishing an algorithm. On the other hand the methods of Euler and Jacobi, and of Poincaré, Hurwitz, Hermite and Charve considered by Voronoĭ, are indeed algorithmic. Nonetheless Voronoĭ, and contemporaneously Minkowski, undertook to compose extensive works abundant in extraordinary difficulties overcome, and proofs of magnificent results, in order to provide even better algorithms. For a mathematician not algorithmically inclined, it is difficult to understand exactly what motivates them. The existence of a set of basic units having been proved, and even a method indicated for finding them by means of a finite number of operations, what further could possibly be involved? On the other hand the algorithmic variety of mathematician understands that, as for most mathematical questions, this represents only the halfway point to a complete answer. It remains to find a regular, and if possible practical, computational procedure, each step of which involves a minimal number of trials, and if possible such that this number is bounded over all steps in the algorithm. The continued fraction algorithm, for example, in some sense involves no "trials" whatsoever. Voronoĭ's algorithm for cubic fields of negative discriminant involves at each step at most 7 trials, and a certain refined version at most 4.

We now turn to our exposition of Voronoĭ's dissertation. We observe first of all that without doubt in this work Voronoĭ framed and elicited his results geometrically. However, at that time in St. Petersburg, geometry, especially in number theory, was out of favour. Markov, the chief opponent at Voronoĭ's defence of his dissertation, would at that time not have passed a dissertation in number theory using geometric methods, and only much later, after an interval of 25 years, had he come round to the extent of recommending that the problems on quadratic forms that he had been unable to solve should be tackled using geometrical methods. Be that as it may, in St. Petersburg in 1896 it was pointless to even think of expounding results in geometric guise. Having obtained his results by geometric means, Voronoĭ was thus compelled to translate his proofs into purely arithmetic terminology, removing all traces of geometry, which is, of course, always possible.

A certain mathematician who knew Voronoĭ well, albeit somewhat later on, related how once, on dropping in on Voronoĭ at home, he saw various geometric models on his desk. On being asked what they were for, Voronoĭ answered: "O, nothing much," and quickly shut them away in a cupboard. There exists a photograph of Voronoĭ seated at his writing desk, on which there stands some kind of complicated wire model obviously made by himself, apparently related to his work on perfect quadratic forms or on parallelohedra.

For this reason it is simplest to expound Voronoĭ's work "On a generalization of the continued fraction algorithm" immediately in geometric terms, thus effectively translating it back into that geometric language in which Voronoĭ originally, it seems clear, conceived it.

The work is divided into three sections, with the following headings:

I. Successive relative minima of systems of covariant forms

$$\omega = X\lambda + X'\mu, \quad \omega' = X\lambda' + X'\mu',$$

with respect to rational integer values of the variables.

II. Successive relative minima of systems of covariant forms

$$\omega = X\lambda + X'\mu + X''\nu, \quad \omega' = X(l + l'i) + X'(m + m'i) + X''(u + u'i),$$

with respect to rational integer values of the variables.

III. Successive relative minima of systems of covariant forms

$$\begin{aligned} \omega &= X\lambda + X'\mu + X''\nu, \\ \omega' &= X\lambda' + X'\mu' + X''\nu', \\ \omega'' &= X\lambda'' + X'\mu'' + X''\nu'', \end{aligned}$$

with respect to rational integer values of the variables.

Section I is, properly speaking, concerned with ordinary continued fractions, but considered from a new point of view which furnishes for Voronoĭ the basis for his generalization of them. Covariant forms

$$\omega = X\lambda + X'\mu, \quad \omega' = X\lambda' + X'\mu',$$

where the variables X, X' are restricted to taking rational integral values, are represented geometrically by the coordinates $x = \omega, y = \omega'$, of the points of the planar lattice[132] $\mathcal{L}$ determined by the frame consisting of the two vectors (λ, λ'), (μ, μ'), assumed linearly independent. We also assume of the lattice $\mathcal{L}$ that apart from the

[132]Denoted by O in the original. *Trans.*

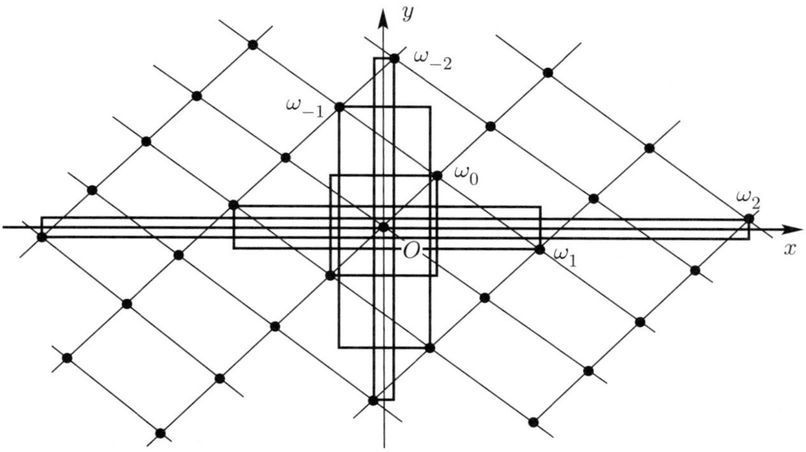

FIGURE 18

origin none of its points is on either the x- or y-axes. Given any point M, we shall call the rectangle centred at the origin O with sides parallel to the x- and y-axes and one vertex at M, the *normal parallelogram (rectangle) determined by* M. In view of the assumption that there are no points of $\mathcal{L}$ other than O on the coordinate axes, if M is a point of the lattice, then the only other point of the lattice on the boundary of the normal rectangle of M is that diametrically opposite M.

A point (ω_0, ω_0') other than O of the lattice $\mathcal{L}$ is called a *relative minimum* if its normal rectangle has no points of $\mathcal{L}$ other than O in its interior. Voronoĭ denotes by (S) the set of all relative minima of the lattice $\mathcal{L}$. We shall now show that the set (S) is infinite. Observe first that there is at least one relative minimum (ω_0, ω_0'), since if (ω, ω') is any point of the lattice other than O, then either it is a relative minimum point, in which case there is nothing left to prove, or not, in which case there is another point $(\neq O)$ of $\mathcal{L}$ in the interior of its normal rectangle. If this new point is a relative minimum, again we have the desired conclusion, while in the contrary case we can in turn find a point $(\neq O)$ of $\mathcal{L}$ in the normal rectangle of the new point, and so on. This process must terminate (in a relative minimum) since each point of $\mathcal{L}$ considered lies in the the normal rectangle of its predecessor, and hence in the normal rectangle of the original point (ω, ω'), which contains only finitely many points of the lattice.

Next, if $\boldsymbol{\omega}_0 = (\omega_0, \omega_0')$ is any relative minimum of $\mathcal{L}$, then (we omit the proof of this) if we stretch the normal rectangle of $\boldsymbol{\omega}_0$ parallel to the x-axis in both the positive and negative directions, while maintaining its height, the boundary of this continuously lengthening rectangle will eventually encounter another point of the lattice $\mathcal{L}$, necessarily as an interior point of a side parallel to the y-axis. The first such point $\boldsymbol{\omega}_1$ encountered (see Figure 18) will be the closest relative minimum to $\boldsymbol{\omega}_0$ in the x-direction, in the sense that there is no relative minimum with x-coordinate between ω_0 and ω_1. Repeating this procedure with the normal rectangle of $\boldsymbol{\omega}_1$, we obtain the next (in the x-direction) relative minimum $\boldsymbol{\omega}_2$, and so on, to infinity.

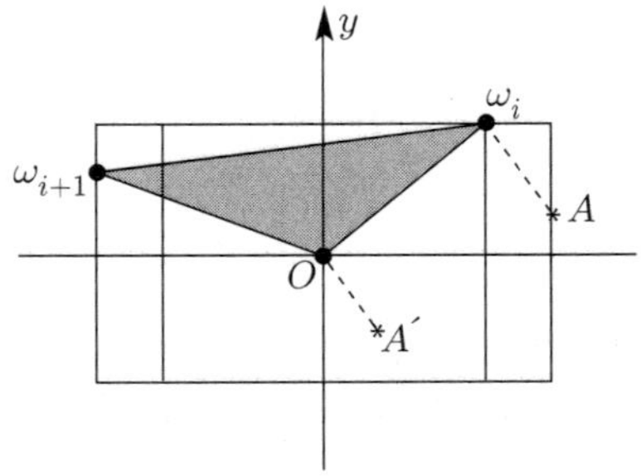

FIGURE 19

Analogously, by stretching the rectangles parallel to the y-axis, one obtains relative minima $\boldsymbol{\omega}_{-1}, \boldsymbol{\omega}_{-2}, \ldots$ with (strictly) decreasing x-coordinates. If need be, one may choose for consideration from each pair of relative minima symmetrically situated relative to the origin the one with positive y-coordinate, i.e., one may restrict attention to just those relative minima in the upper half-plane.[133]

From the recipe just described for finding successive relative minima, it is easy to see that the frame consisting of the two vectors emanating from O with their tips at consecutive relative minimum points $\boldsymbol{\omega}_i, \boldsymbol{\omega}_{i+1}$ (see Figure 19) is for each i a generating frame for the lattice $\mathcal{L}$. For the triangle $O\boldsymbol{\omega}_i\boldsymbol{\omega}_{i+1}$ lies completely in the lengthened normal rectangle of the point $\boldsymbol{\omega}_i$ containing the point $\boldsymbol{\omega}_{i+1}$ on a vertical edge, and this has no points of the lattice (other than O) in its interior; hence the triangle has no points of $\mathcal{L}$ in its interior, from which it follows that it must be fundamental for the lattice.

It is also easily seen that for each i the points $\boldsymbol{\omega}_i$ and $\boldsymbol{\omega}_{i+1}$ must lie on opposite sides of the y-axis. For if $\boldsymbol{\omega}_{i+1}$ coincided with a point A of the vertical side of the lengthened rectangle on the same side of the y-axis as $\boldsymbol{\omega}_i$, then (see Figure 19) the vector $\overrightarrow{\boldsymbol{\omega}_i A}$ when laid off from the origin would have its tip at a point A' inside that extended rectangle, contradicting the fact that, apart from O, that rectangle has no points of the lattice in its interior. Thus successive relative minima lie alternately on one side of the y-axis and then the other.

We shall now reexamine in greater detail the process of locating successive relative minima, but now somewhat differently from Voronoĭ, in order to facilitate the reader's understanding of just how the algorithm described in Section II pertaining to the 3-dimensional case represents a generalization of the continued fraction algorithm, which is what we are at present expounding.

Assume as before that the lattice $\mathcal{L}$ has no points other than O on the coordinate axes, and let M be any *primary* point of the lattice in the upper half-plane, i.e., such that the interior of the line segment OM contains no points of the lattice. We now project the whole lattice $\mathcal{L}$ onto the x-axis parallel to the segment OM (see Figure 20); this results in a linear chain $\mathcal{S}$ of points on the x-axis, since each linear row of points of the planar lattice $\mathcal{L}$ parallel to OM projects parallel to OM onto just one point of the x-axis. In each linear row of points of our lattice parallel to

[133]It can be shown that the chain

$$\ldots, \pm\boldsymbol{\omega}_{-2}, \pm\boldsymbol{\omega}_{-1}, \pm\boldsymbol{\omega}_0, \pm\boldsymbol{\omega}_1, \pm\boldsymbol{\omega}_2, \ldots$$

accounts for all relative minima of the lattice. *Trans.*

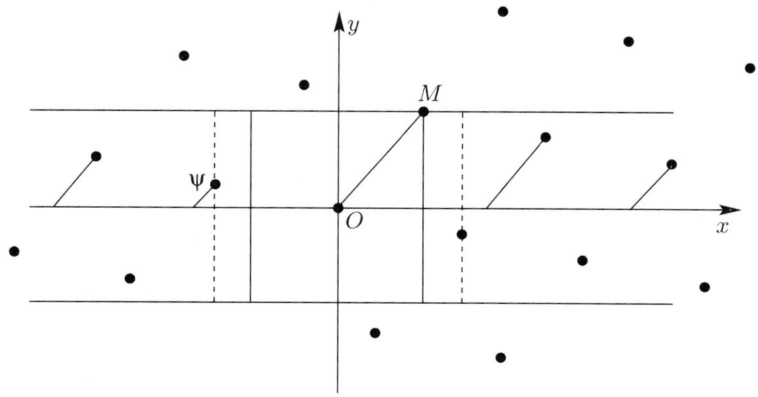

FIGURE 20

OM, and corresponding to a single point of the chain $\mathcal{S}$, consider the point with least positive ordinate; we shall call such a point *lowest* or a *pinhead* of a "pin" protruding parallel to OM from the corresponding point of the row $\mathcal{S}$ on the x-axis (see Figure 20).

It is clear that if M is a relative minimum, then on extending its normal rectangle horizontally in both directions from the y-axis while keeping the height fixed, the first lattice point in the upper half-plane encountered by a vertical side of this rectangle (i.e., the next relative minimum to M in the x-direction having positive ordinate less than that of M) will also be a pinhead, since, in view of the fact that there are no points of the lattice in the interior of OM, all points of the rows of $\mathcal{L}$ parallel to OM that are not pinheads and have positive ordinate, will lie further from the x-axis than the point M. We now seek to characterize that pinhead, the first encountered by the lengthening rectangle.

To this end, consider the point of the chain $\mathcal{S}$ closest to O among those on the opposite side of O from the orthogonal projection of M on the x-axis, and denote by ψ the pinhead of the pin protruding from this point. We have the following geometrical result:

If the point ψ does not lie inside the normal rectangle of M, then M is a relative minimum point of the lattice and ψ is the next relative minimum to M in the x-direction having positive ordinate less than that of M. On the other hand if ψ lies inside the normal rectangle of M, then although M is not a relative minimum, we have found a point of the lattice, namely ψ, inside the normal rectangle of M.

The proof is as follows. If ψ is not inside the normal rectangle of M, then the line of the row of $\mathcal{L}$ containing ψ and parallel to OM, intersects the x-axis outside that rectangle; but then, as is clear from Figure 20, all the other pinheads (except M) lie further from the y-axis than ψ, and consequently the extended rectangle with a vertical edge through ψ (dashed in Figure 20) contains no points of $\mathcal{L}$ other than O in its interior. Hence M is a relative minimum, and ψ is the next relative minimum to M having positive ordinate less than that of M.

If on the other hand it turns out that the point ψ lies inside the normal rectangle of M, we repeat the construction with ψ in place of M, obtaining a point ψ'. If ψ' is inside the normal rectangle of ψ, then it is certainly inside that of M. Since the normal rectangle of M contains only finitely many points of the lattice, this process must terminate in a point $\boldsymbol{\omega}_1$ not lying in the normal rectangle of the preceding point $\boldsymbol{\omega}_0$, so that by the above result we will have found a relative minimum $\boldsymbol{\omega}_0$, together with its neighbouring relative minimum $\boldsymbol{\omega}_1$ in the x-direction with smaller positive ordinate. Repeated applications of this construction and the above result will then yield further successive relative minima $\boldsymbol{\omega}_2, \boldsymbol{\omega}_3, \dots$.

If we wish to take the initial point M to be the point (λ, λ') of the initial frame $(\lambda, \lambda'), (\mu, \mu')$ defining $\mathcal{L}$, then we need to arrange beforehand that $\lambda' > 0$, for instance by replacing the first vector of the frame by its negative if necessary.[134] Assuming, then, that $\lambda' > 0$ we need to ensure that the abscissa of the point of intersection with the x-axis of the line through the point (μ, μ') parallel to the first vector, has sign opposite to that of λ. It is easily calculated that that abscissa is $\frac{1}{\lambda'}\left|\begin{smallmatrix} \mu & \lambda \\ \mu' & \lambda' \end{smallmatrix}\right|$, so that, since $\lambda' > 0$, we need that the sign of $\left|\begin{smallmatrix} \lambda & \mu \\ \lambda' & \mu' \end{smallmatrix}\right|$ coincide with that of λ. If this happens not to be the case, then we replace the second vector by its negative. Next we find an integer δ satisfying $0 < \mu' + \delta\lambda' < \lambda'$, and then the point ψ is given by $(\mu + \delta\lambda, \mu' + \delta\lambda')$, and so on.

Having in this way eventually found two neighbouring relative minima $\boldsymbol{\omega}_0, \boldsymbol{\omega}_1$, the problem of finding the sequence of further successive relative minima reduces to expanding the number $\frac{\omega_1'}{\omega_0'}$ as a continued fraction. It follows easily that these relative minima are in fact just those points which we called "convergent" in our exposition of the geometry of continued fractions.[135]

We shall call two planar figures *similar relative to the given coordinate axes* if one of them can be obtained from the other by means of "stretchings" in the directions of the axes, i.e., by means of any affine transformation sending each axis onto itself (with or without reversal of direction). This means of course that the coordinates of the points of one figure are obtained from the corresponding points of the other by multiplying them by some fixed nonzero numbers τ, τ'. (Voronoĭ does not introduce the term "similar", but we find it convenient.)

Suppose that for a given planar lattice we have such a similarity transformation of the lattice onto itself, with coefficients e, e', which is nontrivial, i.e., at least one of the numbers e, e' is different from 1 in absolute value. Since $|ee'|$ is the factor by which area is changed by the transformation, we must have $|ee'| = 1$ since the fundamental parallelogram of the lattice must be transformed into another such, so that area remains unchanged. Thus such a transformation sending a lattice onto itself is in fact a hyperbolic rotation[136] relative to the coordinate axes as asymptotes (perhaps followed by reflections in them). We shall call such transformations of a lattice *multiplicative automorphisms*. (This term is also not used by Voronoĭ.) Denoting such a transformation by e, the first factor of the pair e, e', we have that $\dots, e^{-3}, e^{-2}, e^{-1}, e^0 = 1, e, e^2, e^3, \dots$ are likewise multiplicative automorphisms. We shall therefore call a lattice having a multiplicative automorphism *periodic*.

It is clear from the very definitions of "relative minimum" and "neighbouring (or adjacent) relative minima", that under any similarity transformation the chain

---

[134]Presumably it has first been arranged that $\lambda' \neq 0$ to conform with the requirement that the lattice have no points other than O on the coordinate axes. *Trans.*

[135]See the Commentary in the chapter on Markov. *Trans.*

[136]This is the definition of a hyperbolic transformation; see the chapter on Markov. *Trans.*

of successive relative minima of a lattice transforms into the chain of successive relative minima of the image lattice, with adjacent relative minima being sent to adjacent relative minima of the image. It is therefore appropriate to call the complete chain of successive relative minima of a periodic lattice *periodic* also.

A frame of the form $\left[\begin{smallmatrix} 1 & \phi \\ 1 & \phi' \end{smallmatrix}\right]$, with first vector $(1,1)$, is said to be *normalized.* Given a frame $\left[\begin{smallmatrix} \lambda & \mu \\ \lambda' & \mu' \end{smallmatrix}\right]$, the process of *normalization* of the frame involves multiplication of the first components of the vectors of the frame by $\tau = \frac{1}{\lambda}$ and the second components by $\tau' = \frac{1}{\lambda'}$.[137]

If the lattice is periodic then the sequence of successive reduced normalized frames

$$\begin{bmatrix} 1 & \phi_1 \\ 1 & \phi_1' \end{bmatrix}, \begin{bmatrix} 1 & \phi_2 \\ 1 & \phi_2' \end{bmatrix}, \ldots, \begin{bmatrix} 1 & \phi_i \\ 1 & \phi_i' \end{bmatrix}, \ldots,$$

obtained by normalizing the members of the sequence of frames

$$\begin{bmatrix} \omega_0 & \omega_1 \\ \omega_0' & \omega_1' \end{bmatrix}, \begin{bmatrix} \omega_1 & \omega_2 \\ \omega_1' & \omega_2' \end{bmatrix}, \ldots, \begin{bmatrix} \omega_{i-1} & \omega_i \\ \omega_{i-1}' & \omega_i' \end{bmatrix}, \ldots,$$

where the $\boldsymbol{\omega}_k$ are as before, is periodic in the usual sense of the word, i.e., for some positive integer n we have for all i

$$\begin{bmatrix} 1 & \phi_{i+n} \\ 1 & \phi_{i+n}' \end{bmatrix} = \begin{bmatrix} 1 & \phi_i \\ 1 & \phi_i' \end{bmatrix}.$$

For, since the lattice is periodic, we must have for some n

$$\begin{bmatrix} \omega_n & \omega_{n+1} \\ \omega_n' & \omega_{n+1}' \end{bmatrix} = \begin{bmatrix} e\omega_0 & e\omega_1 \\ e'\omega_0' & e'\omega_1' \end{bmatrix},$$

where e, e' are the coefficients of a multiplicative automorphism of the lattice, whence we obtain immediately

$$\begin{bmatrix} 1 & \phi_{n+1} \\ 1 & \phi_{n+1}' \end{bmatrix} = \begin{bmatrix} 1 & \phi_1 \\ 1 & \phi_1' \end{bmatrix},$$

and so on.

We have successively

$$\begin{aligned} \phi_1 &= \frac{\omega_1}{\omega_0}, \quad \text{i.e., } \omega_1 = \omega_0\phi_1, \\ \phi_2 &= \frac{\omega_2}{\omega_1}, \quad \text{i.e., } \omega_2 = \omega_1\phi_1 = \omega_0\phi_1\phi_2, \\ \phi_3 &= \frac{\omega_3}{\omega_2}, \quad \text{i.e., } \omega_3 = \omega_2\phi_3 = \omega_0\phi_1\phi_2\phi_3, \\ &\ldots \quad \ldots \ldots \ldots \ldots \ldots \ldots \ldots \ldots \end{aligned}$$

which yields at the nth step $\omega_n = \omega_0\phi_1\phi_2\cdots\phi_n$, and since $\omega_n = e\omega_0$, we infer that $e = \phi_1\phi_2\cdots\phi_n$, and, analogously, $e' = \phi_1'\phi_2'\cdots\phi_n'$. We thus have the following important result:

The coefficients e, e' of a hyperbolic rotation sending a planar lattice onto itself are equal to the products of the respective coordinates of the second vectors of the reduced normalized frames comprising a period.

---

[137] Note that the normalized frame can be identified with the original generating frame of the lattice, if we consider it as resulting merely from a change of units of measurement on each of the coordinate axes. *Trans.*

If e, e' are the coefficients of a hyperbolic rotation of a lattice onto itself and the lattice contains the point $(1,1)$, then it will contain also the point $(1\cdot e, 1\cdot e') = (e, e')$, called a *unit of the lattice* since coordinatewise multiplication of its points by this point maps the lattice onto itself.

It is easy to show that all multiplicative automorphisms e of a lattice are up to sign integer powers of a particular *basic* automorphism E. For, if $\left[\begin{smallmatrix} 1 & \phi_{n+1} \\ 1 & \phi'_{n+1} \end{smallmatrix}\right]$ is the first reduced, normalized frame of the lattice coinciding with $\left[\begin{smallmatrix} 1 & \phi_1 \\ 1 & \phi'_1 \end{smallmatrix}\right]$, then every hyperbolic rotation e must be up to sign an integral power of the one E, say, achieving this transformation of frames, since otherwise there would exist a shorter period; hence $e = \pm E^k$ for some integer k.

In answer to the question as to when a lattice is periodic, Voronoĭ proves the following theorem:

A lattice is periodic if and only if it is similar to one where the coordinates of each vector of a generating frame are conjugate elements[138] *of some quadratic number field, i.e., if the vectors of the frame are* $(\lambda, \lambda'), (\mu, \mu')$, *then* λ *and* μ *should be elements of one and the same quadratic number field, and* λ' *and* μ' *should be their respective conjugates.*

Consequently the continued fraction algorithm affords a method for computing the basic unit of the ring of integers of a quadratic number field.[139]

Voronoĭ gives a further application of the continued fraction algorithm, namely to the question of recognizing whether or not two arbitrarily given planar lattices are similar. Obviously two similar lattices will have identical chains of reduced normalized frames, in the sense that any such frame of one of the lattices will occur as a member of the chain of reduced normalized frames of the other. If the latter chain is not periodic, then the number of computations needed to determine whether a given reduced normalized frame is in that chain is potentially infinite. On the other hand if that chain is periodic, then, once the period n is known, at most n comparisons need be made to resolve the question.

The final application of the continued fraction algorithm given by Voronoĭ, is as follows: Consider a periodic lattice, determined in accordance with the above theorem by a (real) quadratic number field, and the sublattice $\mathcal{I}$ consisting of all points whose coordinates are algebraic integers of this quadratic field. As usual, an *ideal* of the lattice $\mathcal{I}$ is defined to be a sublattice of $\mathcal{I}$ closed under coordinatewise multiplication by points from $\mathcal{I}$. Voronoĭ shows that the largest number h of such sublattices (ideals) not similar to one another (via a hyperbolic rotation relative to the axes) is finite. The number h is called the *ideal class number*, since it is the number of equivalence classes with respect to similarity, or similarity classes, of ideals. (These correspond to the lattices $\Gamma, \Gamma_1, \Gamma_2, \ldots, \Gamma_{h-1}$ contained in the "principal configuration" corresponding to the given number field; see the Commentary in the chapter on Zolotarev.) Thus determining whether or not two given ideals belong to the same class is equivalent to determining whether their lattices are similar. Since we are assuming that the lattices are all periodic, it follows from the above that this question can be decided essentially by computing a reduced

[138]That is, roots of a single rational quadratic polynomial. *Trans.*

[139]Namely the number E defined above. Note also that by Dirichlet's theorem such a basic unit exists if and only if the quadratic number field is real, so that this is, presumably, tacitly assumed in the preceding theorem. *Trans.*

normalized frame of one of the ideals and the period of the chain of such frames of the other.

Although, apart from the approach, Section I of the present work of Voronoĭ contains contains no essentially new results, we have dwelt on it in considerable detail because Sections II and III follow closely the same general pattern. The difference consists in the replacement of the continued fraction algorithm as a means of computing the chain of successive relative minima, by a generalization of that algorithm.

We now examine Section II of Voronoĭ's dissertation, headed, as noted earlier, "Successive relative minima of a system of covariant forms

$$\omega = X\lambda + X'\mu + X''\nu,$$
$$\omega' = X(l + l'i) + X'(m + m'i) + X''(n + n'i),$$

with respect to rational integer values of the variables."

We will identify the xy-plane with the complex plane coordinatized by the complex coordinate $x + iy$, and consider the z-axis as a real vertical axis, so that each point of xyz-space may alternatively be coordinatized by its (real) height z and a complex number $x + iy$, its projection onto the xy-plane. i.e., by the ordered pair $(z, x + iy)$ consisting of one real and one complex coordinate.

Provided that

$$\begin{vmatrix} \lambda & \mu & \nu \\ l & m & n \\ l' & m' & n' \end{vmatrix} \neq 0,$$

which in fact we presuppose, the above system of covariant forms is represented by the lattice generated by the *Voronoĭ frame* consisting of the vectors

$$(\lambda, l + l'i), \ (\mu, m + m'i), \ (\nu, n + n'i),$$

written more compactly as

$$\begin{bmatrix} \lambda & \mu & \nu \\ l + l'i & m + m'i & n + n'i \end{bmatrix}.$$

Voronoĭ also presupposes that:

1° This lattice has no points other than the origin O in the plane $z = 0$; and

2° Any two distinct points of the lattice the same horizontal distance from the z-axis are symmetrically disposed with respect to the origin, i.e., are diametrical opposites (antipodean) with respect to O.

For any point ω in xyz-space, consider the circular cylinder with axis the z-axis, centre at O, and with the circular boundary of its base or lid passing through ω (as in Figure 21). We shall call this the *normal cylinder of the point* ω; it consists of all points whose coordinates z and $x + iy$ are both in modulus less than or equal to those of ω. If ω is a point of the above-defined lattice, then in view of conditions 1° and 2°, apart from the antipodal point $-\omega$ relative to O, which lies on the circular boundary of the base or lid of the normal cylinder of ω according as ω is on the lid or base respectively, there will be no other points of the lattice on the boundary of this cylinder.

Voronoĭ calls a point of the lattice a *relative minimum* if the interior of its normal cylinder contains no points of the lattice other than O. It is easily seen, much as in the above exposition of Section I, that there exists at least one relative minimum $\boldsymbol{\omega}_0$. If we now let the radius r of the normal cylinder of $\boldsymbol{\omega}_0$ increase

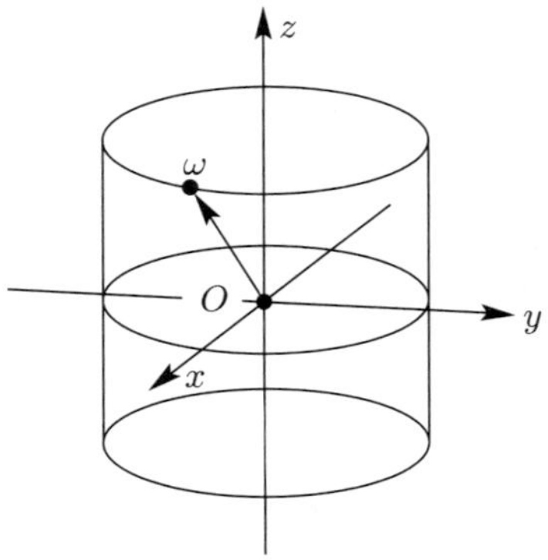

FIGURE 21

continuously, without changing its height, then it is again easy to see that eventually the expanding cylinder will encounter another point $\boldsymbol{\omega}_1$ of the lattice, necessarily at an interior point of the curved vertical side of the cylinder, which will clearly also be a relative minimum, but with a greater value r_1 of r (the radius of the cylinder) than that (r_0) of $\boldsymbol{\omega}_0$, and which is also *adjacent* to $\boldsymbol{\omega}_0$ in the sense that there are no relative minima at radii r strictly between r_0 and r_1. We next repeat this procedure starting with the normal cylinder of $\boldsymbol{\omega}_1$, in order to locate the next relative minimum $\boldsymbol{\omega}_2$, and so on, obtaining in this way an infinite sequence of successive relative minima of increasing r (see Figure 22). Analogously, we will obtain an infinite sequence of successive relative minima with decreasing r if, starting again with the normal cylinder of $\boldsymbol{\omega}_0$, we let its height increase continuously while holding the radius fixed, until the lid encounters another point of the lattice (again a relative minimum), and so on (see Figure 23). These two processes yield a doubly infinite chain

$$\ldots, \boldsymbol{\omega}_{-3}, \boldsymbol{\omega}_{-2}, \boldsymbol{\omega}_{-1}, \boldsymbol{\omega}_0, \boldsymbol{\omega}_1, \boldsymbol{\omega}_2, \boldsymbol{\omega}_3, \ldots$$

of successive relative minima of the lattice. It is not difficult to see that in fact these[140] account for all relative minima of the lattice.

Voronoĭ now proves that the parallelogram determined by the two vectors from the origin to two adjacent relative minima $\boldsymbol{\omega}_i, \boldsymbol{\omega}_{i+1}$, is a fundamental parallelogram for that 2-dimensional lattice consisting of all those points of the 3-dimensional lattice under consideration lying in the plane determined by these two vectors. This follows from the fact that the triangle $O\boldsymbol{\omega}_i\boldsymbol{\omega}_{i+1}$ (see Figure 22) has no points of the lattice in its interior, which follows in turn from the fact that its interior is wholly contained in the interior of the expanded normal cylinder of $\boldsymbol{\omega}_{i+1}$, which contains no points of the lattice.

From this it follows that for the pair of vectors $\overrightarrow{O\boldsymbol{\omega}_i}$, $\overrightarrow{O\boldsymbol{\omega}_{i+1}}$ there are infinitely many choices of a third $\boldsymbol{\nu}_i$ such that the three of them comprise a generating frame of the lattice under consideration. Hence we can find a doubly infinite chain of

[140]And their negatives. *Trans.*

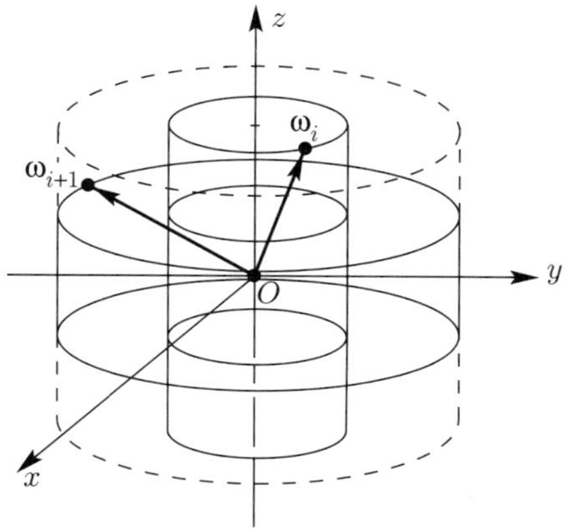

FIGURE 22

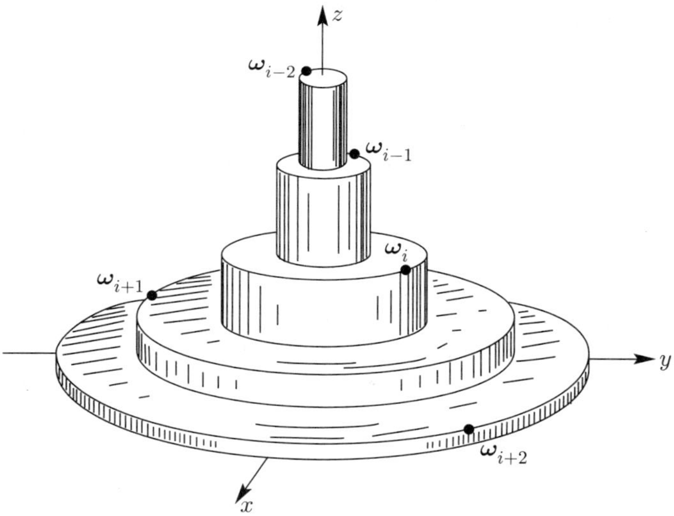

FIGURE 23

frames

$$\ldots, (\boldsymbol{\omega}_{-2}, \boldsymbol{\omega}_{-1}, \boldsymbol{\nu}_{-2}),\ (\boldsymbol{\omega}_{-1}, \boldsymbol{\omega}_0, \boldsymbol{\nu}_{-1}),\ (\boldsymbol{\omega}_0, \boldsymbol{\omega}_1, \boldsymbol{\nu}_0),\\ (\boldsymbol{\omega}_1, \boldsymbol{\omega}_2, \boldsymbol{\nu}_1),\ (\boldsymbol{\omega}_2, \boldsymbol{\omega}_3, \boldsymbol{\nu}_2), \ldots,$$

where in each frame $\boldsymbol{\nu}_i$ can be chosen in infinitely many ways. Voronoĭ chooses these third vectors in a manner specific to each frame, to be defined below, and calls these frames *reduced.*

Analogously to the planar situation described above, we say that two figures in our $(z, x+iy)$-space are *similar* if one can be obtained from the other by coordinatewise multiplication of its points $(z, x+iy)$ by a point (τ, τ'), where τ is some real and τ' some complex number, both nonzero. Thus a figure similar to a given one is obtained by applying to the latter a "stretching" in the z-direction, a radial expansion from the z-axis, and finally a rotation about the z-axis. Under such a similarity transformation the frame

$$\begin{bmatrix} \lambda & \mu & \nu \\ l+l'i & m+m'i & n+n'i \end{bmatrix}$$

is sent to the frame[141]

$$\begin{bmatrix} \tau\lambda & \tau\mu & \tau\nu \\ \tau'(l+l'i) & \tau'(m+m'i) & \tau'(n+n'i). \end{bmatrix}$$

Taking $\tau = \frac{1}{\lambda}$, $\tau' = \frac{1}{l+l'i}$, yields an image frame $\left[\begin{smallmatrix} 1 & \phi & \psi \\ 1 & a+bi & c+di \end{smallmatrix}\right]$, with first vector $(1,1)$, i.e., with Cartesian coordinates $(x,y,z) = (1,0,1)$, which we shall call a *normalized frame*, obtained from the original frame by means of *normalization*. The above doubly infinite chain of frames yields on normalization a chain of reduced normalized frames

$$\ldots, (\mathbf{1}, \boldsymbol{\phi}_{-2}, \boldsymbol{\psi}_{-2}),\ (\mathbf{1}, \boldsymbol{\phi}_{-1}, \boldsymbol{\psi}_{-1}),\ (\mathbf{1}, \boldsymbol{\phi}_0, \boldsymbol{\psi}_0), (\mathbf{1}, \boldsymbol{\phi}_1, \boldsymbol{\psi}_1), (\mathbf{1}, \boldsymbol{\phi}_2, \boldsymbol{\psi}_2), \ldots.$$

As before one verifies that

$$\omega_1 = \omega_0\phi_1,\ \omega_2 = \omega_0\phi_1\phi_2,\ \omega_3 = \omega_0\phi_1\phi_2\phi_3, \ldots, \omega_n = \omega_0\phi_1\phi_2\cdots\phi_n, \ldots,$$

where ω_i and ϕ_i are the real components (i.e., the z-components) of $\boldsymbol{\omega}_i$ and $\boldsymbol{\phi}_i$ respectively.

If the lattice admits a multiplicative automorphism defined by a pair $(\tau, \tau') = (e, e')$, i.e., a similarity sending the lattice onto itself, then as before we call the lattice *periodic*. Voronoĭ shows that a lattice satisfying conditions 1° and 2° is periodic if and only if it is similar to a lattice the coordinates of whose points are conjugate numbers of a cubic field of negative discriminant, i.e., a cubic number field of whose conjugate fields one is real and the other two complex conjugates, or, more precisely:

The above-defined lattice, satisfying conditions 1° *and* 2°*, is periodic if and only if the first (real) components* λ, μ, ν *of the initial frame all belong to the same real cubic number field, and the second components* $l+l'i, m+m'i, n+n'i$ *are respective conjugates of the first components, all from the same conjugate of the cubic number field containing the first components,*[142] *or if the lattice is similar to such a lattice, i.e., differs only by a real factor* τ *of the real components* λ, μ, ν *and a complex factor* τ' *of the complex components* $l+l'i, m+m'i, n+n'i$.

As before, it is easily seen that if the lattice is periodic, then all of its multiplicative automorphisms are up to sign integer powers of a single such automorphism $\mathbf{E} = (E, E')$:

$$\mathbf{e} = \pm\mathbf{E}^m,$$

[141] No longer necessarily a frame of the lattice under consideration. *Trans.*

[142] That is, if and only if there is an irreducible rational polynomial of degree 3 with just one real root α such that the first components λ, μ, ν all lie in $\mathbb{Q}[\alpha]$, while the second components all lie in $\mathbb{Q}[\xi]$ where ξ is a nonreal root, and furthermore if $\lambda = q_0 + q_1\alpha + q_2\alpha^2$ where $q_0, q_1, q_2 \in \mathbb{Q}$, then $l + l'i = q_0 + q_1\xi + q_2\xi^2$, and similarly for $\mu, m+m'i$ and $\nu, n+n'i$. *Trans.*

and further that the chain of reduced normalized frames of the lattice is periodic, and if the period is n, then

$$E = \phi_1\phi_2\cdots\phi_n.$$[143]

Voronoĭ further shows that two such lattices[144] are similar precisely when their chains of reduced normalized frames coincide. Hence if the two lattices are periodic, then they will be similar if and only if any frame of the reduced normalized chain of one of them coincides with some frame in any interval of reduced normalized frames of the other, of length equal to the period; the solution of the problem of determining whether two ideals of a cubic number field are equivalent or not reduces to the solution of this sort of decision problem for lattices.

So far the exposition of Section II has been very similar to that of Section I, i.e., to the 2-dimensional case. It remains to spell out the algorithm which, starting from a given fundamental frame, allows one to calculate a reduced frame of the lattice and then an adjacent reduced frame, so as to be able to compute the chain of successive reduced frames. The eliciting of such an algorithm in fact is the crux of Voronoĭ's dissertation.

This algorithm, proposed by Voronoĭ in Section II, is based on a certain geometrical theorem, the centrepiece of that section, which we now prepare to formulate. Consider as above a 3-dimensional lattice $\mathcal{L}$ satisfying conditions 1° and 2°, and let M be any *primary* point of $\mathcal{L}$, i.e., such that the line segment OM contains no points of $\mathcal{L}$ other than O and M. We now project the lattice $\mathcal{L}$ onto the xy-plane by means of the projection parallel to OM, obtaining thereby a 2-dimensional lattice $\mathcal{S}$ in the xy-plane since each row of points of $\mathcal{L}$ parallel to OM projects onto a single point of the xy-plane.

Analogously to the procedure described above for the 2-dimensional case, we now consider in each row of points of $\mathcal{L}$ parallel to OM that point with least positive height z. We shall call such points *lowest points* or *pinheads* (of "pins" parallel to OM protruding from the points of the lattice $\mathcal{S}$ in the xy-plane). It is clear that if M is a relative minimum of $\mathcal{L}$, then as we allow the radius r of its normal cylinder to increase while keeping its height unchanged, the first point of $\mathcal{L}$ encountered (necessarily by its vertical (curved) side), i.e., the adjacent relative minimum to M in the direction of increasing r, will be another pinhead, since all points other than pinheads of the rows of $\mathcal{L}$ parallel to OM must lie further from the xy-plane than M, in view of the fact that the segment OM has no points of $\mathcal{L}$ in its interior. We now examine this situation in order to see exactly which pinhead this will be.

Consider the hexagon called "Zelling's hexagon", made up of the 6 fundamental acute-angled triangles of the lattice $\mathcal{S}$ having O as a vertex, and choose from these triangles that one containing the negative of the vector from O to the orthogonal projection of M onto the xy-plane; we shall now consider the coordinates of the points of $\mathcal{S}$ with respect to the frame consisting of the two vectors forming the two sides of this triangle incident with O. The seven pinheads of those "pins" protruding from the points $(1,0),(0,1),(-1,0),(0,-1),(1,-1),(-1,1),(1,1)$ written relative to this frame, i.e., from the vertices of Zelling's hexagon and the point $(1,1)$, the fourth point of the fundamental parallelogram determined by the frame $(1,0),(0,1)$,

[143]Here E is the desired generator of the units of the ring of algebraic integers of the cubic number field. *Trans.*

[144]Not necessarily periodic. *Trans.*

will then be called *Voronoĭ pinheads.* Of the seven Voronoĭ pinheads, that having least distance r from the z-axis will be called the *reduced* Voronoĭ pinhead.

We are now ready to formulate the geometric theorem of Voronoĭ in question, analogous to the earlier one in the exposition of Section I:

If the reduced Voronoĭ pinhead N lies outside the normal cylinder of the point M, then M is a relative minimum of the lattice $\mathcal{L}$, and N is its adjacent relative minimum in the direction of increasing r. On the other hand if N lies inside the normal cylinder of M, then although M is not, of course, a relative minimum, we have located a point of $\mathcal{L}$ within the normal cylinder of M.

We omit the rather difficult proof of this theorem. Voronoĭ's proof is contained in §27 of his dissertation; a more geometrical proof is given on pages 211–214 of the book by Delone and Faddeev on irrationalities of the third degree.

On the basis of this remarkable theorem, playing a role analogous to that played by the result stated above in the theory of ordinary continued fractions, the desired algorithm is obtained as follows. Suppose given a generating frame of the lattice in question; then the tips of all three of its vectors will be primary points of the lattice. One now divides the whole lattice coordinatewise by M, the tip of the first vector of this frame, obtaining a normalized lattice $\mathcal{N}$ with normalized frame $(\mathbf{1}, \boldsymbol{\phi}, \boldsymbol{\psi})$.

We next compute the positive binary quadratic form (A, B, C) determined by the frame in the xy-plane obtained by projecting the frame $(\mathbf{1}, \boldsymbol{\phi}, \boldsymbol{\psi})$ onto the xy-plane parallel to its first vector $\overrightarrow{O\mathbf{1}}$.[145] We next Zelling-reduce[146] this form, and choose from the six Zelling frames the one embracing the negative x-semiaxis (which has direction opposite to that of the orthogonal projection of $\overrightarrow{O\mathbf{1}}$ on the xy-plane). We now compute the squares of the distances of the seven Voronoĭ pinheads from the z-axis; it can be shown that if the closest of these to the z-axis is $\boldsymbol{\theta}_g$ and the next closest $\boldsymbol{\theta}_h$, then the triple $(\mathbf{1}, \boldsymbol{\theta}_g, \boldsymbol{\theta}_h)$ is a generating frame for the normalized lattice $\mathcal{N}$, and furthermore, by the above theorem, either $\mathbf{1}$ and $\boldsymbol{\theta}_g$ are relative minima of $\mathcal{N}$ with $\boldsymbol{\theta}_g$ adjacent to $\mathbf{1}$ in the direction of increasing r, or $\boldsymbol{\theta}_g$ is a point of the lattice inside the normal cylinder of $\mathbf{1}$.

We next divide the lattice $\mathcal{N}$ by $\boldsymbol{\theta}_g$, obtaining a second normalized lattice $\mathcal{N}'$, whose frame

$$\begin{bmatrix} \frac{1}{\theta_g} & 1 & \frac{\theta_h}{\theta_g} \\ \frac{1}{\theta'_g} & 1 & \frac{\theta'_h}{\theta'_g} \end{bmatrix}$$

we rewrite as $(\mathbf{1}, \boldsymbol{\phi}', \boldsymbol{\psi}')$, where

$$\boldsymbol{\phi}' = \frac{\boldsymbol{\theta}_h}{\boldsymbol{\theta}_g} = \begin{bmatrix} \theta_h/\theta_g \\ \theta'_h/\theta'_g \end{bmatrix}, \qquad \boldsymbol{\psi}' = \frac{\mathbf{1}}{\boldsymbol{\theta}_g},$$

from which we compute the quadratic form (A', B', C') just as (A, B, C) was calculated from the frame $(\mathbf{1}, \boldsymbol{\phi}, \boldsymbol{\psi})$.

Continuing in this way, we obtain a sequence of lattices $\mathcal{N}, \mathcal{N}', \mathcal{N}'', \ldots$. If the initial point M of the original lattice is not a relative minimum for that lattice, i.e., $\mathbf{1}$ is not a relative minimum of the lattice $\mathcal{N}$, then by the above geometrical theorem the reduced Voronoĭ pinhead $\boldsymbol{\theta}_g$ will lie inside the normal cylinder of the point $\mathbf{1}$. Then in the lattice $\mathcal{N}'$ this pinhead becomes the point $\mathbf{1}$, and if it also fails to be

[145] That is, the vector from O to the point $(x, y, z) = (1, 0, 1)$. *Trans.*

[146] See "The theory of irrationalities of the third degree" by Delone and Faddeev, p. 293 *et seqq. Trans.*

a relative minimum of $\mathcal{N}'$, then again the reduced Voronoĭ pinhead of $\mathcal{N}'$ will lie inside its normal cylinder, and consequently the point of $\mathcal{N}$ it was obtained from via normalization will certainly lie inside the normal cylinder of the point $\mathbf{1}$ of the lattice $\mathcal{N}$, and so on. Consequently after finitely many steps we must arrive at a lattice $\mathcal{N}^{(l)}$ for which the point $\mathbf{1}$ is a relative minimum. Each of the lattices $\mathcal{N}, \mathcal{N}', \mathcal{N}'', \dots$ except for the first is obtained from its predecessor by dividing the latter by a point, i.e., by a similarity; since $\mathcal{N}$ is obtained from the original lattice also in this way, it follows that they are all similar to the original lattice. Hence the relative minimum point $\mathbf{1}$ of the lattice $\mathcal{N}^{(l)}$ is obtained from some relative minimum $\boldsymbol{\omega}_0$ of the lattice $\mathcal{N}$ by dividing by that very point $\boldsymbol{\omega}_0$, so that $\mathcal{N}^{(l)} = \frac{\mathcal{N}}{\boldsymbol{\omega}_0}$. If the lattice $\mathcal{N}$ is periodic, with period n, say, then the interval of lattices $\mathcal{N}^{(l)}, \mathcal{N}^{(l+1)}, \mathcal{N}^{(l+2)}, \dots, \mathcal{N}^{(l+n-1)}$, which may be written as $\mathcal{N}/\boldsymbol{\omega}_0, \mathcal{N}/\boldsymbol{\omega}_1, \mathcal{N}/\boldsymbol{\omega}_2, \dots, \mathcal{N}/\boldsymbol{\omega}_{n-1}$, will repeat.

In the case where $\mathbf{1}$ is a relative minimum of such a lattice, with $\boldsymbol{\theta}_g$ its adjacent relative minimum in the direction of increasing r, we shall call the frame $(1, \boldsymbol{\theta}_g, \boldsymbol{\theta}_h)$ a *Voronoĭ reduced frame.* In view of the uniqueness of the Voronoĭ reduced frame in a lattice having $\mathbf{1}$ as a relative minimum, which follows from the rule for obtaining the points $\boldsymbol{\theta}_g$ and $\boldsymbol{\theta}_h$ in this case, we infer that if the lattice is periodic, then the Voronoĭ reduced frames of the lattices

$$\mathcal{N}^{(l)}, \mathcal{N}^{(l+1)}, \mathcal{N}^{(l+2)}, \dots, \mathcal{N}^{(l+n-1)}$$

will repeat periodically. As soon as we obtain the lattice $\mathcal{N}^{(l+n)}$ having the same Voronoĭ frame as $\mathcal{N}^{(l)}$, by multiplying together the second pairs $\boldsymbol{\theta}_g$ of the Voronoĭ reduced frames of the above interval of lattices, we obtain the fundamental multiplicative automorphism $\mathbf{E}$ of the lattice $\mathcal{N}$. Every multiplicative automorphism will then have the form

$$\mathbf{e} = \pm\mathbf{E}^m$$

for some integer m.

Voronoĭ carries out actual calculations in the case where the frame $(\mathbf{1}, \boldsymbol{\phi}, \boldsymbol{\psi})$ is such that their first components ϕ, ψ are numbers[147] in a certain cubic number field of negative discriminant, i.e., have the form

$$\phi = \frac{k + k'\alpha + k''\alpha^2}{\sigma}, \qquad \psi = \frac{l + l'\alpha + l''\alpha^2}{\sigma},$$

where $k, k', k'', l, l', l'', \sigma$ are prescribed rational integers, and α is a[148] root of an irreducible cubic integer polynomial $x^2 + px + q$. The root α is assumed given accurate to several decimal places. Usually two is enough, since all inequalities of the algorithm are to be solved in whole numbers, so that in practice quite rough approximations of their (irrational) coefficients suffice. These inequalities occur in connection with the reduction of the quadratic form (A, B, C), with the locating of the lowest points of the lattice on the straight lines parallel to $\overrightarrow{O\mathbf{1}}$, with the comparison of the distances r of the seven Voronoĭ pinheads from the z-axis, and so on.

It is in principle possible for the solution to one or more of these inequalities to be unclear, in which case it becomes necessary to calculate α *ad hoc* to a larger number of decimal places.

---

[147]Presumably real. *Trans.*

[148]See the preceding footnote. *Trans.*

In §64 of the book "The theory of irrationalities of the third degree", Delone and Faddeev give a modification of the above algorithm where all the coefficients figuring in the inequalities are also rational integers. This modification relates to the original algorithm of Voronoĭ in much the same way as Gauss' method of adjacent forms (*formae contiguae*) relates to that of Lagrange.

In the case of a cubic field of positive discriminant D, i.e., when all three roots of the integer cubic polynomial defining it are real, the algorithm is completely different.

Consider space coordinatized by means of rectangular Cartesian coordinates x, y, z. The system of covariant forms $\omega, \omega', \omega''$ investigated by Voronoĭ in this situation, is represented, for integer values of the variables X, X', X'', by the points of the lattice $\mathcal{L}$ generated by the frame consisting of the vectors

$$(\lambda, \lambda', \lambda''), \ (\mu, \mu', \mu''), \ (\nu, \nu', \nu'').$$

We make the general assumption that

1°. The lattice $\mathcal{L}$ has no points other than the origin O (where $X = X' = X'' = 0$) on the three coordinate planes.

For each point ω of space the rectangular parallelepiped (box) having one of its vertices at ω, its centre at O, and sides parallel to the coordinate planes, will be called the *normal parallelepiped* of ω. Since the lattice $\mathcal{L}$ is assumed to satisfy condition 1°, the normal parallelepiped of any of its points will have on its boundary only one other point of the lattice, namely the antipodal point $-\boldsymbol{\omega}$, since if there were another, then the vector from it to one or the other of $\pm\boldsymbol{\omega}$ would be parallel to a coordinate plane, and then if this vector were laid off from O its tip would constitute a point of $\mathcal{L}$ in that coordinate plane, violating condition 1°. A point of the lattice is called a *relative minimum* if its normal parallelepiped is "empty", i.e., if the only point of the lattice in its interior is O. It is proved much as in the case of negative discriminant that $\mathcal{L}$ has at least one relative minimum. For comparisons of the dimensions of normal parallelepipeds we shall use the terms "longer" and "shorter" in the x-direction, "wider" and "narrower" in the y-direction, and "higher" and "lower" in the z-direction.

Let $\boldsymbol{\omega}_0$ be a relative minimum of $\mathcal{L}$. Keeping its width and height unchanged, i.e., without changing its y- and z-dimensions, we continuously extend the normal parallelepiped of $\boldsymbol{\omega}_0$ in the directions of both the negative and positive x-semiaxes (see Figure 24). It can be shown that eventually the lengthening parallelepiped will encounter in the interior of a face perpendicular to the x-axis a point $\boldsymbol{\omega}_1$ of the lattice, which will clearly also be a relative minimum of $\mathcal{L}$ since its normal parallelepiped is "empty". Starting now with $\boldsymbol{\omega}_1$, we iterate this procedure, obtaining ultimately an infinite sequence of successive relative minima of $\mathcal{L}$ extending from $\boldsymbol{\omega}_0$ in the x-direction, called an *x-chain of relative minima*, denoted by $(\boldsymbol{\omega}_0)_x$. This shows in particular that there are in fact infinitely many relative minima. Analogously, if we carry out this procedure in the y-direction, i.e., let the width of the normal parallelepiped of $\boldsymbol{\omega}_0$ grow, and iterate for successively encountered relative minima, we shall obtain a y-chain $(\boldsymbol{\omega}_0)_y$ of successive relative minima extending from $\boldsymbol{\omega}_0$ in the y-direction. Finally, by extending successive normal parallelepipeds vertically, we obtain a z-chain $(\boldsymbol{\omega}_0)_z$ of successive relative minima extending from $\boldsymbol{\omega}_0$ in the z-direction.

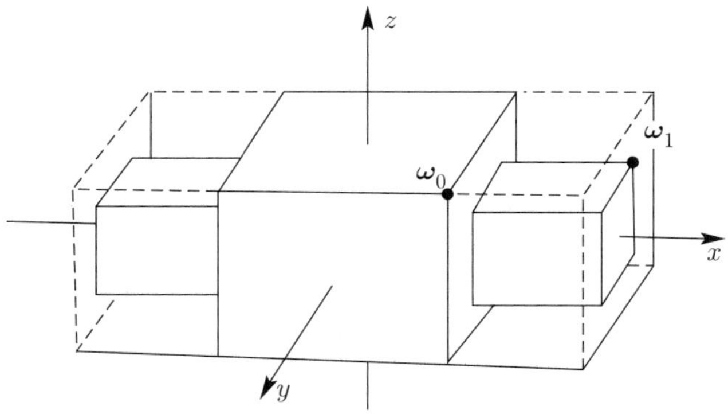

FIGURE 24

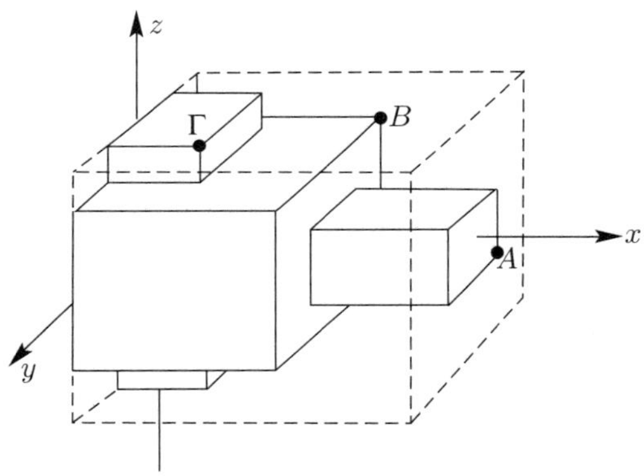

FIGURE 25

So far this procedure is entirely analogous to that involving normal rectangles of relative minima in the situation of a planar lattice considered earlier. However, from this point on significant differences appear. Recall that in the planar case the chain of successive relative minima stretching out from the given relative minimum $\boldsymbol{\omega}_0$ in the x- and y-directions exhausts the relative minima of the lattice; however, in the present spatial case it is not difficult to see that the chains of relative minima extending in the x-, y-, and z-directions from the given particular one $\boldsymbol{\omega}_0$ (together with their negatives) are far from accounting for all relative minima of the lattice. As in the planar case so also in the present spatial situation is it true that corresponding to any given relative minimum there is exactly one other adjacent to it further out in the direction of each axis; however, while in the planar case it is true conversely that any relative minimum is adjacent to a unique relative minimum preceding it in the direction of the relevant axis, this no longer holds in the present 3-dimensional situation, which is more complicated. It can happen that a relative minimum is adjacent in the direction of one of the axes not just to one but to several preceding relative minima, or even to none in that direction.

It is not difficult to construct examples of such situations. In Figure 25 a situation is depicted where the relative minimum A is adjacent in the x-direction

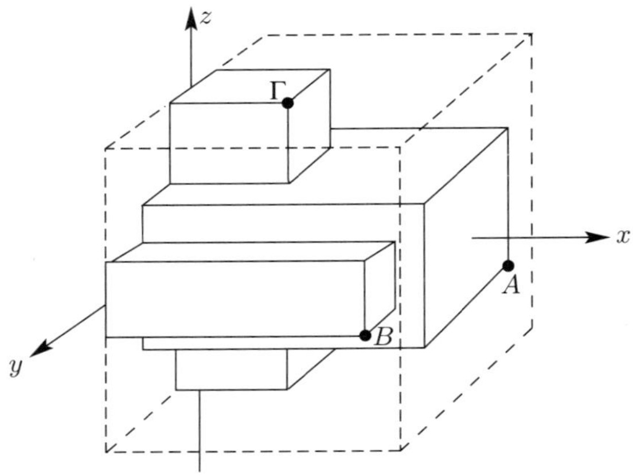

FIGURE 26

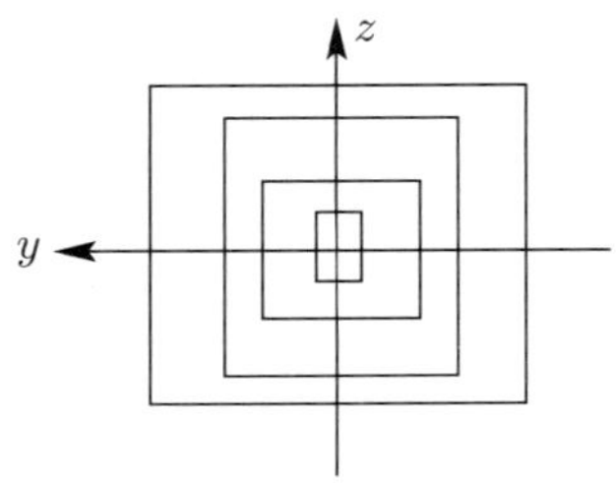

FIGURE 27

to the two distinct preceding relative minima B and Γ, and in Figure 26, that where the relative minimum A is not adjacent in the x-direction to any preceding relative minima. In both diagrams it is assumed that the lattice in question has fundamental parallelepiped $OAB\Gamma$, where A, B, Γ lie on the right-hand, front, and top faces respectively of the dashed rectangular parallelepiped, which has O as the centre of its left-hand face, and also that A, B, Γ have been chosen so that no other points of the lattice in question (generated by the frame $(\overrightarrow{OA}, \overrightarrow{OB}, \overrightarrow{O\Gamma})$) lie in the dashed rectangular parallelepiped, which, as it turns out, can always be arranged. Note also that in both Figures 25, 26, only the the right halves of the normal parallelepipeds of A, B, Γ, i.e., the halves in the direction of the positive x-semiaxis, are sketched in, their other halves being symmetrically placed with respect to the origin O.

If we view successive normal parallelepipeds from somewhere far out on the positive x-semiaxis, then we shall see what looks like concentric rectangles centred at O with sides parallel to the y- and z-axes. Thus the normal parallelepipeds of the points of an x-chain $(\boldsymbol{\omega})_x = (\boldsymbol{\omega}_0, \boldsymbol{\omega}_1, \boldsymbol{\omega}_2, \boldsymbol{\omega}_3, \dots)$, will from such a vantage point look like a nested sequence of rectangles centred at O, as shown in Figure 27. We shall say that one normal parallelepiped *penetrates* another in the direction of a given

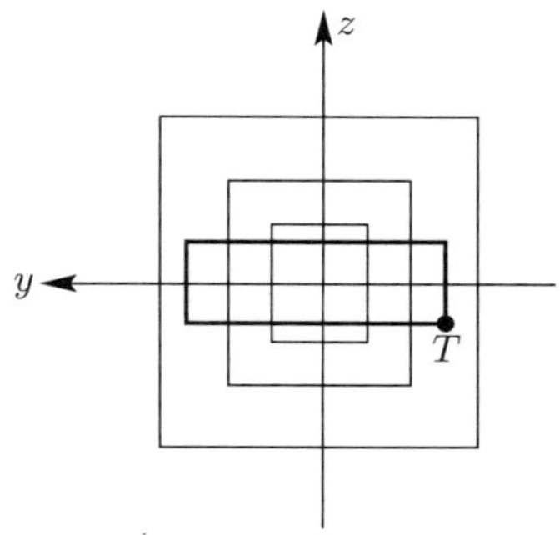

FIGURE 28

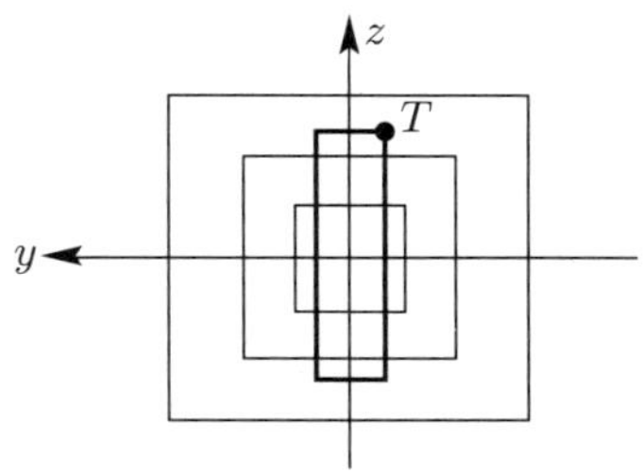

FIGURE 29

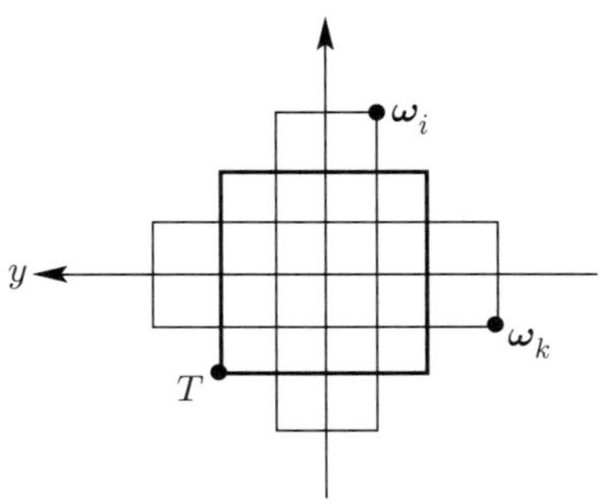

FIGURE 30

axis if it is longer than the second in that direction, but shorter in the directions of the other two axes; and in this case we shall also say that the second *embraces* the first in the given direction. We shall use such terminology also in connection with the rectangles obtained by projecting normal parallelepipeds on a given coordinate plane. For the sake of brevity we shall also extend the term "relative minimum" from the appropriate points of $\mathcal{L}$ to their normal parallelepipeds.

Consider a relative minimum T not in the x-chain $(\boldsymbol{\omega})_x$. We shall say that T is *above* the chain $(\boldsymbol{\omega})_x$ if in the orthogonal projection of (the normal parallelepipeds of) this chain on the yz-plane there are rectangles which penetrate the rectangle corresponding to T in the z-direction (as in Figure 28), and that T is *below* the chain $(\boldsymbol{\omega})_x$ if in the projection of the chain on the yz-plane there are rectangles which embrace that corresponding to T in the z-direction (as in Figure 29). No relative minimum T can lie simultaneously above and below the chain $(\boldsymbol{\omega})_x$ in this sense, since otherwise there would exist relative minima $\boldsymbol{\omega}_k$ and $\boldsymbol{\omega}_i$ in the chain such that the projections of their normal parallelepipeds on the yz-plane did not lie one inside the other (see Figure 30). Similarly, one defines T as being *above* or *below* the chain $(\boldsymbol{\omega})_y$ in terms of the projections of (the normal parallelepipeds of) T and that chain on the xz-plane, and T's being *to the left* or *to the right* of the chain $(\boldsymbol{\omega})_z$ in terms of projections on the xy-plane.

We now consider the possibilities for the situation of an arbitrary relative minimum T with respect to the chain $(\boldsymbol{\omega})_x$ starting at a given relative minimum $\boldsymbol{\omega}_0$. If T does not embrace $\boldsymbol{\omega}_0$ in the x-direction, then there are four possibilities: 1) T coincides with $\boldsymbol{\omega}_0$; 2) T is higher and narrower than $\boldsymbol{\omega}_0$ (i.e., the projections of their normal parallelepipeds on the yz-plane are in this relation to one another), in which case T is above the chain $(\boldsymbol{\omega})_x$; 3) T is lower and wider than $\boldsymbol{\omega}_0$, in which case T is below $(\boldsymbol{\omega})_x$; 4) T is lower and narrower than $\boldsymbol{\omega}_0$, in which case $\boldsymbol{\omega}_0$ embraces T. In the last case, let $\boldsymbol{\omega}_i$ be the longest member of the chain $(\boldsymbol{\omega})_x$ in the x-direction, still shorter than T. If this point is higher and narrower than T, then T is below the chain $(\boldsymbol{\omega})_x$; if $\boldsymbol{\omega}_i$ is lower and wider than T, then T is above $(\boldsymbol{\omega})_x$; and finally if $\boldsymbol{\omega}_i$ is higher and wider than T, then if we lengthen the normal parallelepiped of $\boldsymbol{\omega}_i$ in the x-direction, there will be no point of the chain encountered prior to T, by choice of $\boldsymbol{\omega}_i$ as the longest (in the x-direction) member of the chain shorter than T. Hence in this case T is the adjacent relative minimum to $\boldsymbol{\omega}_i$, i.e. $T = \boldsymbol{\omega}_{i+1}$, so is a member of $(\boldsymbol{\omega})_0$. We have thus proved

THEOREM 1. *If a relative minimum T does not embrace $\boldsymbol{\omega}_0$ in the x-direction, then it is either a member of the chain $(\boldsymbol{\omega})_x$, or else it is above or below $(\boldsymbol{\omega})_x$.*

It is not very difficult to prove also

THEOREM 2. *If a relative minimum T is situated above the chain $(\boldsymbol{\omega})_x$, then its adjacent relative minimum T_1 in the x-direction is either a member of $(\boldsymbol{\omega})_x$ or also above it, and analogously if T is below $(\boldsymbol{\omega})_x$.*

It follows immediately that if a relative minimum T is above the chain $(\boldsymbol{\omega})_x$, then each member of the chain $(T)_x$ is either above $(\boldsymbol{\omega})_x$ or a member of it,[149] and analogously if T is below $(\boldsymbol{\omega})_x$.

These results allow us to extend the definition of T's being above or below the chain $(\boldsymbol{\omega})_x$, to the case where T embraces $\boldsymbol{\omega}_0$ in the x-direction: We shall henceforth understand the expression *T is above the chain $(\boldsymbol{\omega})_x$* to mean that at least one element of the chain $(T)_x$ is above $(\boldsymbol{\omega})_x$ in the former sense (and analogously for the expression *T is below the chain $(\boldsymbol{\omega})_x$*), and, finally, we shall say that *T belongs to the chain $(\boldsymbol{\omega})_x$* if $\boldsymbol{\omega}_0$ is a member of $(T)_x$, so that the chain $(\boldsymbol{\omega})_x$ is part of the chain $(T)_x$.

The following result of Voronoĭ is very important in the sequel. We omit its rather difficult, albeit completely elementary, proof.

THEOREM 3. *If a relative minimum T is situated below $(\boldsymbol{\omega})_x$ and to the right of $(\boldsymbol{\omega})_z$, then the chains $(\boldsymbol{\omega})_x$ and $(T)_z$ have a member in common.*

From this we deduce the following elegant corollary:

[149] In which case the two chains merge from this point on. *Trans.*

THEOREM 4. *If ω_0 and T are any two relative minima of $\mathcal{L}$, then one of the chains $(\omega)_x, (\omega)_y, (\omega)_z$ has a point in common with one of the chains $(T)_x, (T)_y, (T)_z$.*

As Voronoĭ observes, it is curious that none of these theorems requires that the discrete spatial system of points in question be a lattice; it suffices that no two points of the system have a coordinate in common or have a vanishing coordinate,[150] i.e., that the system have no points (other than O) in the coordinate planes, and that the difference of any two points should likewise not yield a point in a coordinate plane.

We shall now show how on the basis of these theorems (especially Theorem 3), if we have an algorithm for finding a relative minimum and computing a chain of relative minima starting from that one and extending in the direction of any axis, then via the periodic lattice arising from a cubic field of positive discriminant we can find a system of fundamental units of that field and decide the fundamental question of the similarity of two periodic spatial lattices. Recall that in the case of periodic planar lattices, and also 3-dimensional periodic lattices of negative discriminant, for deciding these questions it was enough to have an algorithm for obtaining a single relative minimum and computing a sequence of successive relative minima starting from that one; in the first case this was achieved by means of the continued fraction algorithm and in the second via Voronoĭ's algorithm involving normal cylinders. However, in the present case of a 3-dimensional periodic lattice of *positive* discriminant the situation is significantly more complex, and a single algorithm for constructing a chain of successive relative minima is insufficient. Voronoĭ's contribution here consists in his discovery of rather deep geometrical theorems which in conjunction with the aforesaid algorithm for constructing such chains allow the complete resolution of these fundamental questions.

Among the systems of covariant forms considered by Voronoĭ in Section III, a *reduced system of type 1* is defined to be one whose corresponding frame (attached to O) has first vector with its tip at some relative minimum, second vector with its tip at the adjacent relative minimum further in the direction of the x-axis, and the third determined uniquely in a manner depending solely on the first two; *reduced systems of types 2 and 3* are then defined analogously relative to the y- and z-axes respectively. Much as in Section II Voronoĭ now gives in the present case an algorithm for finding reduced systems of covariant forms of types 1, 2 and 3, and for them computing chains of successive reduced systems in the direction of any of the axes. For the sake of brevity we shall omit the description of this algorithm here.

It can be shown using Theorem 3 that in a periodic lattice every x-sequence of successive relative minima will from some point on be periodic in the forward direction, and that then from that point it can be extended backwards with the same periodicity, and similarly of course for y- and z-sequences of successive relative minima. (As noted earlier in connection with Figure 26, in an aperiodic lattice such sequences sometimes admit more than one continuation backwards, and sometimes even none.) Using Theorem 3, it can be further shown that

For a periodic lattice any two periodic doubly infinite chains of successive relative minima along different axes always have a point in common.

[150] Other than O. *Trans.*

From this one obtains the following method for deciding whether two periodic lattices with positive discriminant are similar:

Two periodic lattices $\mathcal{R}$ and $\mathcal{S}$ are similar if and only if one of the lattices obtained from $\mathcal{R}$ by dividing by points of an x-chain of $\mathcal{R}$, coincides with one of the lattices obtained from $\mathcal{S}$ by dividing by the points of a z-chain of $\mathcal{S}$.

Note that this procedure is effective since it suffices to divide each chain by the points of a single repeating interval of their respective periodic chains.

Recall that by Dirichlet's theorem the units of the ring of algebraic integers of a cubic number field of positive discriminant are expressible in terms of a system of two fundamental units ϵ_1 and ϵ_2. Voronoĭ gives the following algorithm for finding such a system: One constructs the x-chain starting from any relative minimum of the corresponding lattice, and then divides the lattice by each point in turn of that chain. Denote by $\mathcal{S}$ the first lattice resulting repeatedly from this procedure, for the first time after $n+1$ successive divisions, say. We may take ϵ_1 to be the ratio of[151] the relative minima of the original lattice; thus starting from the relative minimum $\mathbf{1}$ of $\mathcal{S}$, we will have the repeating interval $\mathbf{1} = \boldsymbol{\omega}_1, \boldsymbol{\omega}_2, \ldots, \boldsymbol{\omega}_n = \boldsymbol{\epsilon}_1$, and the full repeating interval of lattices will be $\mathcal{S} = \frac{\mathcal{S}}{\boldsymbol{\omega}_1}, \frac{\mathcal{S}}{\boldsymbol{\omega}_2}, \ldots \frac{\mathcal{S}}{\boldsymbol{\omega}_n}$. Next, starting from the point $\mathbf{1}$ of the lattice $\mathcal{S}$, one computes in that lattice the z-sequence $\mathbf{1}, \boldsymbol{\psi}_1, \boldsymbol{\psi}_2, \ldots$, and divides $\mathcal{S}$ by each of $\boldsymbol{\psi}_1, \boldsymbol{\psi}_2, \ldots$ in turn, i.e., one computes the lattices $\frac{\mathcal{S}}{\boldsymbol{\psi}_1}, \frac{\mathcal{S}}{\boldsymbol{\psi}_2}, \ldots$. One then finds the first of these lattices coinciding with one of the lattices $\frac{\mathcal{S}}{\boldsymbol{\omega}_1}, \frac{\mathcal{S}}{\boldsymbol{\omega}_2}, \ldots, \frac{\mathcal{S}}{\boldsymbol{\omega}_n}$, obtaining

$$\frac{\mathcal{S}}{\boldsymbol{\psi}_k} = \frac{\mathcal{S}}{\boldsymbol{\omega}_i} \quad \text{for some } \ i,\ k.$$

We may then take $\boldsymbol{\epsilon}_2 = \frac{\boldsymbol{\psi}_k}{\boldsymbol{\omega}_i}$.

Thus Voronoĭ obtains an effective algorithm yielding not just a pair of independent units but resulting directly in a fundamental pair.

We conclude our exposition of Voronoĭ's doctoral dissertation with the following remarks concerning the case just examined.

In the cases of a periodic planar lattice and of a periodic spatial lattice of negative discriminant, the difficulties encountered in spatial lattices of positive discriminant are absent, namely those arising from the fact that doubly infinite chains along all three axes do not exhaust the relative minima of the lattice. Analogous difficulties arise also for arbitrary periodic lattices of dimensions higher than 3. In the case of 3-dimensional periodic lattices of positive discriminant Voronoĭ was able to overcome these difficulties using Theorems 3 and 4 concerning relative minima quoted above, whose proofs required a profound investigation of the mutual disposition of relative minima in the case of positive discriminant. Neither Charve's algorithm nor Minkowski's—indeed no other algorithm—overcomes these difficulties, skirting them instead by means of trial and error in certain situations ; however, the number of trials needed may become arbitrarily large for larger and larger discriminant. On the other hand Voronoĭ by using his theorems on relative minima, provides a purely algorithmic solution of both basic problems by means of an algorithm leading directly to the desired quantities. Voronoĭ gives special emphasis to this aspect of his approach at the end of the preface to his essay,

[151] corresponding coordinates of *Trans.*

ascribing great weight to it. He writes: "It is known that in searching for a fundamental system of units of the cubic field by Dirichlet's method, one has to find the minimum of a certain determinant whose entries are logarithms of the moduli of any two independent units. In our opinion the method we propose for finding a fundamental system of units is of interest, in addition to its practical convenience, in that, in finding a fundamental system of units by this method it turns out to be unnecessary to find the least value of Dirichlet's determinant."

As a matter of fact every other method involves finding by algorithmic means a system of independent units—which need not in general be basic—and then obtaining a basic system from this one by trial and error, in some situations possibly involving a great many trials, and, generally speaking, unboundedly many for increasing discriminant. Thus all earlier authors constructed algorithms only for computing a system of independent units, relying on trial and error to then deduce a basic system, while Voronoĭ, on the strength of the above-described theorems, produces an algorithm leading directly to a basic system. In this lies the essential difference between his work and that of others.

It is curious that in 1939 there appeared a lengthy article by the Hamburg mathematician Bullig, in which he studied in detail precisely the mutual disposition of relative minima of a 3-dimensional lattice of positive discriminant, and, 45 years after Voronoĭ, obtained results similar to his. Bullig remarks in his article that even Voronoĭ had been unable to surmount the difficulties involved. However, this is false; Voronoĭ overcame exactly those difficulties, moreover a half-century earlier, and it is strange that Bullig did not take the trouble to acquaint himself with the famous dissertation of Voronoĭ, one of the greatest arithmeticians of the modern era.[152]

---

[152]It is highly regrettable that in the review of Bullig's work in *Zentralblatt*, which was assigned to Chebotarev, the latter, similarly failing to acquaint himself as behoved him with Voronoĭ's work, and without assessing the essence of the question appropriately, wrote a very unclear phrase failing to establish the priority of our great scholar.

Voronoĭ's Memoir of 1903: "On a Problem from the Theory of Asymptotic Functions"

To Voronoĭ we owe a debt of gratitude not just for his solution of a whole series of problems from the theory of algebraic numbers and the geometry of quadratic forms. In his 1903 memoir entitled "Sur un problème du calcul des fonctions asymptotiques",[153] he achieved a significant breakthrough in a problem of a quite different sort, posed by Dirichlet.[154] This is the "problem of divisors", namely that of estimating the sum

$$S(n) = \tau(1) + \tau(2) + \cdots + \tau(n)$$

for large values of n, where $\tau(k)$ denotes the number of divisors of the integer k. It is easy to see that $S(n)$ is equal to the number of points (x, y) with integer coordinates satisfying

$$x > 0, \quad y > 0, \quad xy \leq n. \tag{1}$$

In other words $S(n)$ is the number of integer points in the curvilinear triangle xOy shown in Figure 31, bounded by the coordinate axes and the hyperbola $xy = n$, with points on the axes excluded. To estimate the size of the set of such points Dirichlet uses the following transformation. One first drops perpendiculars AP, AQ from the point $A = (\sqrt{n}, \sqrt{n})$ of the hyperbola onto the axes, and writes

$$S(n) = (APOy) + (AQOx) - (APOQ), \tag{2}$$

where $(APOy)$ denotes the number of integer points in the quadrilateral $APOy$ excluding those on the sides OP, Oy, and analogously, in obvious notation,

$$(AQOx) = AQOx - OQ - Ox, \quad (APOQ) = APOQ - OP - OQ.$$

It is easily seen that

$$(APOy) = (AQOx) = \sum_{x=1}^{x \leq \sqrt{n}} \left[\frac{n}{x}\right], \qquad (APOQ) = [\sqrt{n}]^2,$$

where $[z]$ denotes the integer part of z. Hence by (2) one has

$$S(n) = 2 \sum_{x=1}^{x \leq \sqrt{n}} \left[\frac{n}{x}\right] - [\sqrt{n}]^2.$$

[153] *J. für die reine und angew. Math. (Crelle)*, Bd. 126, H. 4, pp. 241–282.

[154] "Über die Bestimmung der mittleren Werthe in der Zahlentheorie", Berl. Abh., 1849, pp. 69–74; *Werke* 2, pp. 51–56.

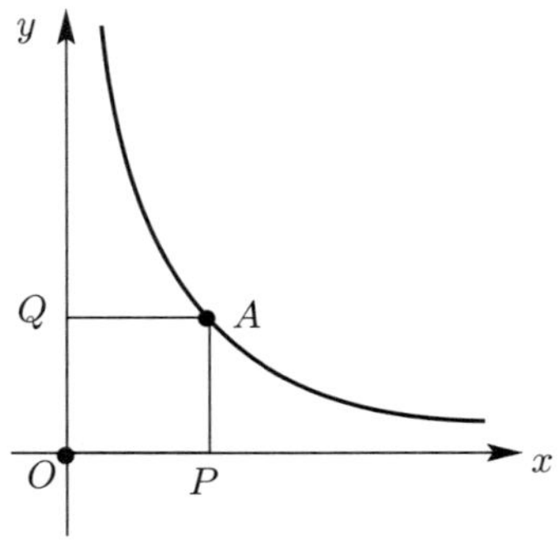

FIGURE 31

Observing that $[z]$ differs from z by less than 1, and invoking the formula

$$\sum_{x=1}^{x \le t} \frac{1}{x} = \ln t + C + \frac{\theta(t)}{t},$$

where $C = 0.57721\ldots$ is Euler's constant, and the quantity $|\theta(t)|$ is bounded as $t \to \infty$, one obtains the following formula for approximating $S(n)$:

$$S(n) = n(\ln n + 2C - 1) + K_n\sqrt{n}, \tag{3}$$

where $|K_n|$ is bounded as $n \to \infty$.

Notwithstanding the simplicity of Dirichlet's approach, further refinement of the formula (3) (i.e., a more accurate estimate of the order of increase of the error term on the right-hand side) turned out to be very difficult. Many mathematicians worked without success on this problem,[155] until in 1903 Voronoĭ managed to obtain, by means of a profound generalization of Dirichlet's method, a substantial sharpening of the formula (3).

Voronoĭ's approach is as follows: Consider all *primitive* integer points (x, y), i.e., with x and y relatively prime, satisfying the conditions (1), and arrange the corresponding fractions $\frac{x}{y}$ in decreasing order (appending the extreme terms $\frac{1}{0}$ and $\frac{0}{1}$) obtaining, say,

$$\frac{a_0}{b_0} = \frac{1}{0} > \frac{a_1}{b_1} > \frac{a_2}{b_2} > \cdots > \frac{a_k}{b_k} > \frac{a_{k+1}}{b_{b+1}} = \frac{0}{1}. \tag{4}$$

This sequence resembles the Farey sequence[156] and like the latter satisfies the equalities

$$a_i b_{i+1} - a_{i+1} b_i = 1, \quad i = 0, 1, \ldots, k. \tag{5}$$

These become easier to see if one rotates a ray fixed at O from the semiaxis Oy to Ox, noting in order the above primitive integer points encountered. However, the sequence (4) can be obtained in another way, very important in what follows,

[155]References to early articles devoted to Dirichlet's problem can be found for instance in the book by P. Bachmann, *Zahlentheorie, II Theil (Die analytische Zahlentheorie, 13 Abschn.)*, Leipzig.

[156]The nth *Farey sequence* consists of the reduced fractions between 0 and 1 with denominator at most n, arranged in decreasing order. *Trans.*

namely by means of interposition (just as the usual Farey series can be constructed). Suppose we have two fractions $\frac{a}{b}, \frac{a'}{b'}$ satisfying the conditions

$$\begin{aligned} &ab' - a'b = 1, \\ (6) \qquad &(a,b) = 1, \quad a \geq 0, \quad b \geq 0, \quad ab \leq n, \\ &(a',b') = 1, \quad a' \geq 0, \quad b' \geq 0, \quad a'b' \leq n. \end{aligned}$$

Consider the fraction (the *mediant* between the given two fractions, in modern terminology)

$$(7) \qquad \frac{a+a'}{b+b'}.$$

It follows from (6) that this fraction is reduced as written, and it clearly lies between $\frac{a}{b}$ and $\frac{a'}{b'}$. If

$$(a+a')(b+b') > n,$$

we discard the fraction (7), otherwise, i.e., if $(a+a')(b+b') \leq n$, we interpose it between $\frac{a}{b}$ and $\frac{a'}{b'}$. Thus starting with the two "fractions" $\frac{1}{0}$ and $\frac{0}{1}$, by means of successive interpositions we obtain the sequence (4). With each fraction $\frac{a_i}{b_i}$ in that sequence we associate the point P_i on the hyperbola $xy = n$ with coordinates

$$P_i\left(\sqrt{\frac{a_i}{b_i}n}, \sqrt{\frac{b_i}{a_i}n}\right), \qquad i = 0, 1, \ldots, k+1.$$

The points $P_0, P_1, \ldots, P_{k+1}$ occur in order along the hyperbola, with the two extreme points $P_0(\infty, 0)$ and $P_{k+1}(0, \infty)$, infinitely distant vertices of the curvilinear triangle xOy (see Figure 32). Take any point P_i with $i = 1, 2, \ldots,$ or k; the fraction corresponding to this point is $\frac{a_i}{b_i}$, which by our alternative construction of the sequence (4) is the mediant between the two fractions $\frac{a_i'}{b_i'}$ and $\frac{a_i''}{b_i''}$ say, adjacent to it in (4), with associated points P_i' and P_i'' say. Consider the tangent lines to the hyperbola at the points $P_0, P_1, \ldots, P_{k+1}$, and denote by Δ_i the (ordinary) triangle bounded by the three tangent lines at P_i', P_i and P_i''. For each $i = 0, 1, \ldots, k$, denote by Γ_i the curvilinear triangle bounded by the tangents at P_i and P_{i+1} and the hyperbola. We shall now show that no two of the $2k+1$ triangles Δ_i, Γ_i overlap, and that they cover the whole triangle xOy.

This is easiest seen by examining again the procedure for generating the sequence (4). We first construct the fraction $\frac{a_2}{b_2}$ (with corresponding point P_2 in Figure 32) as the mediant between the "fractions" $\frac{1}{0}$ (corresponding to $P_0 = (\infty, 0)$) and $\frac{0}{1}$ (corresponding to $P_{k+1} = (0, \infty)$). Having located P_2 on the hyperbola, we now remove the triangle $ABO = \Delta_2$ from the region xOy. Next, having found the mediant P_1 between P_0 and P_2, we remove the triangle $CDA = \Delta_1$, and so on. For general k, once all of the points $P_1, \ldots, P_k$ are located and the corresponding triangles $\Delta_1, \ldots, \Delta_k$ removed, of the original region xOy there will remain just the curvilinear triangles $\Gamma_0 = P_0CP_1, \Gamma_1 = P_1DP_2, \ldots,$ which are also pairwise nonoverlapping. This completes the proof of our claim.

Denote by $|\Delta_i|$ the number of integer points of the form (1) in the triangle Δ_i, excluding points on the tangent lines at P_i' and P_i'', and by $|\Gamma_i|$ the number of points satisfying (1) excluding points on the tangent lines at P_i and P_{i+1}. We then

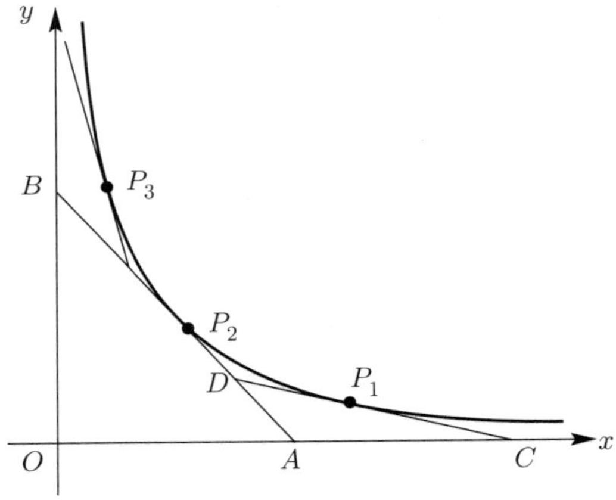

FIGURE 32

have

$$S(n) = \sum_{i=1}^{k} |\Delta_i| + \sum_{i=0}^{k} |\Gamma_i|.$$

After deducing this basic equation, Voronoĭ carries out a detailed estimation of the numbers $|\Gamma_i|$ and $|\Delta_i|$. For the curvilinear triangles Γ_i he uses the above-described approach of Dirichlet, while for the straight-edged triangles Δ_i his estimate depends on the well-known summation formula of Sonin from analysis.[157] By these means, after a rather lengthy and complex computation, Voronoĭ arrives at the following approximating formula:

$$S(n) = n(\ln n + 2C - 1) + \theta_n \sqrt[3]{n} \ln n,$$

where $|\theta_n|$ remains bounded as $n \to \infty$.

In 1906 Sierpiński[158] applied Voronoĭ's method to a problem of similar type (geometrically speaking), namely that of obtaining an approximating formula for the number $A(n)$ of integer points (x, y) in the disc $x^2 + y^2 \leq n$, obtaining the formula

$$A(n) = \pi \cdot n + \theta_n' \sqrt[3]{n},$$

where $|\theta_n'|$ remains bounded as $n \to \infty$.

This article of Voronoĭ served as one of the starting points for the work of the greatest among the contemporary[159] representatives of the St. Petersburg school, Academician I. M. Vinogradov.

[157] For Sonin's formula and further developments concerning Dirichlet's problem, see the succeeding chapter, devoted to Academician Vinogradov.

[158] "Sur un problème du calcul des fonctions asymptotiques", *Prace matematyczno-fizyczne*, vol. 17, 1906.

[159] In 1947. *Trans.*

Voronoĭ's Memoirs on Quadratic Forms

Properties of Perfect Positive Quadratic Forms

We now turn to the extensive research of Voronoĭ published under the general heading "Nouvelles applications des paramètres continus à la théorie des formes quadratiques",[160] and dating from the last years of his life. It was Voronoĭ's intention to publish under this general head a series of memoirs devoted to the solution of various large problems in the theory of positive and indefinite quadratic forms using Hermite's principle (involving continuous parameters). Of these only two were published, both on positive forms: "Sur quelques propriétés des formes quadratiques positive parfaites" (*Crelle*, Bd. 133), and "Recherches sur les paralléloèdres primitifs" (*Crelle*, Bde. 134, 136), the latter appearing only after his death. Unfortunately his further results on positive forms and all of his work on indefinite forms remain unknown.

It is to Hermite that we owe the idea of using continuous parameters in number theory; by means of this method he obtained many important results, one of which is as follows: If

$$f(x_i) = \sum_{i,j=1}^{n} a_{ij}x_ix_j \quad (a_{ij} = a_{ji}) \tag{1}$$

is a positive quadratic form in the n variables $x_1, x_2, \ldots, x_n$, with determinant

$$D(a_{ij}) = \begin{vmatrix} a_{11} & \cdots & a_{1n} \\ \vdots & & \vdots \\ a_{n1} & \cdots & a_{nn} \end{vmatrix}$$

then one can assign integer values $\xi_1, \xi_2, \ldots, \xi_n$ to the variables, not all zero, such that

$$f(\xi_i) \leq \mu(n) \sqrt[n]{D(a_{ij})}, \tag{2}$$

where $\mu(n)$ is a constant depending only on n. (In the sequel we shall understand $\mu(n)$ to denote the smallest possible such constant for a given n.)

As we have seen in an earlier chapter, Korkin and Zolotarev determined the (smallest) value of $\mu(n)$ for $n = 2, 3, 4, 5$. Recall also that they introduced, for any n, the important concept of an "extremal form" ("forme extrême"). We shall now redefine this concept, but in the simplest possible fashion. To this end we set

$$k = \frac{n(n+1)}{2},$$

[160] *J. für die reine und angew. Math. (Crelle)*, Bd. 133, 1908, pp. 97–178; Bd. 134, 1908, pp. 198–287; Bd. 136, 1909, pp. 67–181.

and associate with the form (1) the point in real k-dimensional space E with coordinates a_{ij}, $i \leq j$. Denote by $\mathfrak{P}$ the set of points $(a_{ij}; i \leq j)$ of E for which for $r = 1, 2, \dots, n$, the $r \times r$ submatrices of the symmetric matrix

$$\begin{bmatrix} a_{11} & \cdots & a_{1n} \\ \vdots & & \vdots \\ a_{n1} & \cdots & a_{nn} \end{bmatrix}$$

in the upper left-hand corner have nonnegative determinant.[161] We shall call the region $\mathfrak{P}$ the *positive cone*, since the points in its interior correspond precisely to the positive quadratic forms of order n. Denote by $M(a_{ij})$ the minimum value taken by the form (1) at integral x_i, not all zero, and write

$$\bar{M}(a_{ij}) = \frac{M(a_{ij})}{\sqrt[n]{D(a_{ij})}}. \tag{3}$$

It is clear that this is, where defined, a continuous function of the variables a_{ij}. The extremal forms of Korkin and Zolotarev are just the points inside $\mathfrak{P}$ at which the function (3) has a local maximum. In terms of extremal forms the approach to solving Hermite's problem (namely that of finding $\mu(n)$ for every n) is as follows: One first finds all extremal forms, and then calculates for each of them the value of the function (3); the greatest of these (*maximum maximorum*) is then the desired quantity $\mu(n)$.

As we saw earlier, Korkin and Zolotarev discovered many very important properties of extremal forms. We now recall two of these.

Suppose that (1) is an extremal form, M its minimum, and

$$\begin{matrix} l_{11}, & l_{12}, & \dots, & l_{1n} \\ l_{21}, & l_{22}, & \dots, & l_{2n} \\ \vdots & \vdots & & \vdots \\ l_{\sigma 1}, & l_{\sigma n}, & \dots, & l_{\sigma n} \end{matrix} \tag{4}$$

are all assignments of integer values to $x_1, x_2, \dots, x_n$ yielding the value M of the form; thus we have the equalities

$$f(l_{si}) = M, \quad s = 1, 2, \dots, \sigma. \tag{5}$$

Note that here the two assignments of values $\epsilon l_1, \epsilon l_2, \dots, \epsilon l_n$, $\epsilon = \pm 1$, are not counted as different, and it is understood that only one of them is included in the list (4). In connection with the list (4) and the system of equations (5), Korkin and Zolotarev showed that:

1) The equations (5), which are linear in the coefficients a_{ij} of the form f, determine those coefficients up to a common multiple. It follows immediately that the number of rows in the list (4) must be at least k, i.e., $\sigma \geq \frac{n(n+1)}{2}$.

2) The (absolute value of the) determinant of the matrix made up of any n rows of the list (4) (called a *characteristic determinant* of the form f) is bounded above in terms of n alone.

However, Korkin and Zolotarev did not give a method for computing all extremal forms of a given order n, i.e., they did not provide a practical method

[161]The corresponding form is positive, i.e., takes only positive values at nonzero $(x_1, x_2, \dots, x_n)$, if and only if these determinants are all positive, in which case the matrix is called *positive definite*. *Trans.*

for solving Hermite's problem. This Voronoĭ provides in the first of the above-mentioned memoirs, "Sur quelques propriétés des formes quadratiques positives parfaites". The construction he gives (involving "perfect forms", in French "formes parfaites") not only affords a convenient algorithm for solving Hermite's problem for any n, but is also of great theoretical importance.

We now describe the content of Voronoĭ's memoir. We shall again couch his results directly in geometrical terms, following their interpretation by B. A. Venkov in his article "On the reduction of positive quadratic forms",[162] since this makes for readier comprehension.

Voronoĭ calls a positive quadratic form f *perfect* if it has the property of extremal forms stated in the result 1) of Korkin and Zolotarev above, i.e., if it has a *perfect tableau* (4) of integer values of $x_1, x_2, \ldots, x_n$ at which it attains its minimum (called *representations of the minimum*), in the sense that the corresponding system (5) of linear equations determines its coefficients up to a common multiple. Thus an extremal form is perfect; however, as might be expected the converse is not true in general; for $n \geq 6$ there exist forms that are perfect but not extremal.

In the introductory part of his memoir Voronoĭ reexamines the properties of extremal forms established by Korkin and Zolotarev, nearly all of which turn out to hold also for perfect forms, and finds important new properties. In addition to properties 1) and 2) above, we have the following ones:

3) Among the characteristic determinants of a perfect tableau (4) there always exist nonzero ones.

4) It is easily shown that the number σ of rows of a perfect tableau (4) is bounded above as follows:

$$\sigma \leq 2^n - 1. \tag{6}$$

(This inequality follows very simply from the observation that in the n-dimensional space of the variables x_i, the ellipsoid $f(x_i) = M$ contains in its interior no integer points other than the origin O, while of course the points with coordinates given by the rows of the table (4) lie on its boundary. Two integer points P and Q on the boundary cannot be congruent modulo 2 since otherwise $\frac{P+Q}{2}$ would be an integer point in the interior of the ellipsoid; (6) is now immediate.)

5) As before we call two quadratic forms (and likewise the points corresponding to them in the space E) *equivalent* if one can be changed into the other by means of a unimodular change of the variables x_i (i.e., an integral linear transformation of determinant ± 1); the totality of forms equivalent to a given form is then an *(equivalence) class of forms.* Let f be a perfect form with tableau (4), and

$$x_i' = p_{i1}x_1 + \cdots + p_{in}x_n, \qquad i = 1, 2, \ldots, n,$$

a unimodular transformation transforming f into ϕ; thus

$$\phi(x_i') \equiv f(x_i).$$

It is not difficult to see that then ϕ will likewise be perfect, with tableau obtained from the tableau (4) of f by postmultiplying the latter, considered as a $\sigma \times n$ matrix,

[162] *Izvestiya AN SSSR, Ser. Matem.*, 4, 1940, pp. 37–52.

by

$$\begin{bmatrix} p_{11} & \cdots & p_{1n} \\ \vdots & & \vdots \\ p_{n1} & \cdots & p_{nn} \end{bmatrix}$$

the matrix of the unimodular change of variables, in accordance with the usual multiplication of matrices. The tableaux of f and ϕ will be called *equivalent perfect tableaux*. Hence in particular interchanging any two columns of the tableau (4), or adding to any column any multiple of any other column yields a tableau equivalent to (4). By using these operations together with the inequality (6) and the fact that the characteristic determinants are bounded, one easily infers that for a given n every perfect tableau is equivalent to one of a finite set of well-defined particular perfect tableaux, or, in other words, there exist only finitely many classes of perfect forms of each order n.

Thus, noting that at equivalent forms the function $\bar{M}$ of (3) has the same value, and again that the perfect forms include the extremal ones, we conclude that a method for computing a full set of inequivalent perfect forms would suffice as a practical solution of Hermite's problem.

As the variable point f with coordinates a_{ij}, $i \leq j$, ranges over the cone $\mathfrak{P}$, the point $\bar{f}$ with coordinates

$$(a_{11}, \dots, a_{nn}, 2a_{12}, \dots, 2a_{n-1,n})$$

ranges over a set $\bar{\mathfrak{P}}$, which we shall call the *associated cone*. We call the hyperplane in E defined by the equation

$$p_{11}a_{11} + \cdots + p_{nn}a_{nn} + p_{12}a_{12} + \cdots + p_{n-1,n}a_{n-1,n} = M, \tag{7}$$

where the a_{ij} are the variables and the p_{ij} the coefficients, *elliptic, parabolic, or hyperbolic relative to* $\mathfrak{P}$ according as its normal vector[163] (ϵp_{ij}) is, for $\epsilon = +1$ or -1, directed into the interior of, is in the boundary of, or misses the cone $\bar{\mathfrak{P}}$. It can be shown (see below) that: an elliptic hyperplane relative to $\mathfrak{P}$ either fails to intersect $\mathfrak{P}$ or cuts off a bounded piece of it; in the case $M = 0$ a parabolic hyperplane relative to $\mathfrak{P}$ is tangential to $\mathfrak{P}$; and for any M a hyperbolic such hyperplane cuts $\mathfrak{P}$ into two unbounded pieces. Analogously, a hyperplane (7) is called *elliptic, parabolic, or hyperbolic relative to the cone* $\bar{\mathfrak{P}}$ according as its normal (ϵp_{ij}) is directed into, is on the boundary of, or misses the cone $\mathfrak{P}$, and then these three types of hyperplane relative to $\bar{\mathfrak{P}}$ have the same properties with respect to $\bar{\mathfrak{P}}$ as those relative to $\mathfrak{P}$ have with respect to $\mathfrak{P}$. These properties are readily inferred from the usual decomposition of a positive quadratic form as a sum of squares; in this connection it is convenient to use properties of the scalar product (f, ϕ) introduced by Voronoĭ, defined for

$$f = \sum_1^n a_{ij}x_ix_j \quad \text{and} \quad \phi = \sum_1^n b_{ij}x_ix_j,$$

by

$$\begin{aligned}(f, \phi) &= \sum a_{ij}b_{ij} \\ &= a_{11}b_{11} + \cdots + a_{nn}b_{nn} + 2a_{12}b_{12} + \cdots + 2a_{n-1,n}b_{n-1,n}.\end{aligned}$$

---

[163]Considered attached to O. *Trans.*

If

$$\phi = \sum_{s=1}^{\sigma} \rho_s (l_{s1}x_1 + \cdots + l_{sn}x_n)^2 \tag{8}$$

is any decomposition of ϕ as a sum of squares, then

$$(f, \phi) = \sum_{s=1}^{\sigma} \rho_s f(l_{s1}, \ldots, l_{sn}),$$

whence it is immediate that $(f, \phi) > 0$ for positive forms f, ϕ.

Voronoĭ's algorithm depends also on the study of regions of higher-dimensional spaces defined by systems of linear inequalities, which study originated with Minkowski.[164] Consider the region of k-dimensional space coordinatized by $z_1, \ldots, z_n$, defined by the inequalities

$$y_s(z_i) = \xi_{s1}z_1 + \cdots + \xi_{sk}z_k \geq 0, \quad s = 1, 2, \ldots, m. \tag{9}$$

By analogy with the terminology used in the 3-dimensional situation, we shall call such a region a *gonohedron* ("many-faceted angle"). In turns out to be important to have a criterion for determining whether or not such a region is really k-dimensional, i.e., whether it is contained in some hyperplane or not. Voronoĭ provides such a criterion in the following form:

The region (9) *is k-dimensional if and only if the only nonnegative numbers $\rho_1, \ldots, \rho_m$ ($\rho_i \geq 0$) for which*

$$\sum_{s=1}^{m} \rho_s y_s(z_i) \equiv 0,$$

i.e., vanishes for all z_i, are $\rho_1 = 0, \ldots, \rho_m = 0$.[165]

Next consider any point (z_{i0}) lying on the boundary of the region (9). Since it is on the boundary, on substituting the numbers z_{i0} for the variables z_i, some of the inequalities (9) become equalities, say

$$\xi_{s1}z_{10} + \cdots + \xi_{sk}z_{k0} = 0, \quad s = 1, 2, \ldots, p, \; p < m. \tag{10}$$

If the equations (10) determine the unknowns z_{i0} to within a common multiple, then the ray $(\rho z_{10}, \ldots, \rho z_{k0})$, $0 < \rho < \infty$, will be called an *edge* of the region (9). It is not difficult to see that under the assumption that the inequalities (9) all become equalities only at $(z_1, \ldots, z_k) = (0, \ldots, 0)$, the region defined by those inequalities always has edges, and if

$$(z_{1t}, \ldots, z_{kt}), \quad t = 1, 2, \ldots, q,$$

are points determining all the distinct edges, then the points of the region (9) are precisely those with coordinates of the form

$$z_i = \sum_{t=1}^{q} \rho_t z_{it}, \quad i = 1, 2, \ldots, k, \tag{11}$$

for some nonnegative $\rho_1, \ldots, \rho_q$.

---

[164] H. Minkowski, *Geometrie der Zahlen*, Leipzig und Berlin, 1910, p. 39.

[165] That is, if and only if the normals $(\xi_{s1}, \ldots, \xi_{sk})$, $s = 1, 2, \ldots, m$, to the hyperplanes defined by the corresponding equalities are not linearly dependent with all coefficients ρ_s nonnegative. (Consider for example the case $k = 3, m = 2$.) *Trans.*

We now turn to the geometric elucidation of the concept of a perfect form. Let $(q_1, q_2, \ldots, q_n)$ range over all ordered n-tuples of integers with highest common factor 1, and consider the set $\mathfrak{M}$ of all points of E with coordinates

$$(q_1^2, q_2^2, \ldots, q_n^2, 2q_1q_2, 2q_1q_3, \ldots, 2q_{n-1}q_n),$$

all of which lie on the boundary of the cone $\bar{\mathfrak{P}}$. Denote by Π the convex hull of $\mathfrak{M}$ in E, i.e., the smallest convex subset of E containing $\mathfrak{M}$. In order to grasp Π visually to some extent, we first prove that it is a polyhedron bounded by infinitely many hyperplanes, all elliptic with respect to $\bar{\mathfrak{P}}$. To see this, and also with the further investigation of the properties of Π in view, one needs to carry out a deformation of the set $\mathfrak{M}$ by means of the continuous motion of a hyperplane in E. (It is at this juncture that continuous parameters are brought to bear in Voronoï's method.) We shall describe this deformation in visual form; its precise analytic formulation constitutes the bulk of Voronoï's memoir.[166]

Thus let π denote a variable hyperplane in E, elliptic with respect to $\bar{\mathfrak{P}}$, which we suppose to be situated initially sufficiently close to the origin for the bounded piece of $\bar{\mathfrak{P}}$ that it cuts off to contain no points of $\mathfrak{M}$. We then move π continuously away from O without changing the direction of its normal until one or more points of $\mathfrak{M}$ are encountered. Since π is elliptic with respect to $\bar{\mathfrak{P}}$, the convex hull of the set of points thus encountered will be a bounded polyhedron, which we shall denote by $\Pi(\pi)$.[167] Denoting by s the dimension of the region $\Pi(\pi)$, we have of course $0 \leq s \leq k-1$. If $s < k-1$, then one can further rotate the hyperplane π while keeping the set $\Pi(\pi)$ contained in it throughout, until eventually a further point of $\mathfrak{M}$ is encountered. That this must occur follows from the easily proved fact that if a hyperplane elliptic with respect to $\bar{\mathfrak{P}}$ is rotated about any fixed $(k-2)$-dimensional linear subspace, then prior to becoming parabolic with respect to $\bar{\mathfrak{P}}$ it must encounter a point of $\mathfrak{M}$. (Note in this connection that the rotation can be effected in two different directions.) Thus, denoting the plane in its new, rotated position by π', we obtain a larger polyhedron $\Pi(\pi')$, necessarily of dimension $s' > s$. Iterating this process if necessary we ultimately obtain an elliptic hyperplane π_0 with respect to $\bar{\mathfrak{P}}$, such that the polyhedron $\Pi(\pi_0)$ is $(k-1)$-dimensional, so that π_0 is a hyperplane boundary of the polyhedron Π. Let

$$(12) \qquad (l_{s1}^2, l_{s2}^2, \ldots, l_{sn}^2, 2l_{s1}l_{s2}, 2l_{s1}l_{s3}, \ldots, 2l_{s,n-1}l_{sn}), \; s = 1, 2, \ldots, \sigma,$$

be all of the points of $\mathfrak{M}$ lying on π_0. It is easy to see that these will all be vertices of $\Pi(\pi_0)$. Denoting by a_{ij} the coefficients in the equation of the hyperplane π_0, and by M (> 0) its right-hand side, we arrive back at the equations (5), so that the form $f = \sum a_{ij}x_ix_j$ is a perfect form with the tableau (4) (whence the choice of notation in (12)). We have thus indicated geometrically how the hyperplane faces of the polyhedron Π correspond one-to-one to the perfect forms. Consider the straight line segments joining the origin to the vertices of $\Pi(\pi_0)$, i.e., to the points (12). We call the gonohedron with these line segments as edges the *perfect gonohedron* associated with the form f or the tableau (4). If $\bar{\phi}$ is any point of this

[166] In connection with a different problem this deformation is carried out also in the above-cited work of B. A. Venkov "On the arithmetic group of automorphisms ...".

[167] The basic properties of bounded convex polyhedra in higher-dimensional space (in particular the definition of the vertices) are easily inferred, for example from the properties of gonohedra noted above. Voronoï does this in his second memoir "Recherches sur les paralléloèdres primitifs".

gonohedron, then the corresponding quadratic form ϕ may be written in the form (8) with the coefficients $\rho_s \geq 0$.

The bounded polyhedron $\Pi(\pi_0)$ is, as we have seen, $(k-1)$-dimensional and is therefore bounded on all sides by $(k-2)$-dimensional faces; let Π_0 denote one of these faces. Consider as before an elliptic hyperplane π, variable subject to always containing Π_0, i.e., containing the fixed $(k-2)$-dimensional affine subspace determined by Π_0, which, as noted above, may therefore be rotated in two directions until it becomes parabolic. If, on being rotated in one direction π takes up the position of π_0, then on being rotated in the other direction it will take up the position of an adjacent face π_1. In this way we arrive at a perfect form adjacent to the earlier form f, and to a perfect gonohedron adjacent to the former one.

Let $f_0, f_1, \ldots$ denote all of the perfect forms, and $\nu_0, \nu_1, \ldots$ the associated gonohedra. From our geometrical sketch it can be seen that the latter have pairwise nonintersecting interiors and fill the whole cone $\bar{\mathfrak{P}}$. It is also clear that equivalent perfect forms have equivalent associated gohohedra. (Of course Voronoĭ proves these assertions completely rigorously.)

From the foregoing we obtain the following algorithm for constructing a complete set of inequivalent perfect forms (or, what amounts to the same thing, a complete set of inequivalent gonohedra). First compute a gonohedron and those adjacent to it; write S_0 for a maximal set of inequivalent gonohedra contained among these. Next compute all gonohedra adjacent to those in the set S_0, and from among these together with those in S_0, choose a maximal inequivalent set S_1, and so on. The computation terminates when we no longer obtain new gonohedra, i.e., when $S_{i+1} = S_i$; the set S_i will then be the desired full system of inequivalent gonohedra. (Hermite's easily proved "adjacency principle" is involved here; see also below.)

Note that the complete system of inequivalent gonohedra thus obtained can also be used to provide a solution of the general problem of reduction of positive forms, i.e., of finding in an equivalence class of such forms certain special ones, with coefficients satisfying linear inequalities.

Voronoĭ verifies the practicality of his method by computing full systems of inequivalent perfect forms for $n = 2, 3, 4, 5$.

We remarked earlier that in general a perfect form need not be extremal. Voronoĭ establishes the following exact criterion:

A positive form f is extremal if and only if it is perfect and the point corresponding to the adjoint form $\bar{f}$[168] *lies inside the perfect gonohedron of f.*

At the end of his memoir Voronoĭ investigates the *principal* perfect form

$$f_0 = x_1^2 + x_2^2 + \cdots + x_n^2 + x_1x_2 + x_1x_3 + \cdots + x_{n-1}x_n$$

(which appears in the work of Korkin and Zolotarev) and forms adjacent to it.

This memoir of Voronoĭ was the starting point for the research of B. A. Venkov on the reduction of positive forms.[169]

[168] A form is said to be *adjoint* to f if its symmetric matrix is the adjoint of the matrix of f.

[169] See the article cited above in *Izv. AN*, 1940.

Commentary

The present memoir represents in some sense the culmination of investigations of the circle of questions relating to minima of positive quadratic forms. The problem of finding a form with largest minimum originated with Hermite. In their memoir of 1872, Korkin and Zolotarev solved this problem for $n = 4$. Then in their 1873 memoir they introduced their remarkable method of "decomposition of a form along its minima", allowing them to gain greater insight into their solution in the case $n = 4$, and providing the means for Blichfeldt to find, much later, the largest minima for $n = 6, 7$. In their 1878 memoir Korkin and Zolotarev found all extremal forms for $n = 5$, and obtained several important results on general extremal forms. Minkowski, in a series of articles, elucidated the connection between Hermite's problem and the densest lattice packing of spheres. Finally, Voronoĭ showed that the most elegant version of the theory results if one takes as basic not "forms with largest minimum" and not "extremal forms", but forms which he called "perfect". For each n there are only finitely many inequivalent perfect forms; every extremal form is perfect but not conversely, i.e., the perfect forms properly include the extremal ones; and finally a form with largest minimum is an extremal form with largest minimum among the extremal forms.

Since Voronoĭ gives, for each n, an algorithm for finding in finitely many steps a full set of inequivalent n-variable perfect forms, and also a criterion for deciding if a perfect form is extremal or not, his method allows one to find all n-variable extremal forms, and hence a form with largest minimum, i.e., affords a systematic procedure for solving Hermite's problem.

This procedure, based on his theory of perfect forms, is as follows: One considers first all n-tuples $q_1, q_2, \ldots, q_n$ of integers with highest common factor 1, and then the convex hull in $\left(\frac{n(n+1)}{2}\right)$-dimensional space, of the set of points with coordinates

$$(q_1^2, q_2^2, \ldots, q_n^2, 2q_1q_2, 2q_1q_3, \ldots, 2q_{n-1}q_n),$$

which all lie on the boundary of the cone $\bar{\mathfrak{P}}$. The faces of the polyhedron forming the boundary of this convex hull, then correspond to the perfect forms in n variables.

The actual computation begins with the face corresponding to the "principal" extremal form in n variables, known earlier to Korkin and Zolotarev. One then proceeds to an adjacent face via the algorithm proposed by Voronoĭ, and so on, until one meets with a "layer" all of whose faces are equivalent to those already encountered, at which point a full set of inequivalent perfect forms has been obtained. In order to find the extremal forms among these one uses the characterization of an extremal form as one lying in the gonohedron (many-faceted angle) with its vertex at the origin, having a cross-section coinciding with the face corresponding to the form under consideration. By testing all the perfect forms in this way, one arrives at the extremal ones; and then of course any among these having largest minimum will afford the solution to Hermite's problem.

In order to clarify the method further, we consider it in detail in the simplest case $n = 2$. Thus first we represent each binary quadratic form

$$au^2 + 2buv + cv^2$$

by the point of 3-space with coordinates $x = a, y = b, z = c$. As is well known, the form is positive if and only if

$$\begin{vmatrix} a & b \\ b & c \end{vmatrix} > 0, \quad a > 0,$$

i.e., when the point (a, b, c) representing it lies inside that nappe of the cone $\mathfrak{P}$ of degree 2 with equation

$$xz - y^2 = 0,$$

consisting of those points with positive x-coordinate. This cone has as its axis of symmetry the line $x = z, y = 0$, and the x- and z-axes are two of its generators.

In this case $\mathfrak{M}$ is the set of all points of 3-space with coordinates $(x, y, z) = (q_1^2, 2q_1q_2, q_2^2)$ where (q_1, q_2) ranges over all relatively prime pairs of integers. All of these points lie on the cone $\bar{\mathfrak{P}}$ defined by the equation

$$4xz - y^2 = 0,$$

said earlier to be *associated* with the cone $\mathfrak{P}$. Note that the cone $\bar{\mathfrak{P}}$ can be obtained from the cone $\mathfrak{P}$ by stretching from the xz-plane in the y direction by the factor 2. It is easy to see (as we shall now demonstrate) that a plane in xyz-space intersects the cone $\bar{\mathfrak{P}}$ in an ellipse if and only if its normal, considered as emanating from the origin, lies inside $\mathfrak{P}$. Thus we consider an arbitrary plane

$$x_1x + y_1y + z_1z + C = 0,$$

and compute the equation of the projection on the xy-plane of its intersection with the cone $4xy - y^2 = 0$, by eliminating z from these two equations. This yields $z_1y^2 + 4x_1x^2 + 4y_1xy + 4Cx = 0$. Here the discriminant of the quadratic part of the left-hand side is

$$\begin{vmatrix} 4x_1 & 2y_1 \\ 2y_1 & z_1 \end{vmatrix} = 4(x_1z_1 - y_1^2),$$

which is positive, i.e., the equation is that of an ellipse, if and only if $x_1z_1 - y_1^2 > 0$, i.e., if and only if the vector (x_1, y_1, z_1), which is normal to the plane in question, lies inside the cone $xz - y^2 = 0$.

Returning to the general binary quadratic form $au^2+2buv+cv^2$, we now assume that it is perfect, i.e., positive, and determined by the representations (l_{s1}, l_{s2}), $s = 1, 2, \dots, \sigma$, of its minimum M, i.e., that the coefficients a, b, c are determined (to within a common factor) by the system of linear (with respect to a, b, c) equations

$$al_{s1}^2 + 2bl_{s1}l_{s2} + cl_{s2}^2 = M, \quad s = 1, 2, \dots, \sigma.$$

The plane $ax + by + cz = M$ intersects the cone $\bar{\mathfrak{P}}$ in an ellipse since the form is positive and the plane contains the points $(l_{s1}^2, 2l_{s1}l_{s2}, l_{s2}^2)$, $s = 1, 2, \dots, \sigma$, of the set $\mathfrak{M}$, at least three of which are not collinear. Furthermore in view of the minimality of M, it is readily seen (as we shall now show) that this plane is "supporting" for the set $\mathfrak{M}$,[170] so that it is the plane of a face of the convex hull of the set $\mathfrak{M}$. For, if a point $(q_1^2, 2q_1q_2, q_2^2)$ from $\mathfrak{M}$ lay on that side of the plane containing the vertex of the cone $\mathfrak{P}$, then on substituting its coordinates for x, y, z in $ax + by + cz$, we should obtain a value $aq_1^2 + 2bq_1q_2 + cq_2^2 = N < M$, contradicting the minimality of M.

By arguing in the reverse direction one shows that if $ax + by + cz = M$ is the equation of the plane of a face of the convex hull of $\mathfrak{M}$, then this plane intersects

[170]That is, there are no points of $\mathfrak{M}$ on that side of the plane containing O. *Trans.*

the cone $\bar{\mathfrak{P}}$ in an ellipse, whence it follows that the form $au^2+2buv+cv^2$ is positive, and that those points $(l_{s1}^2, 2l_{s1}l_{s2}, l_{s2}^2)$ of $\mathfrak{M}$ lying on this face afford a representation of the minimum of the form, so that the form is perfect.

Thus real numbers a, b, c qualify as coefficients of a perfect binary quadratic form if and only if they are coefficients of the equation of the plane of some face of the convex hull of the set $\mathfrak{M}$. Herein lies the essence of the *remarkable* central idea of Voronoĭ's memoir on perfect forms.

Investigations Concerning Primitive Parallelohedra

In 1908 and 1909 in Volumes 134 and 136 of Crelle's journal there appeared the substantial last published work of Voronoĭ, entitled "Recherches sur les paralléloèdres primitifs", which, together with the preceding memoir published in vol. 133 of the same journal entitled "Sur quelques propriétiés des formes quadratiques positives parfaites", formed, as already noted, instalments in a series of memoirs under the running head "Nouvelles applications des paramètres continus à la théorie des formes quadratiques".

At the end of this, Voronoĭ's last published memoir, Hensel, the editor of Crelle's journal at that time, appended the following remark: "Just as the first page of this remarkable work had been set, we received the sad news that its author has died. The editor has deployed our best resources in order to ensure that the last work of this too soon deceased scholar be proofread with the utmost accuracy."

The memoir of Voronoĭ on parallelohedra is the substantial result of extensive research, occupying as it does about 200 pages of Crelle's journal. It is about as equally concerned with quadratic forms as with geometry. Although the exposition is couched in analytic language, continual mention is made of the geometric objects being investigated: n-dimensional convex polyhedra and their faces of various dimensions; n-dimensional simplices; simplicial decompositions of n-dimensional space; tesselations of n-space by polyhedra; and so on.

The remarkable problem that Voronoĭ considers in this, the last and perhaps the deepest, of his works, is as follows. Imagine n-dimensional Euclidean space filled with identical nonoverlapping convex polyhedra, placed parallel to one another, sharing only whole faces.[171] For which polyhedra is this possible in general? (Such polyhedra are called *parallelohedra.*)

In the case of 3-dimensional space, this question was first posed and answered by our famous crystallographer Fedorov in his book "The elements of the study of figures" (1885). The same question was considered, but not settled, by Kelvin (*Proc. Roy. Soc.*, 1894; Papers, vol. V, p. 333). In his famous article "Allgemeine Lehrsätze über convexe Polyeder" (1896), Minkowski proved a theorem having as corollary that a parallelohedron must have a centre of symmetry, and showed also that the number of faces[172] of an n-dimensional parallelohedron is at most $2(2^n-1)$. From this it already follows that for each n there are only finitely many topological types of nets[173] of n-parallelohedra. However, the fact that for each n the possible n-parallelohedra are up to topological type finite in number leaves the question as to what they are actually like completely untouched. For instance up to the present

[171] As is well known, Euclidean 3-space can be so tesselated by parallelepipeds and by regular hexagonal prisms.

[172] Of dimension $n-1$. *Trans.*

[173] Presumably complexes, i.e., cell decompositions of n-space. *Trans.*

time (1947) even in the case $n = 5$ not only have they not been found, but there is a lack of any principle on which a method for finding them might be based—and all the more so for larger n.

Voronoĭ imposes on his parallelohedra an additional condition, namely that in the tesselation of space by parallelohedra (according to the conditions already stated), at each vertex there should be incident the smallest possible number of parallelohedra in that number n of dimensions, i.e., $n + 1$. Such tesselations, and also the corresponding parallelohedra, he calls *primitive*. For example cubes and regular hexagonal prisms yield nonprimitive tesselations of 3-space, since in the case of cubes each vertex is incident with 8 cubes, and for such prisms the corresponding number is 6, while the least possible number of polyhedra sharing a vertex in an (allowable) tesselation of 3-space is 4.

Suppose we are given in n-dimensional space an n-dimensional lattice (i.e., a parallelepipedal system of points, in German *Punktgitter*) having the origin O as one of its points. Consider the set of all points of n-space not further from O than from any other lattice point. It is obvious that these points form an n-dimensional convex polyhedron centred at O. If one considers the analogous (congruent) polyhedron centred at each point of the lattice, it is obvious that together these will cover the whole of n-space without overlapping, and it can also be shown that adjacent such polyhedra intersect in whole faces. Hence the totality of such polyhedra forms an (admissible) tesselation of n-space into parallelohedra, which may or may not be primitive. Following Voronoĭ, we call such a tesselation the *Dirichlet tesselation* corresponding to the given lattice. It is completely determined by the lattice, and, conversely, the lattice by it.

The first 60 pages of Voronoĭ's memoir are devoted to proving that every primitive tesselation of n-space is an affine image of some primitive Dirichlet tesselation, thus reducing the study of arbitrary primitive tesselations to the study of Dirichlet tesselations. Since every Dirichlet tesselation is defined in terms of a lattice, there will then arise an associated class of positive quadratic forms. We shall conditionally call this *the first part* of Voronoĭ's memoir. In the remaining 140 pages of his memoir, which we shall call *the second part*, Voronoĭ carries out a detailed investigation of Dirichlet tesselations, ultimately yielding an algorithm for finding, for each n, all such tesselations of n-space. He concludes the memoir by applying his algorithm to the classification of all primitive tesselations of 2-, 3-, and 4-dimensional space.

Notwithstanding the richness of the ideas and the elegance of the results used by Voronoĭ in deriving his method for finding all Dirichlet tesselations in each dimension n, the most impressive part of the memoir remains all the same the first, i.e., the first 60 pages; the reduction of general primitive tesselations to those of Dirichlet type represents the deepest part of the work. In this, the first part, Voronoĭ carries out, in analytic guise, purely geometrical investigations of extraordinary profundity.

We now examine this first part, comprising Chapters I and II. The basic idea here is to exploit "Ampère's principle", whereby a complicated closed contour is replaced by a collection of simple, or in some sense elementary, closed contours.

The first 16 sections of the first part are of a purely preparatory character. Voronoĭ first expounds the general principles associated with the analytic definition of an n-dimensional convex polyhedron in Euclidean n-space in terms of linear inequalities, the determination of its "faces" of various dimensions, and of the n-dimensional many-faceted angles correlated with its vertices; such an angle has as its edges the straight lines through the relevant vertex perpendicular to the faces incident with that vertex. He then goes on to give the analytic definition of a tesselation by means of a parallelohedron, and then proves that such a tesselation has a *translation-invariant lattice*,[174] i.e., that no matter which two parallelohedra of the tesselation one takes, after performing a translation of the space bringing one into the original position of the other, the whole tesselation will have been transformed onto itself. He also shows that the number of parallelohedra of the tesselation sharing a ν-dimensional face is at least $n+1-\nu$, and calls a ν-dimensional face shared by only the minimal number $n+1-\nu$ of parallelohedra a *simple* face of the tesselation. Then if all faces of the tesselation are simple, he calls the tesselation *primitive*. Having made this definition, Voronoĭ writes: "In what follows we shall investigate precisely primitive parallelohedra, and also all nonprimitive ones that can be regarded as limits of primitive ones. I conjecture that every nonprimitive parallelohedron can be regarded in this way (i.e., as a limit of primitives), but have not managed to prove this." In fact this remains unproven up to the present (1947).

In Sections 17 and 18 Voronoĭ makes a few preparatory remarks about primitive parallelohedra. He first observes that it is immediate from the definition that if all vertices of an (admissible) tesselation are simple, then all faces of positive dimensions must automatically be simple. In other words primitivity may be equivalently, but more simply, defined by requiring merely that each vertex of the tesselation is incident with just $n+1$ parallelohedra. He then proves that in a primitive tesselation two parallelohedra can be adjacent only along an $(n-1)$-dimensional face; thus for instance they cannot have only a vertex or an edge in common (unless $n=1$ or 2), etc., which can happen in nonprimitive tesselations. He proves here also that in a primitive tesselation each vertex is incident with precisely $n+1$ edges of the tesselation.

Sections 19 to 44 are devoted to the above-mentioned remarkable deduction using Ampère's principle. It begins as follows: Let x_i, $i=1,2,\ldots,n$, be rectangular Cartesian coordinates for Euclidean n-space, and let

$$a_{0k}+\sum_i a_{ik}x_i \geq 0, \quad k=1,2,\ldots,\sigma,$$

be linear inequalities defining a parallelohedron R of a given tesselation, having, as indicated, σ faces of dimension $n-1$. If $u_1,u_2,\ldots,u_\sigma$ are any positive numbers, then of course the inequalities

$$u_k\left(a_{0k}+\sum_i a_{ik}x_i\right) \geq 0, \quad k=1,2,\ldots,\sigma,$$

also define the parallelohedron R. The equation

$$a_{0k}+\sum_i a_{ik}x_i = 0,$$

[174] Voronoĭ denotes the group of translations of this lattice into itself by G, the coordinates of the corresponding vectors by λ_i, and the vectors themselves by $[\lambda_i]$.

or, equivalently,

$$u_k\left(a_{0k}+\sum_i a_{ik}x_i\right)=0,$$

defines the kth $(n-1)$-dimensional face of the parallelohedron R. Voronoĭ introduces the convention that the equation

$$-\left(a_{0k}+\sum_i a_{ik}x_i\right)=0 \quad \left(\text{or} \quad -u_k\left(a_{0k}+\sum_i a_{ik}x_i\right)=0\right)$$

defines that face not as a face of the parallelohedron R, but as a face of the adjacent parallelohedron R_k having that face in common with R. The convention is, in other words, that a degree-one equation

$$a_{0k}+\sum_i a_{ik}x_i=0$$

defines an $(n-1)$-dimensional face of a parallelohedron of a tesselation if:

1°. This is indeed an equation of the $(n-1)$-dimensional hyperplane containing that face; and

2°. All points (x_i) of the parallelohedron under consideration satisfy the inequality

$$a_{0k}+\sum_i a_{ik}x_i\geq 0.$$

If the tesselation is primitive, then each vertex of a parallelohedron is incident with n of its faces of dimension $n-1$. Let (α_i) be a vertex of a parallelohedron R of the tesselation, and suppose that the $(n-1)$-dimensional faces incident with it are given by the equations

$$a_{0k}+\sum_i a_{ik}x_i=0, \quad k=1,2,\ldots,n,$$

or, equivalently, by the equations

$$\sum_i a_{ik}(x_i-\alpha_i)=0, \quad k=1,2,\ldots,n. \tag{1}$$

Denote by R_k, $k=1,2,\ldots,n$, the n parallelohedra adjacent to R along these faces. Voronoĭ proves that one can normalize the equations of these faces by multiplying them by suitable positive numbers $u_1,u_2,\ldots,u_n$, so that in terms of the normalized equations

$$\sum_i a'_{ik}(x_i-\alpha_i)=0, \quad k=1,2,\ldots,n, \tag{2}$$

one has that the faces of the parallelohedron R incident with the vertex (α_i) are given by them, i.e., by the equations (2), while the n faces of each adjacent parallelohedron R_k, $k=1,2,\ldots,n$, that are incident with (α_i) are given by the n equations[175]

$$\begin{gathered}-\sum_i a'_{ik}(x_i-\alpha_i)=0, \quad \sum_i (a'_{ih}-a'_{ik})(x_i-\alpha_i)=0, \\ h=1,2,\ldots,n; \quad h\neq k.\end{gathered} \tag{3}$$

[175]Consider by way of illustration the tesselation of the plane by identical regular hexagons. *Trans.*

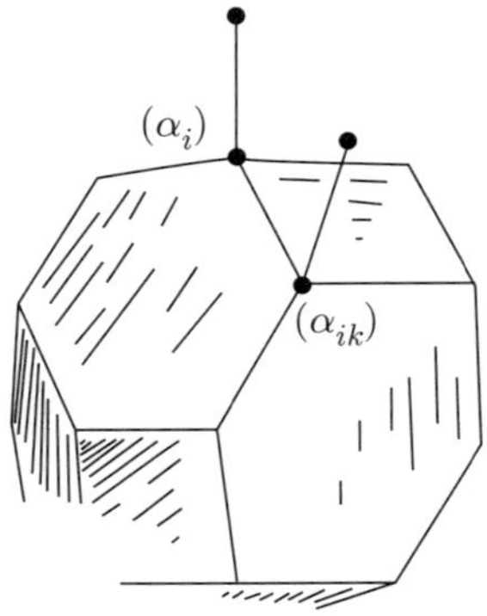

FIGURE 33

Voronoĭ calls the equations (2) and (3) the *canonical equations* of those $(n-1)$-dimensional faces of the parallelohedra $R, R_1, R_2, \ldots, R_n$ incident with the vertex (α_i). We shall say that these equations result from *normalization of the vertex* (α_i).

Voronoĭ now proves the following important theorem:

One can always choose the positive factors $u_1, u_2, \ldots, u_\sigma$ so that after multiplying given equations for all σ $(n-1)$-dimensional faces of a parallelohedron R of the primitive tesselation by them, all vertices of R are normalized.

Voronoĭ calls such normalized inequalities (and the corresponding equations)

$$a'_{0k} + \sum_i a'_{ik} x_i \geq 0, \quad k = 1, 2, \ldots, \sigma,$$

defining the parallelohedron R *canonical.*

He begins the proof of this theorem with the following observation: Let

$$a'_{0k} + \sum_i a'_{ik} x_i = 0, \quad k = 1, 2, \ldots, n,$$

be those among the above σ canonical equations for the faces of the parallelohedron R that are incident with its vertex (α_i), and let (α_{ik}) be a vertex of R adjacent to (α_i) along an edge (see Figure 33).

The vertex (α_{ik}) is incident with (and determined by) $n-1$ of the faces of R incident with the vertex (α_i):

$$a'_{0h} + \sum_i a'_{ih} x_i = 0, \qquad h = 1, 2, \ldots, n; \quad h \neq k,$$

and a further $(n-1)$-dimensional face of R, with equation, say,

$$b_{0k} + \sum_i b_{ik} x_i = 0. \tag{4}$$

Let v_h, $h = 1, 2, \ldots, n$; $h \neq k$, and v_k be positive numbers such that multiplication of the equations (3) and (4) by them normalizes the vertex (α_{ik}); thus the canonical equations of the $(n-1)$-dimensional faces of R incident with (α_{ik}) are:

$$v_h\left(a'_{0h} + \sum_i a'_{ih}x_i\right) = 0, \quad h = 1, 2, \ldots, n;\ h \neq k,$$

$$v_k\left(b_{0k} + \sum_i b_{ik}x_i\right) = 0.$$

One now shows that in fact one may take all $v_h = 1,\ h = 1, 2, \ldots, n;\ h \neq k$. To this end consider for each $h = 1, 2, \ldots, n;\ h \neq k$, and $r = 1, 2, \ldots, n;\ r \neq h, r \neq k$, the canonical equation of the $(n-1)$-dimensional face of the parallelohedron R_h along which it is adjacent to R_r. In view of the result about normalization of a vertex, this canonical equation has the form

$$\sum_i (a'_{ir} - a'_{ih})(x_i - \alpha_i) = 0.$$

On the other hand if we use instead the canonical equations of the faces incident with the point (α_{ik}) to arrive at an equation for that same face of R_h, we obtain

$$\sum_i (v_r a'_{ir} - v_h a'_{ih})(x_i - \alpha_{ik}) = 0.$$

Since these are equations of the same plane we must have

$$v_r a'_{ir} - v_h a'_{ih} = \delta(a'_{ir} - a'_{ih}), \quad i = 1, 2, \ldots, n,$$

for some number δ. Hence $v_r = \delta$, $v_h = \delta$, whence

$$v_r = v_h, \qquad r = 1, 2, \ldots, n; \quad r \neq h, \quad r \neq k.$$

Since the factors v_h, $h = 1, 2, \ldots, n$, are determined only to within a common multiple, we may therefore take

$$v_h = 1, \qquad h = 1, 2, \ldots, n; \quad h \neq k,$$

and for the normalization of the vertex it remains only to choose v_k.

Finally, by applying this process of normalization in turn to the other vertices of the parallelohedron adjacent to (α_i) along an edge, and continuing with these in place of (α_i), and so on, one progressively determines the normalizing factors in all of the inequalities

$$v_k\left(a_{0k} + \sum_i a_{ik}x_i\right) \geq 0, \quad k = 1, 2, \ldots, \sigma,$$

defining the parallelohedron. At first glance it may seem possible that in progressing step by step along edges joining (α_i) to another vertex, different edge paths may lead to different normalizing factors for the faces incident with the latter vertex; however, it turns out that the normalization is in fact independent of the edge path chosen.

In this connection Voronoĭ writes: "The task of proving this is extraordinarily difficult. It is precisely at this juncture in our present investigations that their essentially geometric character reveals itself, and the resulting difficulties can be overcome only by geometric means." There follow 13 pages devoted to the proof of this result. The essence, though not the form, of this proof, is as follows:

The uniqueness of normalization of any vertex of R, i.e., its independence of the choice of edge path (along edges of R) leading to that vertex from the initial vertex, will be established if we can show that on returning to the initial vertex via any closed edge path we end up with the same normalization factors for that vertex as we started with. Voronoĭ shows that any such closed edge path of R can be replaced by a sum of closed edge paths each of which is made up of edges of an $(n-1)$-dimensional face of R. Thus by using Ampère's principle, the problem is reduced to showing that if one starts from any vertex of an $(n-1)$-dimensional face of R, then on traversing any closed edge path of that face one obtains the same normalization factors as one started with; and this is in fact what Voronoĭ establishes, thereby completing the proof of the above theorem.

Next let

$$a_{0k} + \sum_i a_{ik}x_i \geq 0, \quad k = 1, 2, \ldots, \sigma,$$

be canonical inequalities defining a parallelohedron[176] R_0 of the tesselation, and for each $k = 1, 2, \ldots, \sigma$, denote by R_k the parallelohedron adjacent to R_0 sharing the face of R_0 given by the equation

$$a_{0k} + \sum_i a_{ik}x_i = 0.$$

Denoting by (λ_{ik}) the vector defining the parallel transport (i.e., translation) of R_k into the position of R_0, we see that R_k will have as its defining canonical inequalities

$$u_k \left[a_{0h} + \sum_i a_{ih}(x_i + \lambda_{ik}) \right] \geq 0, \quad h = 1, 2, \ldots, \sigma, \tag{5}$$

for suitable positive numbers u_k. Since the $(n-1)$-dimensional face P_k common to R_0 and R_k is defined relative to R_0 by the equation $a_{0k} + \sum_i a_{ik}x_i = 0$, it is, by the above-mentioned convention, defined relative to the parallelohedron R_k by the equation

$$-a_{0k} - \sum_i a_{ik}x_i = 0,$$

together with all multiples of this equation by positive constants u_k. By choosing u_k for each k so that

$$-a_{0k} - \sum_i a_{ik}x_i = u_k \left[a_{0h} + \sum_i a_{ih}(x_i + \lambda_{ik}) \right],$$

where h is the index of the $(n-1)$-face of R_k which it has in common with R_0, we will ensure that the inequality $-a_{0k} - \sum_i a_{ik}x_i \geq 0$ is included among the canonical inequalities (5) defining the parallelohedron R_k.

By iterating this normalization procedure, one can determine canonical inequalities defining each of the parallelohedra $R_1, R_2, \ldots, R_\sigma$ contiguous with R_0, and then for the parallelohedra contiguous with these in turn, and so on. Given any parallelohedron R of the tesselation under examination, one can always find a chain of parallelohedra

$$R_0, R', R'', \ldots, R^{(m)}, R,$$

each having an $(n-1)$-dimensional face in common with its successor, connecting the initial parallelohedron R_0 with R; in the manner just described one derives in succession normalization factors yielding a system of canonical inequalities for each parallelohedron of the chain from those of its predecessor, culminating in canonical

[176] Assumed to have had its vertices all normalized as per the above theorem. *Trans.*

inequalities for R. However, there are of course many such chains connecting R_0 to R, and one needs to show that they all yield the same system of canonical inequalities defining R.

Voronoĭ's proof that this is indeed the case, which takes up 8 pages, is analogous to that of the preceding theorem. Thus one first observes that for the desired conclusion to hold it is necessary and sufficient that if the chain of parallelohedra is closed, i.e., if $R = R_0$, then we end up with the same canonical inequalities of R_0 as we started with. Then Voronoĭ shows that each such closed chain may be considered as a sum of closed chains each of which consists of parallelohedra with a vertex common to all. Ampère's principle thus yields the sufficiency of establishing the desired result only for closed chains of this special sort; and this Voronoĭ does indeed establish.

(Note that Voronoĭ's proofs of both this result and the earlier analogous one do require that the tesselation be primitive.)

Having thus shown that all vertices of the tesselation can be compatibly normalized, Voronoĭ introduces the following linear functions, one for each of the parallelohedra comprising it. Consider a typical vector[177] $(\lambda_i) = (\lambda_1, \lambda_2, \ldots, \lambda_n)$ of the lattice of all parallel transports of the tesselation onto itself, taking an arbitrary given parallelohedron R into the position of an initially specified parallelohedron R_0. The components λ_i then afford an indexing of the parallelohedra of the tesselation. The linear functions Voronoĭ introduces are then as follows:

1. For the initial parallelohedron R_0, which is indexed by $(0, 0, \ldots, 0)$, he sets

$$V(x_1, x_2, \ldots, x_n; 0, 0, \ldots, 0) \equiv 0.$$

2. For each parallelohedron R_k, indexed by $(\lambda_{1k}, \lambda_{2k}, \ldots, \lambda_{nk})$ say, and sharing an $(n-1)$-dimensional face with R_0 with normalized equation

$$a_{0k} + \sum_i a_{ik} x_i = 0,$$

Voronoĭ sets

$$V(x_1, x_2, \ldots, x_n; \lambda_{1k}, \lambda_{2k}, \ldots, \lambda_{nk}) = a_{0k} + \sum_i a_{ik} x_i.$$

3. Finally, if R and R', indexed by (λ_i) and (λ'_i) respectively, share an $(n-1)$-dimensional face with normalized equation (relative to R)

$$a_0 + \sum_i a_i x_i = 0,$$

then he sets

$$\begin{aligned} &V(x_1, x_2, \ldots, x_n; \lambda'_1, \lambda'_2, \ldots, \lambda'_n) \\ &\qquad = V(x_1, x_2, \ldots, x_n; \lambda_1, \lambda_2, \ldots, \lambda_n) + a_0 + \sum_i a_i x_i. \end{aligned}$$

The result on the uniqueness of normalization of the vertices of the tesselation guarantees that these functions are well defined. Voronoĭ calls them the *parallelohedral functions* of the tesselation (which will be denoted by (R)).

The following property of these functions is easily established.

---

[177]See an earlier footnote. *Trans.*

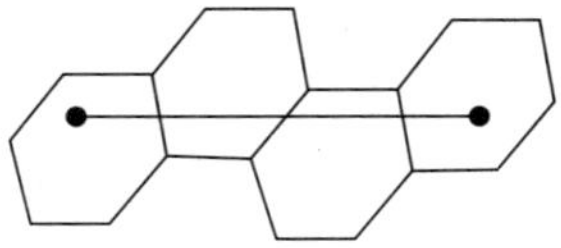

FIGURE 34

Let R and $R^{(0)}$ be any two parallelohedra of the tesselation (R), indexed by $(\lambda_1, \lambda_2, \ldots, \lambda_n)$ and $(\lambda_1^{(0)}, \lambda_2^{(0)}, \ldots, \lambda_n^{(0)})$ respectively. If $(\xi_i^{(0)})$ is any point in the interior of $R^{(0)}$, then

$$V(\xi_1^{(0)}, \xi_2^{(0)}, \ldots, \xi_n^{(0)}; \lambda_1, \lambda_2, \ldots, \lambda_n) > V(\xi_1^{(0)}, \xi_2^{(0)}, \ldots, \xi_n^{(0)}; \lambda_1^{(0)}, \lambda_2^{(0)}, \ldots, \lambda_n^{(0)}).$$

To see this consider the vector $\overrightarrow{\xi^{(0)}\xi}$ from the interior point $(\xi_i^{(0)})$ of the parallelohedron $R^{(0)}$ to an interior point (ξ_i) of R. Let $R^{(0)}, R', \ldots, R^{(m-1)}, R$ be the chain made up of those successive contiguous parallelohedra that this vector meets (see Figure 34). We may suppose that in going from each parallelohedron $R^{(k)}$ of this chain to the next $R^{(k+1)}$, the vector passes through an interior point $(\xi_i^{(k+1)})$ of the $(n-1)$-dimensional face they share, since if this is not the case then it can be arranged by means of an arbitrarily small perturbation of the point (ξ_i) inside R. We write the equation of each of these shared $(n-1)$-faces, normalized with respect to the first of the two parallelohedra $R^{(k)}, R^{(k+1)}$ having it in common, as follows:

$$a_0^{(k)} + \sum_i a_i^{(k)} x_i = 0, \quad k = 1, 2, \ldots, m-1.$$

By definition of the functions V, we have

$$V(x_1, x_2, \ldots, x_n; \lambda_1, \lambda_2, \ldots, \lambda_n)$$
$$= V(x_1, x_2, \ldots, x_n; \lambda_1^{(0)}, \lambda_2^{(0)}, \ldots, \lambda_n^{(0)}) + \sum_{k=0}^{m-1} \left(a_0^{(k)} + \sum_i a_i^{(k)} x_i \right).$$

By choice of the equations for the $(n-1)$-faces shared by successive parallelohedra $R^{(k)}, R^{(k+1)}$, $k = 0, 1, \ldots, m-1$, the point $(\xi_i^{(0)})$ must satisfy the inequalities

$$a_0^{(k)} + \sum_i a_i^{(k)} \xi_i^{(0)} > 0, \quad k = 0, 1, \ldots, m-1,$$

whence in view of the preceding equality we infer that

$$V(\xi_1^{(0)}, \xi_2^{(0)}, \ldots, \xi_n^{(0)}; \lambda_1, \lambda_2, \ldots, \lambda_n) > V(\xi_1^{(0)}, \xi_2^{(0)}, \ldots, \xi_n^{(0)}; \lambda_1^{(0)}, \lambda_2^{(0)}, \ldots, \lambda_n^{(0)}),$$

as required.

"These results," writes Voronoĭ, "open a new path to investigations in the theory of primitive parallelohedra. One may consider the totality (R) of parallelohedra of a primitive tesselation from a new point of view, namely the following:

"Each parallelohedron $R^{(0)}$ of the the tesselation (R), characterized by the vector $(\lambda_i^{(0)})$, consists of the points (x_i) satisfying the inequality

$$V(x_1, x_2, \ldots, x_n; \lambda_1, \lambda_2, \ldots, \lambda_n) \geq V(x_1, x_2, \ldots, x_n; \lambda_1^{(0)}, \lambda_2^{(0)}, \ldots, \lambda_n^{(0)})$$

for all vectors (λ_i) in the group G."

Thus since for the initial parallelohedron R_0 of (R) we have

$$V(x_1, x_2, \ldots, x_n; 0, 0, \ldots, 0) = 0,$$

it follows that that initial parallelohedron is defined by the inequalities

$$V(x_1, x_2, \ldots, x_n; \lambda_1, \lambda_2, \ldots, \lambda_n) \geq 0,$$

where (λ_i) ranges over all vectors in the group G.

We offer, by way of explanation, the following argument, which, while not actually given by Voronoĭ, may well have been crucial to his investigations—at least so it seems to us. Let the points of some lattice Γ be denoted relative to a system of rectangular Cartesian coordinates by (λ_i). For any particular point $(\lambda_i^{(0)})$ of Γ, its Dirichlet region consists precisely of those points (x_i) satisfying the quadratic inequalities

$$\begin{aligned}(x_1 - \lambda_1)^2 + (x_2 - \lambda_2)^2 + \cdots + (x_n - \lambda_n)^2 \\ \geq (x_1 - \lambda_1^{(0)})^2 + (x_2 - \lambda_2^{(0)})^2 + \cdots + (x_n - \lambda_n^{(0)})^2,\end{aligned}$$

which simplify to the linear inequalities

$$\begin{aligned}-2(x_1\lambda_1 + x_2\lambda_2 + \cdots + x_n\lambda_n) + \lambda_1^2 + \lambda_2^2 + \cdots + \lambda_n^2 \\ \geq -2(x_1\lambda_1^{(0)} + x_2\lambda_2^{(0)} + \cdots + x_n\lambda_n^{(0)}) + (\lambda_1^{(0)})^2 + (\lambda_2^{(0)})^2 + \cdots + (\lambda_n^{(0)})^2,\end{aligned}$$

for all points (λ_i) of the lattice Γ. Hence for the Dirichlet tesselation arising from the lattice Γ the linear functions

$$\begin{aligned}D(x_1, x_2, \ldots, x_n; \lambda_1, \lambda_2, \ldots, \lambda_n) \\ = -2(x_1\lambda_1 + x_2\lambda_2 + \cdots + x_n\lambda_n) + \lambda_1^2 + \lambda_2^2 + \cdots + \lambda_n^2,\end{aligned}$$

play the same role as the above functions $V(x_1, x_2, \ldots, x_n; \lambda_1, \lambda_2, \ldots, \lambda_n)$ play in the situation of a general primitive tesselation. Hence if an arbitrary primitive tesselation were an affine image of some Dirichlet tesselation, then we could obtain functions W for it, analogous to the functions V, by applying the appropriate affine transformation to the variables $x_1, x_2, \ldots, x_n$ in the functions D defining that Dirichlet tesselation. It seems obvious that this represents the source of Voronoĭ's idea of introducing the above functions V, with the aim of then showing that they can indeed be obtained from the functions D of some Dirichlet tesselation by means of a suitable affine transformation of the variables $x_1, x_2, \ldots, x_n$.

This is precisely what Voronoĭ proves very elegantly in the final sections of the first part of his memoir, occupying the 12 pages 39–44. However, we must forego examining his proof here. We remark that Voronoĭ incidentally proves that his parallelohedra, at least primitive ones, must have centres of symmetry.

Thus the argumentation of the first 60 pages of Voronoĭ's memoir reduces the problem of classifying all primitive tesselations of n-dimensional space by parallelohedra, to that of classifying Dirichlet regions of certain n-dimensional lattices, i.e., reduces the initial purely geometrical question to one on quadratic forms.

We now turn to the second part of Voronoĭ's memoir, consisting of Chapters III, IV, and V.

Chapter III is entitled:

The definition of parallelohedra by means of positive quadratic forms. Suppose given a positive quadratic form $\phi = \sum\sum a_{ij}x_ix_j$ in the n variables $x_1, x_2, \ldots, x_n$. In Section 45 Voronoĭ considers the set R of all points (α_i) satisfying the inequalities

$$\sum\sum a_{ij}x_ix_j + 2\sum \alpha_i x_i \geq 0, \tag{6}$$

for all integer values of the variables $x_1, x_2, \ldots, x_n$. To motivate this, consider the lattice Γ in n-space with frame E, say,[178] associated in one-to-one fashion with the positive quadratic form ϕ. By definition the Dirichlet region of Γ containing the origin O, is the set of points lying no further from O than from any other point of the lattice, i.e., the set of points whose coordinates (y_i) say, relative to the frame E,[179] satisfy the inequalities

$$\sum\sum a_{ij}(x_i - y_i)(x_j - y_j) - \sum\sum a_{ij}y_iy_j \geq 0,$$

for all points (x_i) of the lattice Γ. These inequalities may be rewritten as

$$\sum\sum a_{ij}x_ix_j - 2\frac{\partial\phi}{\partial x_1}y_1 - 2\frac{\partial\phi}{\partial x_2}y_2 - \cdots - 2\frac{\partial\phi}{\partial x_n}y_n \geq 0,$$

or in the form (6) if one takes

$$\begin{aligned} \alpha_1 &= -2(a_{11}y_1 + a_{12}y_2 + \cdots + a_{1n}y_n), \\ \alpha_2 &= -2(a_{21}y_1 + a_{22}y_2 + \cdots + a_{2n}y_n), \\ &\vdots \qquad\qquad \vdots \\ \alpha_n &= -2(a_{n1}y_1 + a_{n2}y_2 + \cdots + a_{nn}y_n). \end{aligned} \tag{7}$$

Hence if we regard the α_i as the coordinates relative to the frame E^* related to E via the equations (7), then the region R becomes the Dirichlet region about O of the lattice Γ.

It is easy to show that among the infinitely many inequalities (6) there are only a finite number of independent ones. This can be seen geometrically as follows. Each separate[180] inequality (6) defines a half-space containing O, cut off by the $(n-1)$-dimensional subspace (hyperplane) defined in terms of the coordinates α_i relative to the frame E^*, by the equation

$$\sum\sum a_{ij}x_ix_j + 2\sum \alpha_i x_i = 0,$$

which passes through the midpoint of the line segment joining the origin O to the point with coordinates x_i (relative to the frame E) of the lattice Γ, and is perpendicular to that segment. We shall call this hyperplane the *Voronoĭ hyperplane* corresponding to the lattice point (x_i). The (solid) polyhedron R is then the intersection of all such half-spaces, i.e., corresponding to all points (x_i) of the lattice. If we choose $n+1$ points of the lattice Γ so that the corresponding Voronoĭ hyperplanes determine an n-dimensional simplex about O, then only finitely many of the

[178] In the Commentary to the chapter on Korkin, this frame was denoted by $(\mathfrak{A}_1, \mathfrak{A}_2, \ldots, \mathfrak{A}_n)$. *Trans.*

[179] From now on the coordinates x_i and y_i will be understood to refer to the frame E; thus for the points of the lattice Γ they will be integers.

[180] That is, for particular (integer) values of the x_i. *Trans.*

Voronoĭ hyperplanes of other points of the lattice will meet this simplex, all others passing outside it. (Voronoĭ proves this by analytic means.)

In Section 48 Voronoĭ proves the following theorem:

An equation corresponding as above to a point (l_i) *of the lattice* Γ *defines an* $(n-1)$*-face of the polyhedron* R *if and only if not all of the* l_i *are even and the value of the form* $\sum\sum a_{ij}x_ix_j$ *at the points* (l_i) *and* $(-l_i)$ *is strictly less than its value at all other points* (x_i) *of the lattice for which* $x_i \equiv l_i (\bmod 2)$, $i = 1, 2, \ldots, n$.

The number of ordered n-tuples of 0s and 1s other than $(0, 0, \ldots, 0)$ being $2^n - 1$, it follows that there are at most this many pairs of $(n-1)$-faces of R (each face corresponding to (l_i) having an opposite corresponding to $(-l_i)$); hence altogether the number of $(n-1)$-faces of the polyhedron R is at most $2(2^n - 1)$. (This result was found earlier by Minkowski in a somewhat different form.)

In Sections 51–54 Voronoĭ considers the coordinates $l_1, l_2, \ldots, l_n$ relative to the fundamental frame E, of those points of the lattice Γ at the centres of the Dirichlet regions contiguous along $(n-1)$-faces with the Dirichlet region R containing O. Let these points be

$$\pm(l_{i1}), \pm(l_{i2}), \ldots, \pm(l_{i\sigma}), \tag{8}$$

where σ denotes the number of $(n-1)$-faces of R. The following two assertions then hold:

I. *The coordinates* $l_{1k}, l_{2k}, \ldots, l_{nk}$ *of each of these points have, obviously, highest common factor* 1.

II. *If*

$$(p_{i1}), (p_{i2}), \ldots, (p_{in}) \tag{9}$$

are the n *successive minimum points of the quadratic form, i.e.,* (p_{i1}) *is a point of* Γ *closest to* O, (p_{i2}) *a point of* Γ *closest to* O *among those not lying on the line determined by* O *and* (p_{i1}), *and so on, then these points (9) are included among the points (8). From this it follows in particular that all representations of the minimum of the form* $\sum\sum a_{ij}x_ix_j$ *must appear in the sequence (8).*

These two results are simple to prove whether by geometric or algebraic means. Here for instance is a geometric proof of the statement II.

Let $M_1, M_2, \ldots, M_n$ denote the successive minimum points, as defined in that statement. Since M_1 is the first minimum point, its corresponding Voronoĭ hyperplane, meeting the line segment OM_1 at its midpoint, is the only such hyperplane meeting that segment. Part of this hyperplane, therefore, forms an $(n-1)$-face of R.[181] Since this applies also to the antipodal point $-M_1$, the Voronoĭ hyperplanes corresponding to M_1 and $-M_1$ contain between them an $(n-1)$-dimensional layer (or slice), and clearly the next minimum point M_2 must lie inside this layer. The minimality property of M_2 again ensures that there is only one Voronoĭ hyperplane meeting the segment OM_2, namely that corresponding to M_2, meeting the segment OM_2 at its midpoint, so that part of that hyperplane will form an $(n-1)$-face of R, and so on.

In Sections 53 and 54 Voronoĭ proves that the determinant of the matrix with rows any n of the n-tuples from (8) is in absolute value less than $n!$.

[181] Hence M_1 will be at the centre of the Dirichlet region contiguous with R along that face (and similarly for $M_2, M_3, \ldots, M_n$ by the argument that follows). *Trans.*

Finally, in Section 55 he proves that if the Dirichlet region is primitive, then the number of its $(n-1)$-dimensional faces is precisely $2(2^n-1)$.

Chapter IV is entitled:

Various types of primitive parallelohedra. In Section 56 Voronoĭ defines the "type" of a primitive parallelohedron. Suppose a primitive parallelohedron R is defined (as above) by the inequalities

$$\sum\sum a_{ij}l_{ik}l_{jk} + 2\sum \alpha_i l_{ik} \geq 0, \quad k = 1, 2, \ldots, \sigma, \text{ where } \sigma = 2(2^n - 1),$$

and denote its vertices by $(\alpha_{i1}), (\alpha_{i2}), \ldots, (\alpha_{is})$. By the primitivity assumption each of these vertices is incident with precisely $n+1$ parallelohedra of the tesselation, namely R itself and n parallelohedra contiguous with R along those of its $(n-1)$-faces incident with the vertex. Denote by

$$(l_{i1}^{(k)}), (l_{i2}^{(k)}), \ldots, (l_{in}^{(k)}), \quad k = 1, 2, \ldots, s,$$

the centres of the latter n parallelohedra incident with the kth vertex of R. We shall denote this system of n n-tuples briefly by

$$\{l_{ir}^{(k)}\}.$$

Each vertex of the parallelohedron R is characterized by its symbol; thus the symbols

$$\{l_{ir}^{(1)}\}, \{l_{ir}^{(2)}\}, \ldots, \{l_{ir}^{(s)}\}$$

characterize the respective s vertices of the primitive parallelohedron R. If R' is any other n-dimensional primitive parallelohedron,[182] with the same number s of vertices characterized by the same sequence of symbols $\{\ \}$ as those characterizing the vertices of R, Voronoĭ defines R and R' as being *of the same type.*

The centres of the $n+1$ primitive parallelohedra incident with a vertex may be regarded as the vertices of an n-dimensional simplex; Voronoĭ calls this n-simplex the *correlate* of the vertex. The *type of a given primitive tesselation* (R) of n-space by parallelohedra Voronoĭ stipulates as then being given by the totality (L) of simplices correlative to the vertices of (R).

Two simplices transformable one into the other by means of parallel transport are called *congruent.* Voronoĭ proves that the number of distinct incongruent simplices in (L) is finite. He further proves that:

The simplices L cover the whole of n-space, are pairwise nonoverlapping, and intersecting simplices meet in whole faces of dimensions less than n.[183] *Furthermore the totality of their vertices coincides with that of all centres of the parallelohedra of the tesselation* (R).

Voronoĭ continues with a detailed investigation of the number S_ν of ν-dimensional incongruent faces of a parallelohedron R of a given primitive tesselation, ending with a formula giving an upper bound for S_ν.[184] This bound turns out to be exact

---

[182] Clearly not from the same primitive tesselation. *Trans.*

[183] That is, they form a simplicial decomposition of n-space. *Trans.*

[184] In the original the author cites this formula as $S_\nu \leq (n+1-\nu)\Delta^{(n-\nu)}(m^n)_{m=1}$, without specifying the meaning of the symbols appearing. *Trans.*

when certain numbers $\omega_k^{(\mu)}$ are equal to 1, which occurs for $n = 1, 2, 3$ and 4, whence one obtains the following table of values of S_ν:

$$n = 2 : S_0 = 6, \quad S_1 = 6$$
$$n = 3 : S_0 = 24, \quad S_1 = 36, \quad S_2 = 14$$
$$n = 4 : S_0 = 120, \ S_1 = 240, \ S_2 = 150, \ S_3 = 30.$$

Voronoĭ remarks that already in 5 dimensions he encountered primitive parallelohedra for which his bound is not exact.

In Section 69 Voronoĭ defines the concepts "characteristic" and "regulator" of an edge of a primitive polyhedron R, which are important in the sequel. Let $(l_{i0}), (l_{i1}), \dots, (l_{in})$ denote the centres of the $n+1$ parallelohedra incident with a vertex (α_i) of R, or, in other words, the vertices of the simplex L correlative to the vertex (α_i) of R, where the coordinates l_{ir} are relative to the fundamental frame of the lattice Γ. Let (α_{ik}), $k = 1, 2, \dots, n$, be the vertices of R adjacent to (α_i) along the n edges of R incident with that vertex. For each k the set of vertices of the simplex L_k correlative with the vertex (α_{ik}) is obtained by replacing the kth vertex of the correlate of (α_i), namely (l_{ik}), by another point, say (l'_{ik}), since the two simplices L and L_k meet in the $(n-1)$-dimensional face P_k with vertices

$$(l_{ih}), \quad h = 0, 1, 2, \dots, n; \ h \neq k.$$

We shall say that the $(n-1)$-simplex P_k is the *correlate* of the edge $[\alpha_i, \alpha_{ik}]$ of the tesselation (R). Suppose that the hyperplane of which the simplex P_k forms a part has equation

$$\sum p_{ik} x_i = \delta_k,$$

in terms of coordinates x_i relative to the same fundamental frame as before. Since the points (l_{ih}) with $h \neq k$ are on this hyperplane, we have

$$\sum p_{ik} l_{ih} = \delta_k, \quad h = 0, 1, 2, \dots, n; \ h \neq k,$$

whence

$$\sum p_{ik}(l_{ih} - l_{ir}) = 0, \quad h, r = 0, 1, 2, \dots, n; \ h, r \neq k.$$

The latter equations determine the coefficients $p_{1k}, p_{2k}, \dots, p_{nk}$ to within a common multiple; Voronoĭ calls the n-tuple (p_{ik}) the *characteristic* of the edge $[\alpha_i, \alpha_{ik}]$ or face P_k.

Voronoĭ next introduces the important concept of the "regulator" of an edge $[\alpha_i, \alpha_{ik}]$ of the tesselation (R), or of the correlative face P_k of the simplicial complex (L), characterizing in some sense the "length" of the edge. In our commentary on this work below we shall see what the "regulator" is exactly; for now we reveal only that it is given by the formula

$$\rho_k = \sum \sum p_{ij}^{(k)} a_{ij},$$

linear in the coefficients a_{ij} of the original quadratic form $\sum \sum a_{ij} x_i x_j$, where the coefficients $p_{ij}^{(k)} = p_{ji}^{(k)}$ are certain rational numbers obtained using the coordinates $l_{i0}, l_{i1}, l_{i2}, \dots, l_{in}, l'_{ik}$ of the vertices of the two simplices L and L_k correlative to the edge in question.

Section 77 contains the following result, representing the central lemma of the whole of the second part of the memoir, which Voronoĭ had printed in boldface and called "fundamental" (and which we shall likewise call "fundamental").

FUNDAMENTAL THEOREM. *Given any triangulation of n-space into simplices with the property that its vertices form a lattice Γ and that all "stars" (each consisting of all the simplices incident with a vertex) are congruent, i.e., can be obtained from one another by means of parallel transport of the lattice Γ, a necessary and sufficient condition for the simplices of this simplicial complex (L) to be correlates of the vertices of the Dirichlet tesselation (D) determined by the lattice Γ, is that the regulators of all incongruent*[185] *faces of the triangulation be positive.*

A proof (rather different from Voronoĭ's) and generalization of this remarkable theorem will be given in our commentary below.

In Sections 79–84 Voronoĭ considers the deep question of the characterization of a quadratic form in terms of the corresponding characteristics and regulators. We shall not, however, expound this topic here.

Chapter V is entitled:

Properties of the sets of quadratic forms belonging to different types of primitive parallelohedra. This chapter begins with Section 85, where Voronoĭ defines the *region of quadratic forms belonging to a given type of primitive parallelohedron.* Suppose given a type of primitive parallelohedra; by definition this is characterized by a triangulation (L) of n-space by simplices correlative to the vertices of any tesselation of that type. Let $\rho_k = \sum\sum p_{ij}^{(k)} a_{ij}$, $k = 1, 2, \ldots, \tau$, be the regulators of a full set, numbering τ, of incongruent $(n-1)$-faces P_k of the n-simplices of this triangulation. Then Voronoĭ calls the set Δ of all quadratic forms $\sum\sum a_{ij}x_ix_j$ for which

$$\rho_k = \sum\sum p_{ij}^{(k)} a_{ij} \geq 0, \text{ for } k = 1, 2, \ldots, \tau,$$

the *region of quadratic forms corresponding to the given type.*

By the above Fundamental Theorem, for a quadratic form $f = \sum\sum a_{ij}x_ix_j$ to determine a primitive Dirichlet tesselation (D) of the type defined by the triangulation (L), it is necessary and sufficient that f, considered as the point $(a_{11}, a_{22}, \ldots, a_{nn}, a_{12}, \ldots, a_{n-1,n})$, lie in Δ, considered as a region of $(\frac{n(n+1)}{2})$-dimensional space. Voronoĭ defines two forms $\sum\sum a_{ij}x_ix_j$ and $\sum\sum a'_{ij}x_ix_j$ as *belonging to the same region of a type* if the vertices of the respective triangulations (L) and $(L)'$ arising from them have the same coordinates relative to their respective fundamental frames, or, in other words, if the affine transformation transforming the fundamental frame of the second form into that of the first, transforms the triangulation $(L)'$ corresponding to the second form into that corresponding to the first, namely (L).

The region Δ is an $(\frac{n(n+1)}{2})$-dimensional convex many-sided angle, or "gonohedron", to use Fedorov's term. As examples show even for $n = 4$, it can happen that some of the inequalities $\rho_k \geq 0$ follow from the others. Let

$$\rho_1 \geq 0, \rho_2 \geq 0, \ldots, \rho_m \geq 0,$$

be a full set of independent such inequalities, defining the gonohedron Δ; then each of the equations $\rho_k = 0$, $k = 1, 2, \ldots, m$, is an equation of one of its $(\frac{n(n+1)}{2} - 1)$-dimensional faces. Voronoĭ calls such a set of regulators *independent.*

Let $\sum\sum a_{ij}x_ix_j$ be a form corresponding to an interior point of the region Δ; then of course how the simplices of the corresponding triangulation (L) are

[185]Presumably this means simply all faces. *Trans.*

situated with respect to the fundamental frame of the form is determined. We now vary the coefficients of the form so that the corresponding point of Δ moves continuously within Δ until it reaches an interior point of one of its $\left(\frac{n(n+1)}{2} - 1\right)$-dimensional faces; of course at this final point the independent regulator of that face will vanish, and on moving the point through this face the corresponding form will cease belonging to the region of that type. In Sections 87–96 Voronoĭ describes a very remarkable "algorithm for reconstructing a triangulation (L) by simplices" in the case where the point lies within the region Δ, and also the totality $(L)'$ of[186] simplices corresponding to the case when the point representing the form lies in a region Δ' adjacent to Δ along an $\left(\frac{n(n+1)}{2} - 1\right)$-dimensional face. In the course of constructing this algorithm he proves that the regions $\Delta, \Delta', \ldots$ fill the $\left(\frac{n(n+1)}{2}\right)$-dimensional cone in $\left(\frac{n(n+1)}{2}\right)$-dimensional coefficient space, corresponding to the positive forms $\sum\sum a_{ij}x_ix_j$, are nonoverlapping, and intersect in complete $\left(\frac{n(n+1)}{2} - 1\right)$-dimensional faces. Voronoĭ denotes this tesselation of this cone by (Δ).

If we change all forms to equivalent ones by means of an integral linear transformation of determinant ± 1 of the variables $x_1, x_2, \ldots, x_n$, then the coefficient space, as is well known, also transforms linearly and unimodularly. Each region Δ will in this situation either be fixed or sent wholly onto some other such region, corresponding to the transforms of the forms its points represent. We call two such regions *equivalent*. Voronoĭ proves the important result that *the number of inequivalent regions Δ is finite*. The proof relies on the inequality mentioned earlier, established in Section 54, namely $\omega < n!$ where ω is the absolute value of the determinant of the matrix with rows any n of the vectors $(l_{i1}), (l_{i2}), \ldots, (l_{i\sigma})$ from the centre of any parallelohedron to the centres of those adjacent to it.

In Sections 102–104 Voronoĭ shows that the form

$$f = nx_1^2 + nx_2^2 + \cdots + nx_n^2 - 2x_1x_2 - 2x_1x_3 - \cdots - 2x_{n-1}x_n$$

is positive, and that the Dirichlet tesselation corresponding to it is primitive. It can be shown that the lattice Γ corresponding to this form is that of centred n-cubes, i.e., with points the vertices and centres of the n-cubes of the tesselation of n-space by n-cubes. In Sections 102–104 Voronoĭ investigates this form in detail, determining its region Δ, which he calls the *principal region* of the tesselation (Δ).

In Sections 105–110 Voronoĭ finds the regions adjacent to the principal region along $\left(\frac{n(n+1)}{2} - 1\right)$-dimensional faces. It turns out that for $n = 2, 3$ they are equivalent to the principal region, while for $n > 3$ they are not equivalent to the principal region but are equivalent amongst themselves.

Thus having found one region Δ of the tesselation (Δ), given an algorithm for moving to an adjacent one, and proved that (Δ) has only finitely many inequivalent regions, Voronoĭ shows further, applying Hermite's principle, that for any given n his algorithm allows one to find in finitely many steps representatives of all equivalence classes of regions in (Δ). For by means of the algorithm for reconstructing a simplicial decomposition one can find all regions of (Δ) adjacent to the principal region, forming the "first layer". Then one applies the same algorithm to find all regions of (Δ) adjacent to regions of the first layer, forming the "second layer", and so on. Since there are only finitely many inequivalent regions of (Δ), this procedure will eventually culminate in a layer all of whose regions are equivalent to regions of

[186]That is, triangulation by. *Trans.*

earlier layers, or to the principal region (which we consider as forming the "zeroth layer"). It is easy to see that all subsequent layers will then also consist of regions equivalent to regions in earlier layers, for if in earlier layers there are represented h classes of regions, and in the next layer no regions representing new classes appear, then the further regions adjacent to this layer must also be equivalent to regions of earlier layers.[187] It follows that once one has reached a layer consisting of regions all equivalent to regions of earlier layers, all subsequent layers will likewise contain no new inequivalent regions, since otherwise we should have a region of, say, the kth class, where $k \leq h$, represented in an earlier layer, now adjacent to a region of a new $(h+1)$st class, whereas for the representative of the kth class appearing earlier this did not occur. If one applies a suitable unimodular integral transformation of the variables of the corresponding forms so that this region (belonging to the kth class) is sent onto an equivalent region in an earlier layer, then its adjacent region belonging to a putative $(h+1)$st class would be sent to an adjacent region of an earlier layer, which is not possible since by assumption all regions of earlier layers were adjacent to regions belonging only to the first h classes. This represents an application of Hermite's principle.

Thus for any n, knowledge of one region of some type (the "principal" region), together with the algorithm for reconstructing a triangulation, affords Voronoĭ a method for finding a full set of inequivalent regions of types Δ, and hence all topologically distinct primitive parallelohedra R of dimension n.

At the conclusion of his memoir, in Sections 111–118, Voronoĭ finds explicitly all primitive parallelohedra for $n = 2, 3, 4$. It turns out that in the plane there is only one, the regular hexagon, and also in 3-space, namely a 14-sided parallelohedron, while in 4-space there are three, all 30-sided.

Commentary

Voronoĭ's memoir on parallelohedra represents one of the deepest investigations in the geometry of numbers in the world's literature, and the originality of the methods employed in the purely geometrical first part stamps the memoir with the imprint of genius.

As already mentioned, the concept of parallelohedra in 3-space is due to the famous Russian crystallographer Evgraf Stepanovich Fedorov. Voronoĭ exploited this idea in order to penetrate more deeply into the theory of quadratic forms. It cannot be doubted that the consequences of this approach for the theory of quadratic forms, even positive ones, were not exhausted by the work of Voronoĭ. The classification of *all* n-dimensional parallelohedra eluded him; he provided a method for finding only all *primitive* parallelohedra.

Later work relating in one way or another to Voronoĭ's memoir was carried out by B. N. Delone, O. K. Zhitomirskiĭ, and A. D. Aleksandrov.

In his long memoir entitled "Sur la partition régulière de l'espace à 4 dimensions" (*Izvestiya AN SSSR*, 1929), Delone, without the assumption of primitivity, achieves the extremely difficult complete classification of all 4-dimensional parallelohedra affording a so-called "normal" tesselation, i.e., in which adjacent n-dimensional parallelohedra share a full $(n-1)$-dimensional face. He manages to show by means of difficult, purely geometrical arguments, that every 4-dimensional normal tesselation, whether primitive or nonprimitive, is affinely equivalent to some

[187] Since corresponding neighbours of equivalent regions will be equivalent. *Trans.*

Dirichlet tesselation, after which it is easy to actually determine all of these tesselations. It turns out that apart from the three primitive 4-dimensional tesselations found by Voronoĭ, there are further 48 nonprimitive normal tesselations, all representing limiting cases of the primitive ones. Using an approach quite different from that of Voronoĭ, Delone also finds all possible incidence numbers of $(n-2)$- and $(n-3)$-dimensional faces of parallelohedra in n-dimensional normal tesselations.

Delone's methods are not applicable to the problem of finding all 5-dimensional tesselations. Later, in a short article (*Izvestiya AN SSSR*, 1931) Delone shows that in dimension 3, if one removes the assumption of normality, and even the assumption that the parallelohedra are all congruent, retaining only parallelism and the condition that they be similar to one another, then no new tesselations emerge apart from the normal ones. The analogous result in dimension 4 was then proved by A. D. Aleksandrov (*Izvestiya AN SSSR*, 1935). Hence for $n = 1, 2, 3, 4$, we now know all possible (admissible) tesselations, whether primitive or nonprimitive, normal or not normal.

Zhitomirskiĭ followed a different path. In his work "Verschärfung eines Satzes von Voronoï", published in the *Journal of the Leningrad Mathematical Society* in 1931, he gives a direct refinement of Voronoĭ's approach in the first part of his memoir. It turned out to be possible to refine a part of Voronoĭ's method by the introduction of the elegant notion of a special hypersurface, called a "filter" by Zhitomirskiĭ. Equipped with this refinement, he was able to show that Voronoĭ's approach leads to its goal under significantly weaker assumptions than those Voronoĭ made. It turned out that instead of the requirement of "primitivity" of the vertices ("simplicity", in Voronoĭ's terminology—see earlier)—and hence of all faces of higher dimension—it suffices to demand only that the $(n-3)$-dimensional faces be primitive[188] (whence the $(n-2)$-dimensional faces will also necessarily be primitive). If one takes into account the result of Delone on the incidence numbers of $(n-2)$- and $(n-3)$-dimensional faces referred to above, then it turns out to be sufficient in fact to require only that the $(n-2)$-dimensional faces be primitive.

How to approach the problem for $n > 4$ in the case of nonprimitive $(n-2)$-faces remains unclear; in spite of intensive efforts in this direction exerted at various times by Delone, Zhitomirskiĭ, and Aleksandrov, the situation remains unresolved.

There appear to be no other works touching on the question considered by Voronoĭ in the first part of his memoir.[189]

As to the second part of Voronoĭ's memoir, the proofs it gives have been significantly simplified in three articles by Delone: "Sur la sphère vide" (*Proceedings of the International Congress of Mathematicians*, Toronto, 1924); "Sur la sphère vide II" (*Izvestiya AN SSSR*, 1935); and in the survey under the head "The geometry of positive quadratic forms" (Vols. III and IV of *Uspekhi Matem. Nauk"*, 1937 and 1938). The first chapter of the latter article contains a new exposition, with full proofs, of all the results contained in the second part of Voronoĭ's memoir (except for the expression of quadratic forms in terms of regulators and characteristics), and is incomparably easier to read than Voronoĭ's original. It turns out that the result called the "Fundamental Theorem" by Voronoĭ, can be very widely generalized and very simply proved.

---

[188]That is, each such face should be incident with the least possible number of 3-faces, namely 4. *Trans.*

[189]As of 1947. *Trans.*

We begin by setting forth the concepts developed by the author of the present book in his second article on the empty sphere. Let $\mathcal{E}$ denote an arbitrary system of points in n-dimensional Euclidean space subject only to the following two conditions:

1) There exists a fixed positive number r such that for every point of $\mathcal{E}$ the $(n-1)$-sphere[190] of radius r centred at that point contains no other points of $\mathcal{E}$, i.e., the system $\mathcal{E}$ is discrete.

2) There is another fixed positive number R such that every $(n-1)$-sphere of radius R, no matter where it is situated in n-space, contains at least one point of the system $\mathcal{E}$, i.e., n-space does not have arbitrarily large regions empty of points of the system $\mathcal{E}$.

We shall be considering an $(n-1)$-sphere which is allowed to expand, shrink, and move in any manner whatever among the points of the system $\mathcal{E}$ subject only to the condition that it never contain any points of $\mathcal{E}$ in its interior. Delone calls such a sphere *empty*.

In the following four subsections we expound certain general lemmas needed by us in the sequel, concerning the empty sphere and such arbitrary discrete systems of points in n-space.

1. The partition[191] L associated with a given system $\mathcal{E}$ of points of n-space. Take any empty $(n-1)$-sphere, and let its radius increase continuously while keeping its centre fixed, until its surface encounters a point of the system $\mathcal{E}$. This will certainly occur in view of the second of our assumptions, namely that there is no empty sphere of radius exceeding R. We now continue to let the radius of our empty sphere grow by moving its centre away from this point, or, if it encountered several points at once, away from the affine subspace of n-space spanned by those points.[192] By continuing in this way as long as possible while keeping our sphere empty, we convince ourselves that there is an empty $(n-1)$-sphere such that the affine span of the points of the system $\mathcal{E}$ on its surface is the whole of n-space, since at any stage where the points on its surface span an affine subspace of dimension less than n, we can enlarge the sphere by moving its centre away from this affine subspace while keeping it empty—not indefinitely, however, since its radius cannot grow to be larger than R.

We denote by (L) such an empty sphere, i.e., one on whose surface lie points of $\mathcal{E}$ with affine span the whole of n-space;[193] the number of such points must be at least $n+1$, but in general there may of course be more. The convex hull of the set of all points of $\mathcal{E}$ on the surface of the sphere (L) will be a convex polyhedron inscribed in that sphere with its vertices at those points; we denote this polyhedron by L.

LEMMA. *All such polyhedra L of the system $\mathcal{E}$ are pairwise nonoverlapping and fill the whole of n-space, and are furthermore contiguous along complete $(n-1)$-dimensional faces. This partition of n-space by the polyhedra L is uniquely determined by the point system $\mathcal{E}$, and, conversely, uniquely determines it.*

---

[190] Or the n-ball. *Trans.*

[191] The terms "partition" and "tesselation" are used synonymously. *Trans.*

[192] That is, the smallest affine subspace (linear subspace + a constant vector) containing these points. *Trans.*

[193] In the original this is expressed by the phrase "forming an n-dimensional complex of points of $\mathcal{E}$". *Trans.*

To see this, let, to begin with, L_1 and L_2 be any two polyhedra corresponding to any two empty spheres (L_1) and (L_2). If these spheres do not overlap then obviously the polyhedra L_1 and L_2 will also not overlap. On the other hand if the spheres do overlap, then the vertices of the polyhedron L_1 lie in that region of the surface of the sphere (L_1) not contained in the interior of the sphere (L_2) since the latter sphere is empty, and *vice versa.* Hence the vertices of the polyhedron L_1 and those of L_2 lie on opposite sides of the chordal $(n-1)$-dimensional hyperplane determined by the intersection of the surfaces of the corresponding spheres, except that some may lie on that hyperplane, so that the polyhedra L_1 and L_2, as the convex hulls of these respective sets of vertices, must also lie on opposite sides of that hyperplane.

Consider now any $(n-1)$-dimensional face of the polyhedron L_1. (We shall henceforth understand the word "face" to refer to an $(n-1)$-face.) We move the centre of the sphere (L_1) circumscribed about the polygon in the direction perpendicular to this face, i.e., in the perpendicular direction outwards from the centre of the polyhedron L_1 through the centre of the face in question, at the same time allowing the radius of the sphere to change so that it continues to pass through the vertices of the face being considered.[194] This changing sphere immediately moves away from all other vertices of the polyhedron L_1, and ultimately encounters some point or points of the system $\mathcal{E}$ lying outside that polyhedron. The convex hull of the union of these point(s) with the set of vertices of the face of L_1 being considered, is another polyhedron L_2 say, adjacent to L_1 along the whole of that face. It follows that the polyhedra L arising from the system $\mathcal{E}$ fill the whole of n-space, since if there were empty regions between them, there would be faces or parts of faces of some such polyhedra, namely those bounding these interstitial regions, not wholly shared by other such polyhedra.

The partition of n-space by the polyhedra L is uniquely determined by the system $\mathcal{E}$, since if $\mathcal{E}$ gave rise to a partition by means of other polyhedra L', then some polyhedron L would have a common interior point with some distinct polyhedron L', which we have shown to be impossible.

For the sake of brevity we shall call the partition of n-space by polyhedra L arising from the point system $\mathcal{E}$ simply the *partition* L.

2. The fundamental lemma on partitions. Suppose that on the one hand we are given a point system $\mathcal{E}$ in n-space, and on the other a partition of n-space by nonoverlapping polyhedra T filling the whole of n-space, with the properties that adjacent polyhedra intersect in complete faces, and the set of all vertices of the polyhedra T coincides with the system $\mathcal{E}$. We then ask: What further conditions must the partition T satisfy for it to be the partition L corresponding to the system $\mathcal{E}$? Obvious necessary conditions are of course that each polyhedron T should be convex and inscribable in a sphere; however, these conditions are satisfied by any simplicial decomposition with vertices the point system $\mathcal{E}$, and there is a nondenumerable infinity of such decompositions, moreover essentially distinct, so that these conditions are not sufficient.

LEMMA. *Necessary and sufficient conditions for a partition T related to a point system $\mathcal{E}$ as above, to be that system's partition L, are as follows:*

1) *All polyhedra T should be convex and inscribable in spheres; and*

[194] Doubtless its radius initially decreases. *Trans.*

2) *for each pair of polyhedra of the partition T sharing an $(n-1)$-dimensional face, neither should be contained in the sphere circumscribed about the other.*

The necessity of conditions 1) and 2) being obvious, we need prove only their sufficiency. Let T_1 denote any polyhedron of the partition T, (T_1) its circumscribed sphere, and A any point of the system $\mathcal{E}$ that is not a vertex of this polyhedron. Choose any point B inside the polyhedron T_1 and join it by a straight line segment BA to A. (Note that of course B cannot be a point of $\mathcal{E}$.) It may of course happen that the line segment BA, in going from one polyhedron (namely T_1) of the partition T to an adjacent one, passes not through an interior point of their common $(n-1)$-face, but through a common face of smaller dimension (for example through a shared vertex, or edge, etc.); in this case we may clearly arrange that this does not occur by moving B slightly within the polyhedron T_1. We shall assume that this has been done. Denote by $T_1, T_2, \dots, T_k$ the polyhedra of the partition T that the line segment BA intersects going in the direction from B to A; thus the point A is a vertex of the polyhedron T_k. Consider now the hyperplanes determined by the $(n-1)$-dimensional faces shared by successive polyhedra in this chain; we shall call that side of each of these hyperplanes facing towards the point A *upper*. Denote by $(T_1), (T_2), \dots, (T_k)$ the corresponding circumscribed spheres; pairs of successive such spheres intersect, and the chordal hyperplanes of their intersections are the aforementioned hyperplanes. It is easy to show that for any two intersecting spheres the "degree"[195] of a point outside their chordal hyperplane will be less relative to that sphere whose free part, i.e., not belonging to the intersection, lies on the same side of the chordal hyperplane as the point in question. Applying this to each pair $(T_i), (T_{i+1})$ of successive spheres in the above chain, and assuming condition 2) holds, we shall have that a part of (T_{i+1}) lies above the chordal hyperplane of the intersection of (T_i) and (T_{i+1}), and therefore that the degree of the point A relative to the sphere (T_{i+1}) is less than that relative to (T_i). Hence if we denote the degrees of A relative to the spheres of the above chain by $\tau_1, \tau_2, \dots, \tau_k$ respectively, we have

$$\tau_1 > \tau_2 > \cdots > \tau_{k-1} > \tau_k.$$

However, $\tau_k = 0$ since the point A lies on the sphere (T_k); hence $\tau_1 > 0$, whence A lies outside the sphere (T_1). Since T_1 was any polyhedron of the partition T, and A any point of the system $\mathcal{E}$ other than a vertex of T_1, it follows that under conditions 1) and 2) all spheres (T) are empty, i.e., the partition T coincides with the partition L.

3. The partition V associated with a given point system $\mathcal{E}$. Given a point system $\mathcal{E}$ of n-space, and any point O of $\mathcal{E}$, we call the polyhedron V consisting of all points of n-space not further from O than from any other point of $\mathcal{E}$ the *Voronoĭ* (or *Dirichlet*) *region of the point* O. The Voronoĭ region V of a point O may be constructed as follows: Join the point O to every point of the system $\mathcal{E}$ by means of straight line segments, and consider the hyperplanes ($(n-1)$-dimensional affine subspaces) perpendicular to these line segments and passing through their midpoints (which we shall call *Voronoĭ hyperplanes of the system $\mathcal{E}$ with respect to the point* O); then the n-dimensional convex polyhedron that these hyperplanes delineate about O will be the Voronoĭ region V of O relative to the system $\mathcal{E}$.

[195]The *degree* of a point relative to a sphere is defined as the square of the length of the tangent line from this point to the sphere.

LEMMA. *The polyhedra V are nonoverlapping, fill the whole of n-space, and are adjacent along complete $(n-1)$-dimensional faces. The partition of n-space by the Voronoĭ regions V is uniquely determined by the system $\mathcal{E}$.*

The fact that the polyhedra V of all the points of $\mathcal{E}$ fill n-space without overlapping is clear, since for every point of n-space there is some point of $\mathcal{E}$ as close to it as any other point of $\mathcal{E}$, and of course no point of n-space can be simultaneously closer to each of two points of $\mathcal{E}$ than to the other. Essentially this argument also shows that the partition of n-space by the polyhedra V of the system $\mathcal{E}$ is uniquely determined by that system, since each point of n-space, if not on the boundary of a polyhedron V, is in the interior of the Voronoĭ region of the unique point of $\mathcal{E}$ closest to it. That adjacent polyhedra share a full $(n-1)$-dimensional face can be seen as follows: If a polyhedron V of a point A of the system $\mathcal{E}$ were adjacent along parts of some $(n-1)$-face with the polyhedra of two points B and C of $\mathcal{E}$, then these two points would have to lie on the one and the same line segment starting from A with the $(n-1)$-face in question as perpendicular bisector, and ending at both B and C, so that B and C cannot in fact be distinct.

We shall write briefly *partition* V for the partition of n-space by the Voronoĭ regions corresponding to a system $\mathcal{E}$.

In the next section we provide a geometric explanation of what is at foot in Voronoĭ's lemma asserting for each n the finiteness of the number of inequivalent types of partitions L and V of n-space in the case of point systems $\mathcal{E}$ that are lattices.

4. The finiteness of the number of types of partitions L and V for lattices of prescribed dimension. From now on we shall assume that the system $\mathcal{E}$ is an n-dimensional lattice in n-dimensional Euclidean space. Obviously lattices satisfy the basic general conditions 1) and 2) we impose on all our point systems. Let L_1 be any polyhedron of the partition L of the lattice, and consider any $n+1$ vertices of L_1 determining an n-dimensional simplex; this will then be a simplex of the lattice $\mathcal{E}$.[196] It is not difficult to show that from this it follows that the volume of the parallelepiped determined by this simplex is equal to the volume v_0 of a fundamental parallelepiped of the lattice or to some integer multiple of it; hence the volume of the simplex itself, and therefore certainly of the polyhedron L_1, is greater than or equal to $\frac{v_0}{n!}$. Consider now the star consisting of all polyhedra L of the lattice $\mathcal{E}$ incident with a given point O of the lattice $\mathcal{E}$; these will in fact correspond to the vertices of the polyhedron V defined by the point O. Let N denote the number of polyhedra in this star.

We now shrink this star of polyhedra L linearly by a factor of one half by means of a homothety centred at O; all vertices of those polyhedra except for O will then lie on the boundary of the polyhedron V. This can be seen as follows: Observe first that every vertex other than O of the original polyhedra forming this star, is the centre of some polyhedron V' adjacent to the polyhedron V of the point O, along a face of some dimension. However, the midpoint between O and the centre of such a polyhedron V' is a point of their shared face, namely its centre, since the lattice $\mathcal{E}$ has this midpoint as a centre of symmetry (in fact the midpoint between any two points of a lattice is a centre of symmetry of the lattice), so that this midpoint is

[196] That is, of a simplicial decomposition of n-space with vertices the points of $\mathcal{E}$. *Trans.*

the centre of symmetry for the figure made up of the two polyhedra V adjacent along this face.

Thus all vertices except O of the polyhedra L comprising the star will, after the aforementioned homothety has been applied, lie on the boundary of the polyhedron V, and hence, as convex hulls of their vertices, these contracted polyhedra will then be contained in the polyhedron V. It follows that the volume of the star contracted linearly by the factor $\frac{1}{2}$ is at most that of the polyhedron V, which is equal to v_0, the volume of a fundamental parallelepiped of $\mathcal{E}$. We infer that

$$\frac{v_0}{n!2^n}N < v_0,$$

whence $N < n!2^n$. Thus the number of polyhedra L comprising a star of the partition L associated with an n-dimensional lattice $\mathcal{E}$, is bounded in terms of the dimension n alone of that lattice. Since the polyhedra L forming the star correspond one-to-one to the vertices of the polyhedron V, it follows that the number of these vertices is likewise bounded in terms of n alone. Since the number of distinct topological types of complexes in the form of a polyhedron of given dimension n with a bounded number of vertices is clearly finite, and in the case of a lattice the topological type of the polyhedron V fully determines the type of the partition of n-space by the polyhedra V, and hence also that of the partition by the polyhedra L, we have the following

THEOREM. *The number of distinct topological types of partitions L and V arising from a lattice of dimension n, is bounded in terms of n.*

We shall show below that for $n = 1, 2, 3$ there is essentially only one topological type of each of the partitions L and V, while for $n = 4$ there are three, all other types of partition being limiting cases of these. It is unknown how many distinct types of partition there are for $n = 5$, and all the more so for larger n.

We now consider Voronoĭ's further results in the light of the hollow sphere.

5. Regions of types. We shall now examine the question as to where in coefficient space those quadratic forms are situated that correspond to n-frames[197] associated in a particular manner with a star of polyhedra of lattices having partition L of a given type. In greater detail the question is as follows: Let (a_{ij}) be a point of the space of coefficients of positive n-ary quadratic forms; as explained earlier each such point corresponds to an n-frame, and hence determines a lattice. We shall assume that this lattice is "general".[198] Its partition L will then be of a definite topological type, and the n-frame of (a_{ij}) will be associated in a perfectly definite manner with a star of this partition. We now let the point (a_{ij}) move from its original position in coefficient space; its lattice will then also vary continuously, but the type of the corresponding partition L will for small perturbations of the point (a_{ij}) from its original position remain the same, since for all inhomologous pairs of contiguous simplices of the partition L, adjacent along $(n-1)$-faces, the condition that no vertex of either simplex of each such pair lie inside the sphere circumscribed about the other will continue to hold.

The partition of n-space into simplices, with a star of which the n-frame of (a_{ij}) continues to be associated in the prescribed manner throughout the movement of

[197]In the original the expression "n-sider" is used here and later. *Trans.*

[198]A lattice is called *general* if all of its polyhedra L are simplices. It can be shown that a lattice that is not general is always a limit of general lattices.

this point, will, by the above Fundamental Lemma on Partitions, remain throughout a partition L of the lattice. Let L_1, L_2 be, as above, a pair of inhomologous adjacent simplices, with vertices $(l_i^{(1)}), (l_i^{(2)}), \ldots, (l_i^{(n+1)})$, and $(l_i^{(2)}), (l_i^{(3)}), \ldots, (l_i^{(n+1)})$, $(l_i^{(n+2)})$ $(i = 1, 2, \ldots, n)$, respectively, where the coordinates $l_i^{(m)}$ are relative to the n-frame of (a_{ij}). If the centre of the sphere (L_2) has coordinates ξ_i relative to the same frame, and radius r_2, then for any point (t_i) (always with coordinates taken relative to the same n-frame) on the sphere (L_2) we have

$$f(t_i - \xi_i) = r_2^2, \quad m = 2, 3, \ldots, n+2,$$

where f denotes the quadratic form $\sum\sum a_{ij}x_ix_j$. This can be rewritten as

$$f(t_i) - \frac{\partial f}{\partial t_1}\xi_1 - \cdots - \frac{\partial f}{\partial t_n}\xi_n + f(\xi_i) - r_2^2 = 0.$$

This will hold in particular for the vertices of the simplex L_2:

$$f(l_i^{(m)}) - \frac{\partial f}{\partial l_i^{(m)}}\xi_1 - \cdots - \frac{\partial f}{\partial l_n^{(m)}} + f(\xi_i) - r_2^2 = 0, \;\; m = 2, 3, \ldots, n+2.$$

Hence the sphere (L_2) has equation

$$\text{(1)} \qquad \begin{vmatrix} f(t_i) & \frac{\partial f}{\partial t_1} & \frac{\partial f}{\partial t_2} & \cdots & \frac{\partial f}{\partial t_n} & 1 \\ f(l_i^{(2)}) & \frac{\partial f}{\partial l_1^{(2)}} & \frac{\partial f}{\partial l_2^{(2)}} & \cdots & \frac{\partial f}{\partial l_n^{(2)}} & 1 \\ \vdots & \vdots & \vdots & \vdots & \vdots & \vdots \\ f(l_i^{(n+2)}) & \frac{\partial f}{\partial l_1^{(n+2)}} & \frac{\partial f}{\partial l_2^{(n+2)}} & \cdots & \frac{\partial f}{\partial l_n^{(n+2)}} & 1 \end{vmatrix} = 0.$$

If we reorder the vertices of the simplex L_2 so that the minor of the entry $f(t_i)$ in (1) is positive, we obtain on replacing (t_i) by $(l_i^{(1)})$ a precise condition for the vertex $(l_i^{(1)})$ to lie outside the sphere (L_2), namely

$$\text{(2)} \qquad \begin{vmatrix} f(l_i^{(1)}) & \frac{\partial f}{\partial l_1^{(1)}} & \frac{\partial f}{\partial l_2^{(1)}} & \cdots & \frac{\partial f}{\partial l_n^{(1)}} & 1 \\ f(l_i^{(2)}) & \frac{\partial f}{\partial l_1^{(2)}} & \frac{\partial f}{\partial l_2^{(2)}} & \cdots & \frac{\partial f}{\partial l_n^{(2)}} & 1 \\ \vdots & \vdots & \vdots & \vdots & \vdots & \vdots \\ f(l_i^{(n+2)}) & \frac{\partial f}{\partial l_1^{(n+2)}} & \frac{\partial f}{\partial l_2^{(n+2)}} & \cdots & \frac{\partial f}{\partial l_n^{(n+2)}} & 1 \end{vmatrix} > 0,$$

with an analogous condition for the vertex $(l_i^{(n+2)})$ of L_2 to lie outside the sphere (L_1). The inequality (2) may be rewritten as

$$\text{(3)} \qquad (-1)^n \begin{vmatrix} 1 & 0 & 0 & \cdots & 0 \\ 0 & 1 & 0 & \cdots & 0 \\ 0 & 0 & a_{11} & \cdots & a_{1n} \\ 0 & 0 & a_{21} & \cdots & a_{2n} \\ \vdots & \vdots & \vdots & \vdots & \vdots \\ 0 & 0 & a_{n1} & \cdots & a_{nn} \end{vmatrix} \times \begin{vmatrix} f(l_i^{(1)}) & l_1^{(1)} & \cdots & l_n^{(1)} & 1 \\ f(l_i^{(2)}) & l_1^{(2)} & \cdots & l_n^{(2)} & 1 \\ \vdots & \vdots & \vdots & \vdots & \vdots \\ f(l_i^{(n+2)}) & l_1^{(n+2)} & \cdots & l_n^{(n+2)} & 1 \end{vmatrix} > 0.$$

Since the first determinant in this product, as the discriminant of the form f, is positive, the condition (2) simplifies to

$$(4)\qquad \rho = (-1)^n \sum\sum a_{ij} \begin{vmatrix} l_i^{(1)} l_j^{(1)} & l_1^{(1)} & l_2^{(1)} & \cdots & l_n^{(1)} & 1 \\ \vdots & \vdots & \vdots & \vdots & \vdots & \vdots \\ l_i^{(n+2)} l_j^{(n+2)} & l_1^{(n+2)} & l_2^{(n+2)} & \cdots & l_n^{(n+2)} & 1 \end{vmatrix} > 0.$$

Note that this inequality is linear in the coefficients of the form f. Let $\rho_1, \rho_2, \dots, \rho_\kappa$ be the linear expressions corresponding to the left-hand side of (4) for a full set, numbering κ say, of inhomologous pairs of simplices of the partition, with each pair adjacent along an $(n-1)$-face; a necessary and sufficient condition for this partition to be the partition L of the corresponding lattice is therefore that

$$(5)\qquad \rho_1 > 0, \rho_2 > 0, \dots, \rho_\kappa > 0.$$

Voronoĭ calls the quantities $\rho_1, \rho_2, \dots, \rho_\kappa$, linear in the a_{ij}, the *regulators* of the respective pairs of simplices.

We thus have the following

THEOREM. *A region of the space of coefficients a_{ij} corresponding to those positive forms whose lattices yield partitions L of a single topological type, and with n-frames each corresponding in the prescribed manner to a star of the corresponding partition, forms a convex gonohedron (many-faceted angle) given by (5), with vertex at the origin, which is, moreover, $\left(\frac{n(n+1)}{2}\right)$-dimensional (by virtue of the above discussion concerning small perturbations of the point (a_{ij}) from its initial position).*

For any two distinct such gonohedra, no interior point of one can lie in the other, since each such point (a_{ij}) determines a simplicial decomposition L defining the topological type of that decomposition (partition), and an n-frame associated in the prescribed manner with a star of this partition, i.e., the point completely determines the gonohedron belonging to that type. Conversely, provided it corresponds to a general lattice (see the footnote above) each point of the coefficient space of positive forms is an interior point of a completely determinate such gonohedron. Any point (a_{ij}) corresponding to a "special" lattice (i.e., not general), has points corresponding to general lattices arbitrarily close to it, i.e., points corresponding to special lattices are boundary points of the gonohedra, whose interiors consist of general points. Conversely, for any point on the boundary of such a gonohedron, at least one of the inequalities (5) becomes an equality, and hence for at least one pair of adjacent simplices their circumscribed spheres coincide, yielding a hollow sphere with more than $n+1$ points of the lattice on its boundary, so that the corresponding partition L is not simplicial, and the lattice is therefore special. We thus have the following

THEOREM. *The gonohedra of general types fill the whole region of positive forms and are nonoverlapping. Moreover their interior points correspond to general lattices, and their boundary points to special lattices.*

Consider now any unimodular integral transformation of the variables x_i, applied to all positive forms $\sum\sum a_{ij}x_ix_j$. The forms lying in a region belonging to a type will then be transformed to equivalent ones, corresponding to lattices congruent to those of their preimages, but with transformed fundamental frames (however, obviously as before associated in the uniquely prescribed manner with stars of their lattices). The region belonging to the type of the transforms of the forms of the given type will be said to be *equivalent* to the region belonging to that given type. Conversely it is clear that if the topological types of two partitions L belonging to two different regions, coincide, then the difference must lie in the manner in which the fundamental frames are associated with stars of these partitions. It follows that every unimodular integral transformation of forms $\sum\sum a_{ij}x_ix_j$ permutes the set of gonohedra belonging to types, sending each gonohedron to one equivalent to it. Since, as we have seen, the number of distinct topological types of partitions L arising from lattices of given dimension n is bounded in terms of n alone, it follows that the number of inequivalent regions of types is likewise bounded.

We shall now elucidate by means of the concept of the hollow sphere what exactly is involved in Voronoĭ's algorithm for reconstructing simplices.

6. The algorithm for finding the regions of types adjacent to a given region of a type. Imagine an interior point of some region Δ belonging to a given type, moving continuously within Δ till it reaches an interior point M of one of $\left(\frac{n(n+1)}{2}-1\right)$-dimensional faces of the region. Each interior point of such a face corresponds to a lattice where the spheres circumscribing the simplices L comprising an adjacent pair associated with that face of Δ, coincide. This will be the case for all those inhomologous pairs of adjacent simplices whose regulator ρ is such that the equation $\rho=0$ defines the $\left(\frac{n(n+1)}{2}-1\right)$-dimensional plane of the face in question, i.e., those pairs whose regulators, considered as linear expressions in the coefficients a_{ij} of the forms, are proportional to one another.

Consider one such pair of simplices L corresponding to the $\left(\frac{n(n+1)}{2}-1\right)$-dimensional face of Δ in question, with the point M in its interior. Denote by L_0 and L^0 these simplices, by $l_{i1}, l_{i2}, \dots, l_{in}$ $i=1,2,\dots,n$, the coordinates, relative to the fundamental n-frame of the lattice, of the vertices of their shared face, and by (l_{i0}) and (l'_i) the $(n+1)$st vertices of L_0 and L^0 respectively, which we shall call the *opposite* vertices of the pair. Thus all $n+2$ vertices $(l_{i0}), (l_{i1}), (l_{i2}), \dots, (l_{in}), (l'_i)$ lie on the boundary of the single empty sphere (L_0). We have the following

THEOREM. *With M as above an interior point of an $\left(\frac{n(n+1)}{2}-1\right)$-dimensional face of a region Δ belonging to some type, and L_0, L^0 an adjacent pair of simplices of the corresponding lattice E whose circumscribed spheres coincide, there are no points of that lattice on the boundary of that sphere (L_0) other than the $n+2$ vertices of the pair of simplices.*

For the proof, suppose there is a point (l''_i) of the lattice E on the boundary of the sphere (L_0) different from the $n+2$ vertices of the pair of simplices in question. The point (l'_i) lies outside or on the boundary of the sphere (L_0) if and only if

$$\rho' = (-1)^n \sum\sum a_{ij} \begin{vmatrix} l_{i0}\cdot l_{j0} & l_{10} & \cdots & l_{n0} & 1 \\ \vdots & \vdots & \vdots & \vdots & \vdots \\ l_{in}\cdot l_{jn} & l_{1n} & \cdots & l_{nn} & 1 \\ l'_i\cdot l'_j & l'_1 & \cdots & l'_n & 1 \end{vmatrix} \geq 0, \tag{6}$$

and similarly the point (l_i'') lies outside or on the boundary of (L_0) if and only if

$$(7)\qquad \rho'' = (-1)^n \sum\sum a_{ij} \begin{vmatrix} l_{i0}\cdot l_{j0} & l_{10} & \cdots & l_{n0} & 1 \\ \vdots & \vdots & \vdots & \vdots & \vdots \\ l_{in}\cdot l_{jn} & l_{1n} & \cdots & l_{nn} & 1 \\ l_i''\cdot l_j'' & l_1'' & \cdots & l_n'' & 1 \end{vmatrix} \geq 0.$$

For the coordinates a_{ij} of the point M, in view of the assumption that the point (l_i'') lies on the sphere (L_0), both of the inequalities (6) and (7) become equalities. The coefficients of the a_{ij} in the left-hand side of (6) cannot all vanish, since otherwise the $n+2$ points $(l_{i0}), (l_{i1}), \ldots, (l_{in}), (l_i')$ would remain cospherical under arbitrary affine transformations (produced by changing the coefficients a_{ij}), which is clearly impossible. This argument of course also applies to (7). Hence (6) and (7) changed to equalities, define in the coefficient space $\left(\frac{n(n+2)}{2} - 1\right)$-dimensional planes (i.e., hyperplanes) passing through the point M.

We shall show that under the assumption that the point (l_i'') is different from the $n+2$ vertices of the simplices L_0 and L^0, these two hyperplanes are distinct, i.e., the ordered $\left(\frac{n(n+1)}{2}\right)$-tuples of their determinantal coefficients are not proportional. To this end, in place of the fundamental n-frame of the lattice we choose the new coordinate n-frame consisting of the vectors $\mathfrak{b}_1, \mathfrak{b}_2, \ldots, \mathfrak{b}_n$ from the point (l_{i0}) to the points $(l_{i1}), (l_{i2}), \ldots, (l_{in})$ respectively, i.e., the vectors defined by the edges of the simplex L_0 leading from the vertex (l_{i0}); this n-frame need not be fundamental for the lattice E, but for our purposes this is immaterial. With respect to this frame the points $(l_{i0}), (l_{i1}), \ldots, (l_{in})$ (in the old coordinates) have respective coordinates

$$\left.\begin{array}{c} (0,0,0,\ldots,0) \\ (1,0,0,\ldots,0) \\ (0,1,0,\ldots,0) \\ \cdots\cdots\cdots \\ (0,0,0,\ldots,1) \end{array}\right\} (n+1).$$

If we denote the new coordinates of the points (l_i') and (l_i'') by s_i' and s_i'' respectively, then in the equations for the hyperplanes obtained from (6) and (7), the determinantal coefficients of the new variables $b_{ij} = \mathfrak{b}_i \cdot \mathfrak{b}_j$ [199] become respectively $s_i's_j'$ and $s_i''s_j''$ for $i \neq j$, and $\begin{vmatrix} (s_i')^2 & s_i' \\ 1 & 1 \end{vmatrix}$ and $\begin{vmatrix} (s_i'')^2 & s_i'' \\ 1 & 1 \end{vmatrix}$ for $i = j$. These hyperplanes coincide if and only if the coefficients in their equations are proportional, i.e., if and only if

$$(8)\qquad s_i's_j' = \lambda s_i''s_j'', \quad \text{for } i \neq j,$$

$$(9)\qquad (s_i')^2 - s_i' = \lambda((s_i'')^2 - s_i''), \text{ for } i = 1, 2, \ldots, n,$$

for some nonzero proportionality factor λ.

Observe that at least two of the coordinates s_i' must be nonzero, since if they were all zero, the point (l_i') $(=(s_i')$ in the new coordinates) would coincide with (l_{i0}), since this point is $(0, 0, \ldots, 0)$ in the new coordinates, and if just one of the s_i' were nonzero, then (s_i') would be a multiple of one of the other n-tuples in the above list, and then would have to coincide with that n-tuple since it lies on the sphere (L_0). Thus if m denotes the number of nonzero s_i', then $m \geq 2$. Clearly we may without loss of generality assume that $s_1', s_2', \ldots, s_m' \neq 0$; $s_{m+1}' = \cdots = s_n' = 0$.

[199] This is the dot product; see the Commentary in the chapter on Korkin. *Trans.*

Then from (8) it follows that likewise $s''_1, s''_2, \ldots, s''_m \neq 0$; $s''_{m+1} = \cdots = s''_n = 0$. For any k, l satisfying $1 \leq k < l \leq m$, we have from (8) that

$$\lambda = \frac{s'_k s'_l}{s''_k s''_l}.$$

Replacing this expression for λ in (9) for $i = k$ and $i = l$, we obtain

$$s'_l s''_k + (1 - s'_k) s''_l = s'_l,$$
$$(1 - s'_l) s''_k + s'_k s''_l = s'_k.$$

Regarding this as a linear system of equations in the unknowns s''_k and s''_l, we have that the determinant of its coefficient matrix is $s'_k + s'_l - 1$. If this is not zero, then the solution of the system is $s''_k = s'_k, s''_l = s'_l$, whence $\lambda = 1$. It then follows from (8) with $i = k$ and j varying, that $s''_i = s'_i$ for all $i = 1, 2, \ldots, n$, i.e., the points (s'_i) and (s''_i) coincide, contradicting our initial assumption.

There remains the case that for every k, l satisfying $1 \leq k < l \leq m$, we have $s'_k + s'_l - 1 = 0$. In this case $s'_l = 1 - s'_k$, whence, replacing $1 - s'_k$ by s'_l in the first equation of the above system, we obtain $s'_l s''_k + s'_l s''_l = s'_l$, giving $s''_k + s''_l - 1 = 0$. If $m = 2$, then the equality $s'_1 + s'_2 = 1$ implies that the point $(s'_i) = (s'_1, s'_2, 0, \ldots, 0)$ lies on the line segment joining the tips of the vectors $\mathfrak{b}_1$ and $\mathfrak{b}_2$, so that in order for it to lie on the boundary of the sphere (L_0) it would have to coincide with one of those endpoints, contrary to assumption. If $m > 2$ then from the equalities $s'_k + s'_l = 1$ and $s''_k + s''_l = 1$ for all k, l satisfying $1 \leq k < l \leq m$, it follows easily that $s'_1 = s'_2 = \cdots = s'_m = \frac{1}{2}$, and $s''_1 = s''_2 = \cdots = s''_m = \frac{1}{2}$, and again the points (s'_i) and (s''_i) coincide. We conclude that if the point (l''_i) is a point of the sphere (L_0) different from the $n + 2$ vertices of the pair of simplices L_0, L^0, then the $\left(\frac{n(n+1)}{2} - 1\right)$-dimensional hyperplanes in coefficient space defined by (6) and (7) (changed to equalities) having M as a point in common, do not coincide, but merely intersect. However, this contradicts the assumption that M is an interior point of the face in question of the region Δ.

THEOREM. *If M is an interior point of a face of a region Δ, belonging to some type, then certain empty spheres (L) of the partitions corresponding to the points of the region Δ become identified as M is approached from within Δ. Those spheres arising in this way from the merging of two empty spheres and circumscribed about an adjacent pair of simplices L_0 and L^0, result from the merging of precisely those pairs of spheres circumscribed about adjacent simplices L_0 and L^0 which correspond to the face of the region Δ in question, together with those circumscribing "supplementary" simplices each having as one of its vertices one of the diametrically opposite vertices of such an adjacent pair, and as base opposite this vertex one of the faces of the other simplex of the pair visible from that vertex, i.e., not hidden behind that simplex. Taken together all these simplices form a convex polyhedron K with $n + 2$ vertices, inscribed in the sphere in question.*

To see this, observe first that each of the supplementary simplices is a simplex L of the partition, since for every face of the simplex L_0 there is a simplex L^* of the partition, adjacent along that face, and if that face is one visible from the far vertex of L^0, then the supplementary simplex T^* with that as a vertex and the face of L_0 in question as base opposite this vertex must be the simplex L^*. For, if the vertex of L^* not in the face of L_0 in question were outside the sphere L_0, then since the simplices L^* and T^* have their base in common, the vertex of T^* opposite that base would lie inside the sphere (L^*), which would then not be empty; hence the vertex of L^* not in the face it shares with L_0 must lie on the empty sphere (L_0), and then by the preceding theorem must actually coincide with a vertex of L_0. For the converse, note first that since by the preceding theorem there are no points of the lattice on the boundary of the sphere (L_0) other than the $n+2$ vertices of the adjacent pair of simplices L_0, L^0, any simplex whose sphere coincides with (L_0) must have as its vertices some $n+1$ of these. From this, and the obvious fact that the simplices L_0 and L^0 together with all their supplementary simplices form the convex hull K of these $n+2$ vertices, we infer that these simplices together account for all simplices L whose spheres coincide with the sphere (L_0).

REMARK. It may happen that a single hyperplane in coefficient space corresponds to mergings of spheres (L) corresponding to different inhomologous pairs of adjacent simplices of the partition L, not forming part of the same K. Hence it may occur (but does not for $n = 2, 3$ or 4) that there are several inhomologous such polyhedra K corresponding to a single interior point M of a given face of the region Δ. However, we can find all of these by examining all inhomologous pairs of adjacent simplices, which are finite in number, and which all form part of the star of the point O,[200] moreover in $n+2$ ways, corresponding to the different vertices of the pair. Let $K_1, K_2, \dots, K_\tau$ comprise a full set of inhomologous such polyhedra. The partition corresponding to the point M is not simplicial, but contains polyhedra made up of simplices L coming from partitions corresponding to interior points of the region Δ, some of which are combined into the polyhedron K_1, some others into K_2, and so on. Thus $K_1, K_2, \dots, K_\tau$ form a full set of inhomologous nonsimplicial polyhedra L of the partition corresponding to the point M.

THEOREM. *Consider a point M' in a neighbourhood of M and on the exterior side relative to Δ of the $\left(\frac{n(n+1)}{2} - 1\right)$-dimensional face of the region Δ on which M lies, and suppose that M' is sufficiently close to M that no further pairs of empty spheres coincide in passing from M to M' apart from those that coincided on reaching M through interior points of the region Δ. Then all simplices L of the initial partition (i.e., corresponding to M) remain simplices L in passing to the new partition corresponding to M', and the nonsimplicial polyhedra K of the initial partition again break up into simplices L' of the new partition, but in a manner different from that in which each K was formed of simplices coming from the partition corresponding to an interior point of Δ as it approached M. In fact each polyhedron K decomposes into simplices having (in the notation used in the proof of the preceding theorem) the point $S' = (s'_i)$ as a vertex*[201] *and as respective bases those faces of the simplex L_0 not visible from the point S'. All of these simplices will have the line segment OS' as an edge. After this reconstruction*[202] *each polyhedron K becomes decomposed into simplices of the partition corresponding to the point M' of the region Δ' adjacent to the region Δ in the vicinity of the point M.*

For the proof consider to begin with the following necessary and sufficient condition for the vertex $S' = (s'_i)$ of the simplex L^0 not to lie in the interior of the

[200] Here $O = (0, 0, \dots, 0)$ in the new coordinates. *Trans.*

[201] The s'_i and also L_0, L^0, etc. are to be considered as varying with a point moving from the interior of Δ to M and beyond to M'. *Trans.*

[202] Or (simplicial) redecomposition. *Trans.*

sphere circumscribing the simplex L_0:

$$\begin{vmatrix} f(s_i') & s_1' & s_2' & \cdots & s_n' & 1 \\ f(1,0,\dots,0) & 1 & 0 & \cdots & 0 & 1 \\ f(0,1,\dots,0) & 0 & 1 & \cdots & 0 & 1 \\ \vdots & \vdots & \vdots & \vdots & \vdots & \vdots \\ f(0,0,\dots,1) & 0 & 0 & \cdots & 1 & 1 \\ f(0,0,\cdots,0) & 0 & 0 & \cdots & 0 & 1 \end{vmatrix} \geq 0.$$

If we denote the minor of the bottom right-hand entry by d, this condition simplifies to

$$d \geq 0.$$

At M' the point (s_i') has gone inside the sphere (L_0),[203] so that at M' we have $d < 0$. However, the face P_0 of the simplex L_0 opposite its vertex O is visible from the point (s_i') from the outside, since it is the face shared by the simplices L_0 and L^0. Of the remaining n faces of L_0 at least two are not visible from (s_i') (since if n of the $n+1$ faces of L_0 were visible from S', then the vertex incident with these n faces would lie inside the polyhedron K, and therefore certainly inside the sphere (L_0), which is not possible), but generally speaking it could happen that all n of these faces are not visible from S'. Each of these faces has as its vertices O $(= (0,0,\dots,0))$ together with some $n-1$ of the points

$$(1,0,\dots,0),(0,1,\dots,0),\dots,(0,0,\dots,1),$$

and of course is determined by these. Denoting by P_k that face not having the kth of the latter points as a vertex, we clearly have that P_k is not visible from (s_i') precisely if the kth coordinate s_k' of (s_i') is positive.

We next show that under the condition $d < 0$, the spheres S_k circumscribing the simplices with (s_i') as a vertex in common and bases opposite this vertex the faces P_k with $s_k' > 0$, are such that the kth one, S_k, does not contain the remaining $(n+2)$nd vertex of the polyhedron K, i.e., the vertex $(0,0,\dots,0,1,0,\dots,0)$ where the kth coordinate is 1. This follows from the fact that since the minor of the entry $f(0,0,\dots,0,1,0,\dots,0)$ of the submatrix whose determinant is the minor d, is s_k', the condition for the vertex $(0,0,\dots,0,1,0,\dots,0)$ to lie outside the sphere circumscribing the simplex without this point as vertex is precisely $d < 0$, the condition we are assuming holds. Thus the sphere circumscribing each of these simplices does not contain that vertex of K different from the $n+1$ vertices of the simplex it circumscribes, and by taking M' sufficiently close to M for these spheres to differ by arbitrarily little from (L_0), we can ensure also that they are empty. It follows that these new simplices are the simplices L' of the lattice corresponding to the point M', concluding the proof of the theorem.

We give a further explanation of the situation. Each ray emanating from the point (l_i') and meeting the simplex L_0, enters it through some face visible from the point (l_i') and exits through a face not visible from that point. We call the first kind of face *upper* and faces of the second kind *lower*, relative to the point (l_i'). In general there may be in addition faces of the simplex L_0 which when extended pass through the point (l_i'); we call such faces *vertical* relative to (l_i'). The preceding

[203] Presumably because the region Δ' containing M' belongs to a different topological type from Δ; if (s_i') had gone back outside (L_0), the partition corresponding to M' would have the same type as that corresponding to M. *Trans.*

theorem then tells us the following: The simplex L_0 together with those simplices with shared vertex (l_i') and faces opposite this vertex the upper faces of L_0 (which include the simplex L^0) form the initial simplicial decomposition of the polyhedron K.[204] On the other hand the simplices sharing the vertex (l_i') with bases the lower faces of the simplex L_0 form the new decomposition of the polyhedron K.

THEOREM. *Regions of types Δ are adjacent along complete $\left(\frac{n(n+1)}{2} - 1\right)$-dimensional faces.*

To see this, note first that for different points of a single $\left(\frac{n(n+1)}{2} - 1\right)$-dimensional face of a region Δ, the polyhedra K corresponding to merged spheres (L) will be the same, since they are characterized by those hyperplanes determining noncontainment arising from various inhomologous pairs of adjacent simplices of the partition L of the type to which Δ belongs, and these coincide with the hyperplane of the face in question. On the other hand under an affine transformation of the lattice, the upper faces of the simplex L_0 relative to the point (l_i') remain upper, lower faces lower, and vertical faces vertical, in every polyhedron K. Hence by the preceding theorem, in moving to a point M' close to M and lying on the side away from Δ of the face in question, all polyhedra K undergo the same simplicial reconstruction independently of the position of the point M in the interior of that face of Δ.

[204]That is, corresponding to points in Δ. *Trans.*

Ivan Matveevich Vinogradov (1891–1983)

Ivan Matveevich VINOGRADOV

Ivan Matveevich Vinogradov was born on the 14th September (new style), 1891. His father, Matveĭ Avraamyevich, was priest in the graveyard church at Milolyub, a village in the Velikie Luki district, Pskov province. The boy showed talent at drawing, and his parents, instead of enrolling him in a religious school, as was customary for the clergy, sent him (in 1903) to a secular nonclassical secondary school in the district centre, Velikie Luki, to which his father moved from Milolyub with his family on obtaining a living as priest in the Pokrovskaya church.

In 1910 Vinogradov completed his studies at the school and in the same year enrolled in the physics and mathematics faculty of St. Petersburg University, from which he graduated in 1914. In 1915 he wrote his first work, on the sum of the values of the Legendre symbol, and was kept on at the university to be trained for the professoriat. From 1918 to 1920 Ivan Matveevich lived in Perm, where at first he was docent and then professor at Perm University giving lectures in number theory.

In November 1920 Vinogradov returned to St. Petersburg to take up the position of professor at the Polytechnic Institute, and then, at the university, that of docent, and, from 1925, professor and head of the department of number theory and probability theory. In January 1929 Ivan Matveevich was elected a full member of the Academy of Sciences. From 1930 to 1933 he served as director of the Demographic Institute of the Academy of Sciences, and from 1932 as director of the Mathematics Institute of the Academy. Vinogradov was a highly respected member of many domestic and foreign scientific societies: in 1942 he became a member of the Royal Society of London, in 1946 a member of the Paris Academy of Sciences and the US National Academy, etc.

Among the most powerful of contemporary[205] mathematicians, Vinogradov devoted himself to the analytic theory of numbers. This branch of science, originating in the 19th century, has by now become very highly developed; in western Europe such prominent mathematicians as Hardy, Littlewood, van der Corput, and Heilbronn were working[206] on problems from analytic number theory. These scholars all followed essentially the path pioneered by earlier mathematicians, in that they applied the highly evolved methods of analysis—chiefly the theory of functions of a complex variable—to analytic functions related in one way or another to the whole numbers.

Vinogradov introduced new methods into this area of mathematics, which one might call elementary in the sense that they make minimal use of analytic means; in his approach the centre of gravity of the subject is transferred back to number

[205]As of 1947. *Trans.*

[206]In the original the present tense is used here and elsewhere, as was appropriate at the time of writing (1947). *Trans.*

theory, which thus recaptures its arithmetic essence while preserving the flexibility and generality of analysis. The power of Vinogradov's methods and the depth of his understanding of the essence of this new theory of numbers, can be gauged from the fact that the results of his research on a wide range of questions are approximately on the same level as those of the above-mentioned western scholars, while his work on a few particular problems (those of Waring and Goldbach) is far superior to theirs.

However, it must not be thought that Vinogradov's methods are in some sense isolated from the rest of number theory; on the contrary, the proof of one of his most outstanding theorems, settling a conjecture closely related to that of Goldbach, synthesizes his final results on the estimation of trigonometric sums with results of western mathematicians on the distribution of primes in arithmetic progressions.

In terms both of the style and methodology of his work, Vinogradov must be included in the St. Petersburg school. Chebyshev, Voronoĭ and Vinogradov are the three greatest arithmeticians of that school.

Analytic number theory was in significant part the creation of the St. Petersburg school. The first important results in this area are to be found in the works of Euler.[207] The next big step forward in analytic number theory was made by Chebyshev in his ingenious work on the distribution of the primes, serving as the starting point for later work of Riemann and Hadamard. In his article on the hyperbola, Voronoĭ provided the first example of an arithmetic approach involving partitioning an interval of integration into elementary segments. Finally, Vinogradov invented a remarkable new arithmetic approach to problems of analytic number theory (the arithmetic of inequalities), allowing him to make extraordinary discoveries constituting some of the glories of Soviet science.

In 1941 Ivan Matveevich was awarded the Stalin prize first class for his book "A new method in the analytic theory of numbers", and in 1945 by decree of the Presidium of the Supreme Soviet of the USSR was accorded the title of Hero of Socialist Labour.

[207]Dirichlet might perhaps have been mentioned next! *Trans.*

Works of Vinogradov from the First Period of His Mathematical Activity

In addressing the task of expounding Vinogradov's main mathematical achievements, it is appropriate for us to make the following remark. As noted above, the analytic theory of numbers is at the time of writing (1947) one of the most active and productive areas of research in western European science. For this reason in its origins the work of Vinogradov is closely linked to that of prominent contemporary foreign mathematicians, such as Hardy, Littlewood, van der Corput, and others. Since our goal is to acquaint the reader with the scientific character of Vinogradov's work, we shall omit from our exposition the details of the relevant work of other mathematicians, confining ourselves to giving exact references to the literature. Proofs of auxiliary lemmas utilizing standard analytic means will likewise be omitted.

We shall use the following notation:

$\theta, \theta', \dots$ numbers satisfying $|\theta| \leq 1,\ |\theta'| \leq 1, \dots$;

ϵ an arbitrarily small but fixed positive number;

$[z]$ the integer part of the real number z, i.e., the integer defined by the inequalities $[z] \leq z < [z] + 1$;

$\{z\}$ the fractional part of z, i.e., the difference $z - [z]$.

(z) the distance from z to the nearest integer;

(a, b) the highest common factor of the integers a, b.

Finally, if $x, y, \dots$ denote the main independent variables figuring in some question of interest, and A and $B > 0$ quantities depending on $x, y, \dots$, then the formula

$$A = O(B) \text{ or } A \ll B$$

means that for all indicated values of the variables $x, y, \dots$, one has

$$|A| < KB$$

for some K independent of $x, y, \dots$. The quantity K may be an absolute constant or depend on some other parameters; in situations where confusion might arise the dependence of K will be indicated explicitly.

The earliest works of Vinogradov relate to problems in analytic number theory posed by 19th century mathematicians; these problems were chiefly concerned with computing the error of asymptotic approximating formulae involving functions of the integers. Two such problems were mentioned in the preceding summary of Voronoĭ's work, namely Dirichlet's problem of divisors (whose solution involves estimating the number of integer points in the triangle formed by the coordinate axes and the hyperbola $xy = n$), and the problem of estimating the number $A(n)$ of

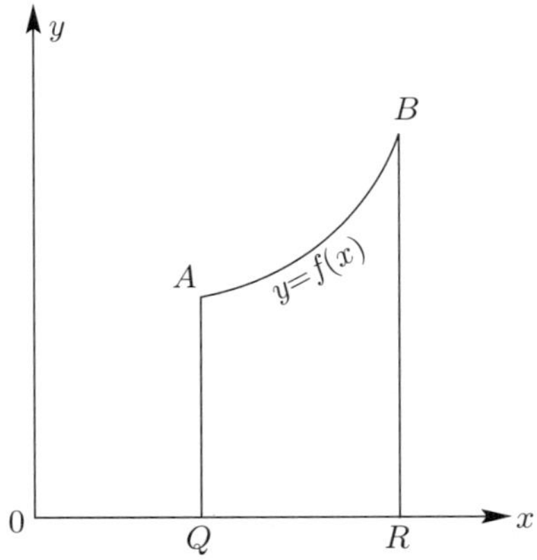

FIGURE 35

integer points in the disc $x^2+y^2 \leq n$. In these two cases Voronoĭ's method yielded the following asymptotic formulae:

$$\tau(1)+\tau(2)+\cdots+\tau(n) = n(\ln n + 2C - 1) + O(\sqrt[3]{n}\ln n),$$
$$A(n) = \pi n + O(\sqrt[3]{n}),$$

where $\tau(k)$ is the number of divisors of the integer k, and C is Euler's constant. For all its elegance Voronoĭ's method (described in the preceding chapter) is rather complex, and closely bound up with special properties of the particular curve in question (the hyperbola). Vinogradov set himself the goal of finding a simple method applicable to a wide class of curves, yet yielding for the problem considered by Voronoĭ the same error estimate as was obtained by him.

The problem of determining the number of integer points in an arbitrary planar region reduces easily to the case where the region has the form $ABRQ$ indicated in Figure 35, i.e., is bounded by the x-axis, two straight-line segments parallel to the y-axis, and a curve (AB in Figure 35) with equation $y = f(x)$, lying entirely above the x-axis. Obviously the number of integer points in this region (excluding those on those parts of the boundary defined by $x = Q$ and $y = 0$) is given by the sum

$$T = \sum_{x>Q}^{x\leq R} [f(x)],$$

taken over all integer x between the indicated limits. Since the sum

$$\sum_{x>Q}^{x\leq R} f(x)$$

can be calculated using known formulae from analysis (for instance the Euler–Maclaurin formula), it follows that the problem of computing T can be replaced by that of computing the sum

$$\sum_{x>Q}^{x\leq R} \{f(x)\}. \tag{1}$$

The early work of Vinogradov contains methods for computing such sums under various assumptions concerning the function $f(x)$. The first remarkable general method for solving problems of this type is given by Vinogradov in his article "A new method for obtaining asymptotic expressions for arithmetic functions" (*Izv. AN SSSR*, 1917). The method for computing a sum of type (1) set out in this paper depends on three lemmas.

LEMMA 1. *Given any real numbers $\tau \geq 1$ and α, there exists a reduced fraction $\frac{Y}{X}$ satisfying*

$$\left|\alpha - \frac{Y}{X}\right| < \frac{1}{\tau X}, \text{ with } 0 < X \leq \tau.$$

This is classical, due to Dirichlet; its proof may be found in any sufficiently detailed course in number theory.[208]

LEMMA 2. *Let X, Y, M be any whole numbers with $X > 0$ and $(X, Y) = 1$, and let $P(z)$ be a function with the property that the difference between its largest and smallest values at the numbers*

$$z = M+1, M+2, \ldots, M+X,$$

does not exceed a given number c. Then

$$\left|\sum_{z=M+1}^{M+X} \left\{\frac{Yz + P(z)}{X}\right\} - \frac{1}{2}X\right| \leq c + \frac{1}{2}.$$

We omit the proof of this simple lemma[209] also, since in the sequel a more complicated lemma of this type will be given with proof (Lemma 6).

LEMMA 3. *Suppose given a sequence of positive integers*

$$X_1, X_2, \ldots, X_n, \tag{2}$$

less than or equal to a number τ. Then, after excluding fewer than τ of the last terms of this sequence, the sequence of remaining terms may without changing their order be subdivided into subsequences satisfying the following condition: The number m of members of each such subsequence

$$X_{\alpha+1}, X_{\alpha+2}, \ldots, X_{\alpha+m}, \tag{3}$$

is equal to the largest of the members $X_{\alpha+i}$ of the subsequence.

PROOF. If $n < \tau$ the assertion is trivial. Assuming $n \geq \tau$, we consider the first X_1 terms of the sequence (2):

$$X_1, X_2, \ldots, X_{X_1}. \tag{4}$$

Let X' denote the largest of these integers; then in particular $X' \geq X_1$. If $X' = X_1$, then the subsequence (4) fulfils the condition of the lemma, and we have found the desired first subsequence. If on the other hand $X' > X_1$, we consider the subsequence of the first X' terms of the sequence (2):

$$X_1, X_2, \ldots, X_{X'}, \tag{5}$$

[208] See, for example: B. A. Venkov, "Elementary number theory" (ONTI, Moscow, 1937; English transl., Wolters-Noordhoff, Groningen, 1970), p. 34.

[209] I. M. Vinogradov, "Elements of number theory", 4th ed. (OGIZ, Moscow-Leningrad, 1944; English transl. of 5th ed., Dover, New York, 1954), Problem 3a (in Chapter III), p. 39.

and let X'' denote the largest of these, so that in particular $X'' \geq X'$. If $X'' = X'$ then again the sequence (5) will be the desired first subsequence. If $X'' > X'$ we continue. Since the increasing sequence $X_1 < X' < X'' < \cdots$ must terminate, we will sooner or later arrive at the first of the desired subsequences. We then iterate this procedure with the remainder of the given sequence, obtaining the second of the desired subsequences, and so on, until there are fewer than τ terms of the sequence (2) left. □

REMARK. For any subsequence (3) satisfying the condition of the lemma we have

$$1 \leq \sum_{i=1}^{m} \frac{1}{X_{\alpha+i}}.$$

Summing these inequalities over all such subsequences, we infer that the number of such subsequences is at most

$$\sum_{k=1}^{n} \frac{1}{X_k}.$$

THEOREM. *With reference to Figure 35, suppose that $A > 1$, $Q < R$, and the function $f(x)$ is such that it has a second derivative in the interval $Q \leq x \leq R$, satisfying*

$$\frac{1}{kA} \leq |f''(x)| \leq \frac{1}{A},$$

for some number $k \geq 1$. Then the following formula holds:

$$\sum_{x>Q}^{R} \{f(x)\} = \frac{1}{2}(R - Q) + H,$$

where

$$H \ll k\left(\frac{R-Q}{A} + 1\right)(A \ln A)^{\frac{2}{3}}.$$

PROOF. 1°. Choose τ arbitrarily subject to

$$1 \leq \tau \leq \sqrt{A}.$$

By Lemma 1 for each

$$x = [Q] + 1, [Q] + 2, \ldots, [R],$$

there exists a pair of relatively prime integers $X = X(x), Y = Y(x)$, satisfying the conditions

$$-\frac{1}{\tau} < Xf'(x) - Y < \frac{1}{\tau}, \quad 0 < X \leq \tau.$$

Consider $X(x)$ at these values of x, i.e., the sequence

$$X([Q] + 1), X([Q] + 2), \ldots, X([R]), \tag{6}$$

and subdivide it into subsequences in accordance with Lemma 3. Let one such subsequence be

$$X(\alpha_s + 1), X(\alpha_s + 2), \ldots, X(\alpha_s + n_s);$$

thus among its terms there must be one equal to n_s, say $X(x_s)$. Hence

$$f'(x_s) = \frac{Y(x_s)}{n_s} + \frac{\theta}{\tau n_s}, \quad 0 < n_s \leq \tau. \tag{7}$$

Now consider that part of the sum

$$\sum_{x>Q}^{R}\{f(x)\}$$

corresponding to this subsequence, namely

$$\Omega_s = \sum_{x=\alpha_s+1}^{\alpha_s+n_s}\{f(x)\}.$$

Define z by $x = x_s + z$. By Taylor's theorem

$$f(x_s+z) = f(x_s) + zf'(x_s) + \frac{z^2}{2}f''(x_s+\theta z).$$

Observing that for the range of x of present interest we have $|z| < n_s$, and invoking (7) and the hypothesis concerning $f''(x)$, one readily obtains

$$\Omega_s = \sum_{z=\alpha_s-x_s+1}^{\alpha_s-x_s+n_s}\left\{\frac{zY(x_s)+n_sf(x_s)+\frac{\theta z}{\tau}+\theta'\frac{n_s^3}{2A}}{n_s}\right\}.$$

We now apply Lemma 2 to this sum, with

$$X = n_s,\ Y = Y(x_s),\ P(z) = n_sf(x_s) + \frac{\theta z}{\tau} + \frac{\theta' n_s^3}{2A},\ c = \frac{n_s}{\tau} + \frac{n_s^3}{2A},$$

obtaining

$$\left|\Omega_s - \frac{1}{2}n_s\right| \leq \frac{n_s}{\tau} + \frac{n_s^3}{2A} + \frac{1}{2}.$$

Adding the analogous inequalities corresponding to all subsequences, together with the obvious inequalities

$$\left|\{f(x)\} - \frac{1}{2}\right| \leq \frac{1}{2},$$

one for each value of x in (6) not belonging to any of those subsequences, and replacing $[R]-[Q]$ by $R-Q+\theta$, we obtain

$$\sum_{x>Q}^{R}\{f(x)\} - \frac{1}{2}(R-Q) = H,$$

where

$$|H| \leq \sum_s\left(\frac{n_s}{\tau} + \frac{n_s^3}{2A} + \frac{1}{2}\right) + \frac{\tau}{2} + \frac{1}{2},$$

with summation $\sum_s$ over all subsequences. Taking into account the second inequality of (7) and the remark following the proof of Lemma 3 concerning the number of subsequences, we obtain

$$\begin{aligned}\sum_s\left(\frac{n_s}{\tau} + \frac{n_s^3}{2A} + \frac{1}{2}\right) &\leq \left(\frac{1}{\tau} + \frac{\tau^2}{2A}\right)\sum_s n_s + \sum_s\frac{1}{2}\\ &\leq \left(\frac{1}{\tau} + \frac{\tau^2}{2A}\right)(R-Q+1) + \frac{1}{2}\sum\frac{1}{X},\end{aligned}$$

where the last summation is over all numbers in the sequence (6). From this and the initial assumption $1 \le \tau \le \sqrt{A}$, we infer that

$$H \ll \left(\frac{1}{\tau} + \frac{\tau^2}{2A}\right)(R - Q) + \tau + \frac{1}{2}\sum \frac{1}{X}. \tag{8}$$

2°. Since

$$f'(x+1) - f'(x) = f''(x+\theta), \quad 0 < \theta < 1,$$

the difference δ between any two successive terms of the sequence

$$f'([Q]+1), f'([Q]+2), \ldots, f'([R]), \tag{9}$$

satisfies $\frac{1}{kA} \le \delta < \frac{1}{A}$. Hence if K denotes the least number in this sequence, then the largest will be at most $K + \frac{R-Q}{A}$, and furthermore the number of terms of the sequence (9) contained in an interval $[U, V]$, $U < V$, will be at most $kA(V-U)+1$. These observations facilitate the estimation of the sum

$$\sum \frac{1}{X}$$

on the right-hand side of (8). To this end we need to estimate the number of times any value of X is repeated in the sequence (6). Given X, any associated Y must satisfy the inequalities[210]

$$XK - \frac{1}{\tau} < Y < X\left(K + \frac{R-Q}{A}\right) + \frac{1}{\tau},$$

and the number of such Y is at most

$$\frac{R-Q}{A}X + 3.$$

The condition governing the choice of X, Y, also entails that for a term of the sequence (9) to be associated with a particular pair X, Y, it must satisfy

$$\frac{Y - \frac{1}{\tau}}{X} < f'(x) < \frac{Y + \frac{1}{\tau}}{X},$$

whence the number of such terms is at most

$$kA\frac{2}{\tau X} + 1 \ll \frac{kA}{\tau X}.$$

Hence the number ξ of terms of the sequence (6) equal to X, satisfies

$$\xi \ll \frac{kA}{\tau X}\left(\frac{R-Q}{A}X + 3\right) = k\left(\frac{R-Q}{\tau} + \frac{3A}{\tau X}\right).$$

Applying this to the sum $\sum \frac{1}{X}$, we obtain

$$\sum \frac{1}{X} \ll \sum_{X=1}^{\tau} k\left(\frac{R-Q}{\tau} + \frac{3A}{\tau X}\right)\frac{1}{X} \ll k\frac{R-Q}{\tau}\ln A + \frac{kA}{\tau}, \tag{10}$$

where we have used the well-known formula

$$\sum_{X=1}^{\tau} \frac{1}{X} \ll \ln \tau.$$

[210] By the condition imposed at the beginning of the proof; see the displayed inequalities preceding (6). *Trans.*

For $A \leq 4$, the theorem now follows immediately from (8) and (10). For $A > 4$, the formulae (8) and (10) yield

$$H \ll \frac{\tau^2}{A}(R-Q) + \frac{kA}{\tau} + k\frac{R-Q}{\tau}\ln A,$$

and taking

$$\tau = (A\ln A)^{\frac{1}{3}},$$

we obtain

$$H \ll k\left(\frac{R-Q}{A}+1\right)(A\ln A)^{\frac{2}{3}},$$

completing the proof of the theorem. □

Note that later Vinogradov significantly simplified the proof of this theorem, to the extent of giving it as an exercise in the latest edition of his course in number theory.[211]

As mentioned above, from this estimate for the sum

$$\sum_{x>Q}^{R}\{f(x)\},$$

it is easy to obtain an estimate for the number T of integer points inside the curve $ABRQ$ by using the summation formula due to N. Ya. Sonin. To this end we write

$$\rho(x) = \frac{1}{2} - \{x\}, \quad \sigma(x) = \int_0^x \rho(x)dx.$$

It is easy to see that for all x

$$|\rho(x)| \leq \frac{1}{2}, \qquad |\sigma(x)| \leq \frac{1}{8}.$$

In this notation Sonin's formula is as follows:

$$\sum_{x>Q}^{x\leq R} f(x) = \int_Q^R f(x)dx + \rho(R)f(R) - \rho(Q)f(Q)$$
$$- \sigma(R)f'(R) + \sigma(Q)f'(Q) + \int_Q^R \sigma(x)f''(x)dx.$$

Let μ denote the largest value of $|f'(x)|$ in the interval

(11) $$Q \leq x \leq R.$$

We continue to assume that $f(x)$ satisfies the conditions imposed in the theorem; hence in particular $f''(x)$ does not change sign in the interval (11), and Sonin's formula yields

$$\sum_{x>Q}^{R} f(x) = \int_Q^R f(x)dx + \rho(R)f(R) - \rho(Q)f(Q) + \frac{\theta}{2}\mu.$$

Combining this with the estimate of $\sum_{x>Q}^{R}\{f(x)\}$ given by the theorem, we obtain

(12) $$T = S + \rho(R)f(R) - \rho(Q)f(Q) - \frac{1}{2}(R-Q) + H,$$

---

[211] I. M. Vinogradov, loc. cit., Exercise 5a (in Chapter III), p. 40.

where

$$S = \int_Q^R f(x)dx,$$

the area of the region $QRBA$, and now

$$H \ll \mu + k\left(\frac{R-Q}{A} + 1\right)(A \ln A)^{\frac{2}{3}}.$$

We now assume that on the arc AB of the graph of $y = f(x)$, the radius of curvature

$$\rho = \frac{(1+(y')^2)^{3/2}}{|y''|} \tag{13}$$

does not vanish or become infinite, and that the ratio of its largest value r to its smallest value r' is less than or equal to a prescribed number σ. From this it follows first of all that $f'(x)$ is monotonic on AB, whence we infer that the arc AB can be subdivided into three subarcs on each of which either $|f'(x)| \leq 1$ or $|f'(x)| \geq 1$. Since the second of these cases becomes the first if we interchange the coordinate axes, we may suppose without loss of generality that

$$|f'(x)| \leq 1.$$

In this case the formula (13) gives

$$\frac{1}{r} \leq |y''| \leq \frac{2^{3/2}\sigma}{r},$$

which has the form of the chief hypothesis of the theorem, so that the formula (12) applies. Writing y_0' and y_1' for the values of $f'(x)$ at the respective endpoints of the arc AB, we have

$$2 \geq |y_1' - y_0'| = \int_Q^R |y''|dx \geq \frac{1}{r}(R-Q).$$

Using this it is not difficult to see that the final term H in the formula (12) satisfies

$$H \ll \sigma^{\frac{4}{3}}(r \ln r)^{\frac{2}{3}}.$$

Note finally that any bounded planar region having a simple closed curve as boundary can be considered as the difference of two regions of the form $QRBA$.[212] Hence we have the following

THEOREM. *Let Ω be a planar region of area S bounded by a curve satisfying the condition that the ratio of the greatest to the smallest values of its radius of curvature is less than a prescribed constant. Then the number T of integer points contained in Ω satisfies*

$$T = S + O\left(r^{\frac{2}{3}}(\ln r)^{\frac{2}{3}}\right).$$

In this formulation the theorem is very convenient for calculating the number of integer points in regions bounded by varying curves, in particular in the situation where the boundary curve expands unrestrictedly while remaining similar to some fixed curve. This is exactly the case in the problem of estimating the number $A(n)$

---

[212] This seems to assume that the region in question is also convex, or at least with boundary in some sense not very convoluted. *Trans.*

of integer points in the disc $x^2 + y^2 \leq n$, where the above theorem with $\tau = \sqrt{n}$, gives

$$A(n) = \pi n + O\left(n^{\frac{1}{3}}(\ln n)^{\frac{2}{3}}\right).$$

For the problem of divisors the first theorem yields, after some calculation,[213]

$$\tau(1) + \cdots + \tau(n) = n(\ln n + 2C - 1) + O\left(n^{\frac{1}{3}}(\ln n)^2\right).$$

Thus notwithstanding the generality of the above theorems, the error estimates they afford are not significantly worse than those obtained using Voronoĭ's method.

We conclude with some remarks on one of the above problems. Write

$$A(n) = \pi n + K(n),$$

and denote by ϑ the greatest lower bound of the set of those exponents $\alpha > 0$ for which

$$K(n) \ll n^{\alpha} \quad (n \to \infty).$$

The precise value of ϑ remains unknown up to the present (1947). In 1916 Hardy proved[214] that

$$\vartheta \geq \frac{1}{4}.$$

The result obtained using Voronoĭ's method or the first of the above theorems of Vinogradov shows that

$$\vartheta \leq \frac{1}{3}.$$

The outstanding Dutch mathematician van der Corput devoted several articles to this problem; in the last of these[215] he proves that

$$\vartheta \leq \frac{27}{82}.$$

Finally, in 1932 Vinogradov returned to this problem and proved[216] that

$$\vartheta \leq \frac{17}{53}.$$

For easy comparison we list the various estimates in order:

$$\frac{1}{4} < \frac{17}{53} < \frac{27}{82} < \frac{37}{112} < \frac{1}{3}.$$

The work of van der Corput and Vinogradov is very complicated; a relatively simple proof of the inequalities

$$\frac{1}{4} \leq \vartheta \leq \frac{37}{112}$$

can be found in Landau's book.[217]

It is appropriate to mention another classical problem which attracted the attention of Vinogradov in the early years of his mathematical activity, namely the estimation of the average value of the integer-valued function $h(-\Delta))$ giving the number of inequivalent binary quadratic forms

$$ax^2 + 2bxy + cy^2, \ a, b, c \text{ integers}, \ a > 0,$$

---

[213] I. M. Vinogradov, loc. cit., Exercise 6B (in Chapter III), p. 40.

[214] *Proc. Lond. Math. Soc.* (2) 15, 1916, pp. 1–25.

[215] *Math. Ann.*, 98, 1928, pp. 697–716; 100, 1928, p. 480.

[216] "On the number of integer points inside a circle", *Izv. AN SSSR*, 1932.

[217] E. Landau, "Vorlesungen über Zahlentheorie", Bd. II.

with prescribed negative discriminant $b^2 - ac = -\Delta$, and satisfying the condition that $a, 2b, c$ have highest common factor 1. The function $h(-\Delta)$ had been studied already by Gauss, and remains to this day of great significance in the theory of numbers. Gauss conjectured[218] that the sum

$$h(-1) + h(-2) + \cdots + h(-n)$$

should be asymptotic to

$$\frac{4\pi}{21\alpha} n^{\frac{3}{2}},$$

where

$$\alpha = \sum_{k=1}^{\infty} \frac{1}{k^3},$$

and this was proved by Lipschitz and Mertens in 1865. In 1912 Landau[219] obtained the more accurate asymptotic formula

$$h(-1) + h(-2) + \cdots + h(-n) = \frac{4\pi}{21\alpha} n^{\frac{3}{2}} - \frac{2}{\pi^2} n + O\left(n^{\frac{5}{6}} \ln n\right). \tag{14}$$

This formula (but with improved error term $O(n^{\frac{5}{6}}(\ln n)^{\frac{2}{3}}))$ was reproved by Vinogradov at the end of the 1917 memoir discussed above. In that same year he devoted a special article[220] to Gauss' problem, in which he reduces the error term to $O(n^{\frac{3}{4}}(\ln n)^2)$. Finally, powerful methods invented by Vinogradov later allowed the reduction of the order to $O(n^{0\cdot 7+\epsilon})$.[221]

A large number of Vinogradov's works are concerned with the distribution of residues or nonresidues of a given degree, or of primitive roots, relative to moduli, etc., in arithmetic progressions or in intervals of prescribed length. Classical number theory, whence interest in these quantities originated, is silent on the question of their distribution, for example on the question as to what the least positive quadratic nonresidue or the least primitive root modulo a (large) prime might be. Up until the appearance of Vinogradov's work on such questions there was no satisfactory way of answering them. Just one result of this kind, and that very partial, was obtained by Gauss[222] in the course of proving the law of quadratic reciprocity, namely the result that given any prime of the form $8n + 1$, there exists a prime $q < 2\sqrt{p} + 1$, for which p will be a quadratic nonresidue. By means of very simple estimates of trigonometric sums Vinogradov was able to establish a whole complex of theorems of this type. Here are the statements of a few of these:

1) *The least positive quadratic nonresidue relative to any prime modulus p is less than $p^{\frac{1}{2\sqrt{e}}}(\ln p)^2$ (where e is the natural logarithmic base).*

2) *The least positive primitive root modulo any prime p is less than*

$$2^{2k}\sqrt{p}\ln p,$$

where k is the number of distinct prime divisors of $p - 1$.

3) *Let p be any prime, and c a divisor of $p - 1$. Partition the set of (nonzero)* residues mod *p into c subsets, where each subset consists of the residues congruent*

[218] "Disquisitiones arithmeticae", Art. 302.

[219] *Sitzungsberichte d. K. Akad. d. Wissenschaften in Wien*, 1912.

[220] "On the average value of the number of essential forms with negative discriminant", *Soobshch. Khar'kov. Matem. Obshch. (Communications of the Kharkov Math. Soc.)*, 1918.

[221] "On certain new problems of number theory", *DAN SSSR*, vol. III, No. 1, 1934.

[222] "Disquisitiones arithmeticae", Art. 125.

to $ax^c (\bmod\ p)$ *for a fixed (nonzero)* a *(so that in particular the residues expressible as* c*th powers will form one subset).*[223] *Then in any arithmetic progression*

$$\beta, \alpha+\beta, 2\alpha+\beta, \ldots, (L-1)\alpha+\beta, \ \textit{with}\ 0 < L < p,$$

where the common difference α *is not divisible by* p*, the number of terms belonging to any single subset of the partition has the form*

$$\frac{L}{c} + O(\sqrt{p}\ln p).$$

We shall not give the proofs of these results here, since they have achieved classical status and form part of the standard material of detailed courses in number theory, along with the theorems of Fermat, Euler, and Lagrange.[224]

[223] Where x ranges over the set $\{1, 2, \ldots, p-1\}$ of nonzero residues mod p; thus the subsets in question are just the cosets of the subgroup of index c of the multiplicative group of the field of p elements. *Trans.*

[224] Landau, "Vorlesungen über Zahlentheorie", Bd. II; Venkov, "Elementary number theory", Ch. III.

Waring's Problem

In 1770 Edward Waring made the following assertion:[225]

For each positive integer $n \geq 2$, there exists a positive integer $r = r(n)$ such that every integer $N > 0$ can be expressed in the form

$$N = x_1^n + x_2^n + \cdots + x_r^n, \tag{15}$$

for some integers $x_i \geq 0$.

This assertion, remaining unproven for the whole of the 19th century, is remarkable for providing the impetus for the creation of powerful new methods in analytic number theory.

The first general proof of the assertion was given by Hilbert in 1909. His proof is of a special character and gives values for the desired number r that are rather too large. Although many mathematicians (Hausdorff, Remak, Frobenius) studied Hilbert's approach with a view to reducing the value of r, their efforts were largely fruitless, and as of the present time (1947). Hilbert's approach has been abandoned.

The next step forward in the history of Waring's problem was made in a series of important articles by Hardy and Littlewood; from 1920 to 1925 these scholars published 6 memoirs under the general head "Some Problems of Partitio Numerorum", in which they developed a new general method for solving problems of additive type in number theory (i.e., problems involving the expression of integers as sums of terms of a certain form). In particular their memoirs I, II, IV, and VI are devoted to Waring's problem.

Before describing the results of Hardy and Littlewood we introduce some notation. Note first that it suffices to solve Waring's problem for some sufficiently large $N = N_0$, since then an equation of the form (15) (albeit with a large number of terms) can be written down for all $N > 0$ by using $0 = 0^n, 1 = 1^n$. Let $G(n)$ denote the least value of r for which (15) holds for all $N \geq N_0(n)$; thus $N_0(n)$ is such that every $N \geq N_0$ can be expressed in the form $x_1^n + \cdots + x_{r'}^n$, where $r' = G(n)$, but there are infinitely many integers $N > 0$ not expressible in the form $x_1^n + \cdots + x_{r'-1}^n$ for integral $x_i \geq 0$. The investigation of the number r in Waring's problem leads to the problem of finding upper and lower bounds (depending on n) for the quantity $G(n)$. It is easy to show[226] that for $n \geq 2$ we must have

$$G(n) > n. \tag{16}$$

Hardy and Littlewood investigate, for given n and r, the number I_N of representations of N in the form (15), and for

$$r \geq (n-2) \cdot 2^{n-1} + 5 \text{ (and } n > 2\text{)}, \tag{17}$$

[225] "Meditationes algebraicae", Cambridge, 1770, pp. 204–205.

[226] Landau, "Vorlesungen über Zahlentheorie", Bd. I, p. 359.

obtain the asymptotic formula

$$I_N = \frac{[\Gamma(1+\nu)]^r}{\Gamma(r\nu)} N^{r\nu-1}\mathfrak{S} + O(N^{r\nu-1-c}), \tag{18}$$

where $c = c(n)$ is a positive constant,

$$\nu = \frac{1}{n}, \tag{19}$$

and $\mathfrak{S} = \mathfrak{S}(n, r, N)$ is a so-called "singular series" (see below). The latter quantity $\mathfrak{S}$ is bounded below independently of N:

$$\mathfrak{S} \geq c_1 > 0, \quad c_1 = c_1(n, r).$$

Formula (18) implies that $I_N > 0$ for all sufficiently large N, i.e., that Waring's assertion is valid, and furthermore yields the bound

$$G(n) \leq (n-2) \cdot 2^{n-1} + 5. \tag{20}$$

The work of Hardy and Littlewood has been elaborated on and supplemented by many prominent researchers, of which mention should be made of Landau and Kamke.

The first of Vinogradov's published articles on Waring's problem dates from 1924. He subsequently returned to the problem several times, until in 1934 he arrived at an original idea allowing a significant reduction in the bound for $G(n)$. Vinogradov's idea was also applicable to other problems and became in his hands the kernel of a remarkable new method, which by virtue of its simplicity and precision represented a great advance on all that had up till then been achieved in analytic number theory. In 1937 he gave a systematic exposition of this method in his book "A new method in analytic number theory". In connection with Waring's problem the method yields

$$G(n) < 6n(\ln n + 1) \text{ for } n \geq 16, \tag{21}$$

and the asymptotic formula (18) under the assumption

$$r > \left(1 + \frac{\ln \ln n}{\ln n}\right) n^3 \ln n.$$

If one takes into account that $\ln n$ increases more slowly than any power n^ϵ of n, it becomes clear how close the bound (21) must be to the actual value of $G(n)$.

For small values of n the function $G(n)$ has been investigated separately. Thus for instance Landau showed in 1927 that $G(3) \leq 8$; Esterman, Davenport, and Heilbronn proved in 1936 that $G(4) \leq 17$; and of course $G(2) = 4$, since by the well-known theorem of Lagrange every integer $N > 0$ is representable as a sum of four squares, while on the other hand it is easy to see that no integer of the form $8k + 7$ is representable as a sum of three squares.

Thus in the wake of Vinogradov's work we may reckon Waring's long-outstanding problem finally settled.

We mentioned above the notable simplicity of Vinogradov's method. The Hardy–Littlewood approach, which is based on the representation of I_N by a coefficient of a certain power series and on the well-known Cauchy integral formula for the coefficients of such a series, involves highly intricate calculations, even in Landau's simplified exposition. By contrast Vinogradov's method utilizes finite sums, avoids complex function theory, and manages with relatively simple calculations.

His method may therefore be regarded as coming fully within the tradition of the St. Petersburg school whereby deep results are obtained by simple means.

Since it is our wish to communicate mainly the essence of Vinogradov's approach to Waring's problem, we shall not discuss the derivation of the asymptotic formula (18), but confine ourselves to showing how his method is used to obtain a bound for $G(n)$, following the exposition in his 1937 book, but with certain simplifications resulting in the following somewhat less sharp bound than (21):

$$G(n) \leq 6n \ln n + 11n.$$

Thus let a, q be positive integers with $(a, q) = 1$, and set

$$S_{aq} = \sum_{x=0}^{q-1} e^{2\pi i \frac{a}{q} x^n}. \tag{22}$$

Then for positive integral N and r, set

$$A(q) = A(q, N, r, n) = q^{-r} \sum_a S_{aq}^r e^{-2\pi i \frac{a}{q} N}, \tag{23}$$

where a ranges over the nonzero residues modulo q.[227] It can be shown that

$$S_{aq} \ll q^{1-\nu}, \quad A(q) \ll q^{1-r\nu}, \tag{24}$$

where ν is as in (19). From this it is immediate that for $r \geq 2n+1$, the series

$$\mathfrak{S} = \mathfrak{S}(N, r, n) = \sum_{q=1}^{\infty} A(q) \tag{25}$$

converges absolutely. The sum $\mathfrak{S}$ of the series is real and positive, and for $r \geq 4n$, satisfies

$$\mathfrak{S} \gg 1. \tag{26}$$

We omit the proofs of the formulae (24) and (26), which are due essentially to Hardy and Littlewood.[228]

Note that the constant implicit in each use of the symbol $\ll$ will both here and in the sequel be understood to depend on n and r; and since towards the conclusion of this exposition we shall take $r = 4n$, we may in effect assume such constants depend only on n.

In his work on Waring's problem and other additive problems Vinogradov takes as starting point the very simple integral

$$\int_0^1 e^{2\pi i \alpha m} d\alpha = \begin{cases} 1 & \text{if } m = 0, \\ 0 & \text{if } m \neq 0, \end{cases} \tag{27}$$

where m is an arbitrary integer. We write $P = [N^\nu]$, and observe that if N could be expressed in the form (15) for integral $x_i > 0$, then it would follow that $x_i < P$. In (27) we now replace the integer m by the difference

$$(x_1^n + \cdots + x_r^n) - N,$$

---

[227] That is, over the set $\{1, \ldots, q-1\}$. *Trans.*

[228] Landau, "Vorlesungen über Zahlentheorie", Bd. I, VI Teil, Kap. 2; Vinogradov, "A new method in analytic number theory", Ch. II.

and sum the resulting equations as $x_1, \dots, x_r$ range independently over the numbers $1, 2, \dots, P$. The right-hand side of this sum will then be equal to the number W_N of representations of N of the form (15), whence

$$(28) \qquad W_N = \sum_{x_1=1}^{P} \cdots \sum_{x_r=1}^{P} \int_0^1 e^{2\pi i\alpha(x_1^n+\cdots+x_r^n-N)}\,d\alpha = \int_0^1 L_\alpha^r e^{-2\pi i\alpha N}\,d\alpha,$$

where

$$L_\alpha = \sum_{x=1}^{P} e^{2\pi i\alpha x^n}.$$

We now apply Lemma 1 ("Dirichlet's lemma") from the preceding section: Writing

$$(29) \qquad \tau = 2nP^{n-1},$$

we have by that lemma that for each real number α in the interval $(0,1)$, there exist integers a, q with highest common factor 1 satisfying the conditions

$$(30) \qquad -\frac{1}{q\tau} < z < \frac{1}{q\tau}, \ 0 < q \le \tau, \quad \text{where } z = \alpha - \frac{a}{q}.$$

Consider the interval of values of α given by (30) (i.e., $\frac{a}{q} - \frac{1}{q\tau} < \alpha < \frac{a}{q} + \frac{1}{q\tau}$) determined by some fraction $\frac{a}{q}$ satisfying $q \le P^{1-\nu}$. If $q > 1$ and $0 < a < q$, that interval is contained in the interval $(0,1)$, since

$$0 \le \frac{a-1}{q} < \frac{a}{q} - \frac{1}{q\tau}, \qquad \frac{a}{q} + \frac{1}{q\tau} < \frac{a+1}{q} \le 1;$$

and if $a < 0$ or $a > q$ a similar argument shows that the interval in (30) is entirely outside the interval $(0,1)$. Finally, in the cases $q = 1, a = 0$ and $q = 1, a = 1$, the interval $(0,1)$ contains only the following initial and terminal segments respectively, of the interval of values of α in (30):

$$(31) \qquad \alpha = \frac{0}{1} + z, \ 0 < \alpha < \frac{1}{\tau}; \qquad \alpha = \frac{1}{1} + z, \ 1 - \frac{1}{\tau} < \alpha < 1.$$

Note that in the crucial formula (28) the integrand has period 1; hence integrating that integrand over the second interval in (31) will produce the same value as integrating it over the first, so that in the second case we can use the same formula for W_N as in the first, namely

$$(32) \qquad W_N = \int_{-\frac{1}{\tau}}^{1-\frac{1}{\tau}} L_\alpha^r e^{-2\pi i\alpha N}\,d\alpha.$$

Intervals of values of α of the form given in (30) under the conditions

$$0 < q < P^{1-\nu}, \quad 0 \le a < q$$

(with equality between a and q allowed in the second condition if $q = 1$) will be called *intervals of class I*. They are all contained in the new interval of integration $\left(-\frac{1}{\tau}, 1 - \frac{1}{\tau}\right)$.[229] It is not difficult to verify that any two distinct class I intervals (i.e., corresponding to distinct fractions $\frac{a}{q}$) do not overlap. For otherwise we should have

$$\alpha = \frac{a}{q} + z = \frac{a_1}{q_1} + z_1, \text{ with } |z| \le \frac{1}{q\tau}, \ |z_1| \le \frac{1}{q_1\tau},$$

[229] Or can be replaced by intervals so contained. *Trans.*

for distinct fractions $\frac{a}{q}$ and $\frac{a_1}{q_1}$, whence

$$\left|\frac{a}{q} - \frac{a_1}{q_1}\right| \leq \frac{1}{q\tau} + \frac{1}{q_1\tau}.$$

On the other hand, since these fractions are distinct we have

$$\left|\frac{a}{q} - \frac{a_1}{q_1}\right| \geq \frac{1}{qq_1}.$$

Hence

$$\frac{1}{qq_1} \leq \frac{1}{q\tau} + \frac{1}{q_1\tau}, \text{ whence } \tau \leq q + q_1,$$

i.e., $2nP^{n-1} \leq 2P^{1-\nu}$, which is impossible.

We shall call the subintervals of $\left(-\frac{1}{\tau}, 1 - \frac{1}{\tau}\right)$ remaining after all class I intervals have been removed *intervals of class II.* From the foregoing it follows that every number α in a class II interval has the form

$$\alpha = \frac{a}{q} + z,\ |z| \leq \frac{1}{q\tau},\ P^{1-\nu} < q \leq \tau.$$

Corresponding to this subdivision of the interval of integration into subintervals of the two types, the quantity W_N decomposes as a two-term sum

$$W_N = W_N^{(I)} + W_N^{(II)},$$

where $W_N^{(I)}$ is the sum of integrals with integrand as in (32) over each class I interval, and $W_N^{(II)}$ the analogous sum of integrals over all class II intervals. Denoting by H_{aq} the integral with integrand as in (32) over the class I interval given by (30), we have

$$W_N^{(I)} = \sum H_{aq},$$

where the summation is taken over $q = 1, 2, \ldots, [P^{1-\nu}]$, and, for each such q, over $a = 0, 1, \ldots, q-1$. We begin by obtaining an asymptotic formula for $W_N^{(I)}$. For this we first need

An asymptotic formula for L_α. We shall now estimate the quantity

$$L_\alpha = \sum_{x=1}^{P} e^{2\pi i \alpha x^n}$$

for values of α in a class I interval given by (30). For this purpose we shall use the following lemma of van der Corput, easily deduced using the theory of Fourier series.[230]

LEMMA 4. *Let M, N be integers with $M < N$, and $f(x)$ a function defined on the interval $M \leq x \leq N$ satisfying the conditions*

$$0 \leq f'(x) \leq \frac{1}{2},\ f''(x) \geq 0.$$

Then for either sign $\eta = \pm 1$, we have

$$\sum_{z=M}^{N} e^{2\pi i \eta f(z)} = \int_M^N e^{2\pi i \eta f(x)} dx + 2\theta.$$

[230]Vinogradov, "A new method in analytic number theory", Chapter I, Lemma 15.

In the sum L_α the summation index x takes on the values $1, 2, \ldots, P$; dividing these x by q, we have $x = qt + s$, where now s ranges over the integers $0, 1, \ldots, q-1$, and for each such s the quotient t takes on all integer values in the interval

$$-sq^{-1} < t < (P - s)q^{-1}. \tag{33}$$

Hence we may rewrite L_α in the form

$$\begin{aligned} L_\alpha &= \sum_{s=0}^{q-1} \sum_t e^{2\pi i \left(\frac{a}{q} + z\right)(qt+s)^n} \\ &= \sum_{s=0}^{q-1} \sum_t e^{2\pi i \left[\frac{a}{q} s^n + z(qt+s)^n\right]} = \sum_{s=0}^{q-1} e^{2\pi i \frac{a}{q} s^n} D_s, \end{aligned} \tag{34}$$

where

$$D_s = \sum_t e^{2\pi i z (qt+s)^n}.$$

In order to estimate the latter sum, we apply Lemma 4 with

$$\begin{gathered} \eta = \operatorname{sign} z, \quad f(t) = |z|(qt+s)^n, \\ M = [-sq^{-1}] + 1, \qquad N = [(P-s)q^{-1}], \end{gathered}$$

noting that in the interval (33) the function $f(t)$ satisfies the conditions of the lemma since

$$f'(t) = n|z|q(qt+s)^{n-1} \leq \frac{nP^{n-1}}{\tau} = \frac{1}{2}, \; f''(t) \geq 0.$$

Hence by that lemma we have

$$D_s = \int_M^N e^{2\pi i z (qt+s)^n} \, dt + 2\theta.$$

Noting that M and N differ from the endpoints of the interval (33) by at most 1, and changing back to the variable x, we then have

$$D_s = \int_{-sq^{-1}}^{(P-s)q^{-1}} e^{2\pi i z (qt+s)^n} \, dt + 4\theta' = \frac{1}{q} \int_0^P e^{2\pi i z x^n} \, dx + 4\theta'.$$

Finally, since $P = [N^\nu]$ differs from N^ν by less than 1, we have

$$D_s = \frac{1}{q} \int_0^{N^\nu} e^{2\pi i z x^n} \, dx + 5\theta''.$$

Substituting this expression for D_s in (34), and recalling the notation (22), we obtain

$$L_\alpha = \phi \frac{S_{aq}}{q} + 5q\theta, \text{ where } \phi = \int_0^{N^\nu} e^{2\pi i z x^n} \, dx. \tag{35}$$

We are now ready to obtain an

Asymptotic formula for $W_N^{(I)}$. In order to estimate the integral ϕ that we have arrived at, we use the following lemma.[231]

LEMMA 5. *Suppose given $Q \geq 1, n > 1, z$ a real number, and*

$$I = \int_0^Q e^{2\pi i z x^n} dx;$$

then

$$|I| \leq Q \text{ for } |z| \leq Q^{-n},$$
$$|I| \leq \sqrt{2}|z|^{-\nu} \text{ for } |z| > Q^{-n}.$$

Applying this lemma to the integral ϕ, we obtain

$$(36) \qquad |\phi| \leq Z, \text{ where } \begin{cases} Z = N^{\nu} & \text{for } |z| \leq N^{-1}, \\ Z = \sqrt{2}|z|^{-\nu} & \text{for } |z| > N^{-1}. \end{cases}$$

By the first estimate in (24) it then follows that

$$\left|\phi \frac{S_{aq}}{q}\right| \ll Zq^{-\nu},$$

whence from (35) we have

$$(37) \qquad \left|L_\alpha^r - \left(\phi \frac{S_{aq}}{q}\right)^r\right| \ll (Zq^{-\nu} + q)^r - (Zq^{-\nu})^r.$$

Noting that

$$\frac{q}{Zq^{-\nu}} = \frac{q^{1+\nu}}{N^\nu} \leq \frac{P^{(1-\nu)(1+\nu)}}{P} < 1 \text{ for } |z| \leq N^{-1},$$
$$\frac{q}{Zq^{-\nu}} \ll q^{1+\nu}|z|^\nu \ll \frac{P^{1-\nu}}{(P^{n-1})^\nu} = 1 \text{ for } |z| > N^{-1},$$

we obtain from (37)

$$\left|L_\alpha^r - \left(\phi \frac{S_{aq}}{q}\right)^r\right| \ll (Zq^{-\nu})^{r-1} q = Z^{r-1} q^{1-\nu(r-1)}.$$

From now on we shall assume that

$$r \geq 4n,$$

so that $1 - \nu(r-1) < -1$, whence

$$L_\alpha^r e^{-2\pi i \alpha N} = \phi^r \left(\frac{S_{aq}}{q}\right)^r e^{-2\pi i (\frac{a}{q} + z)N} + O(q^{-1} Z^{r-1}).$$

Integration with respect to z from $-\frac{1}{q\tau}$ to $+\frac{1}{q\tau}$ yields

$$\begin{aligned} H_{aq} &= \int_{-\frac{1}{q\tau}}^{+\frac{1}{q\tau}} L_\alpha^r e^{-2\pi i \alpha N} dz \\ &= R_q \left(\frac{S_{aq}}{q}\right)^r e^{-2\pi i \frac{a}{q} N} + O\left(q^{-1} \int_{-\frac{1}{q\tau}}^{+\frac{1}{q\tau}} Z^{r-1} dz\right), \end{aligned}$$

[231] Vinogradov, "A new method in analytic number theory", Chapter I, Lemma 16.

where

$$R_q = \int_{-\frac{1}{q\tau}}^{+\frac{1}{q\tau}} \phi^r e^{-2\pi i z N} dz.$$

For sufficiently large N we shall have

$$N^{-1} < \frac{1}{P^{1-\nu}\tau} \leq \frac{1}{q\tau},$$

whence, taking into account the definition of the quantity Z in (36), we obtain

$$\begin{aligned}\int_{-\frac{1}{q\tau}}^{\frac{1}{q\tau}} Z^{r-1} dz &= 2\int_0^{\frac{1}{q\tau}} Z^{r-1} dz \\ &\ll \int_0^{N^{-1}} N^{\nu(r-1)} dz + \int_{N^{-1}}^{\frac{1}{q\tau}} z^{-\nu(r-1)} dz \\ &\ll N^{-\nu} + z^{(-\nu(r-1)+1)}\Big|_{1/(q\tau)}^{N^{-1}} \ll P^{r-n-1}.\end{aligned}$$

Hence

$$H_{aq} = R_q \left(\frac{S_{aq}}{q}\right)^r e^{-2\pi i \frac{a}{q} N} + O(q^{-1}P^{r-n-1}).$$

We wish now to replace the integral R_q in this formula by the integral over the whole line, namely

$$R = \int_{-\infty}^{\infty} \phi^r e^{-2\pi i z N} dz.$$

For this purpose we need to estimate the quantity to be added to the above expression for H_{aq} to achieve this, namely

$$\Delta = \left(\frac{S_{aq}}{q}\right)^r e^{-2\pi i \frac{a}{q} N} \left\{ \int_{\frac{1}{q\tau}}^{\infty} \phi^r e^{-2\pi i z N} dz + \int_{-\infty}^{-\frac{1}{q\tau}} \phi^r e^{-2\pi i z N} dz \right\}.$$

Using (24) and (36) we estimate this as follows:

$$\begin{aligned}\Delta &\ll q^{-\nu r} \int_{\frac{1}{q\tau}}^{\infty} Z^r dz \ll q^{-\nu r} \int_{\frac{1}{q\tau}}^{\infty} z^{-\nu r} dz \ll q^{-\nu r} \frac{1}{(q\tau)^{-\nu r+1}} \\ &= q^{-1}\tau^{\nu r-1} \ll q^{-1}P^{r-\nu r-n+1} \ll q^{-1}P^{r-n-1}.\end{aligned}$$

Hence

$$H_{aq} = R\left(\frac{S_{aq}}{q}\right)^r e^{-2\pi i \frac{a}{q} N} + O(q^{-1}P^{r-n-1}).$$

Summing over all a corresponding to a given q, and bringing in the notation introduced in (23), we obtain

$$\sum_a H_{aq} = RA(q) + O(P^{r-n-1}).$$

Summation over all $q = 1, 2, \ldots, [P^{1-\nu}]$ then yields

$$W_N^{(I)} = R \sum_{q=1}^{P^{1-\nu}} A(q) + O(P^{r-n-1}). \tag{38}$$

We have yet to bring the expression

$$R = \int_{-\infty}^{\infty} e^{-2\pi izN} \left(\int_0^{N^\nu} e^{2\pi izx^n} dx \right)^r dz$$

into appropriate form. To this end in the integral in parentheses we make the substitution $x = N^\nu u$, and then in the outer integral $zN = t$, obtaining

$$R = \int_{-\infty}^{\infty} e^{-2\pi izN} N^{\nu r} \left(\int_0^{1} e^{2\pi izNu^n} du \right)^r dz = N^{\nu r-1} Q,$$

where

$$Q = \int_{-\infty}^{\infty} e^{-2\pi it} \left(\int_0^{1} e^{2\pi itu^n} du \right)^r dt,$$

which is independent of N. The quantity Q may be evaluated by means of an additional summation of the formula (38) over N,[232] or by direct integration using properties of the "Dirichlet integral";[233] in any case the end result is

$$Q = \frac{[\Gamma(1+\nu)]^r}{\Gamma(r\nu)}.$$

Substitution of the resulting formula for R in (38) yields

$$W_N^{(I)} = N^{\nu r-1} \frac{[\Gamma(1+\nu)]^r}{\Gamma(r\nu)} \sum_{q=1}^{P^{1-\nu}} A(q) + O(P^{r-n-\nu}). \tag{39}$$

Taking into account the estimate of $A(q)$ in (24) together with the assumption $r \geq 4n$, we have

$$\sum_{q>P^{1-\nu}} A(q) \ll \sum_{q>P^{1-\nu}} \frac{1}{q^{r\nu-1}} \leq \sum_{q>P^{1-\nu}} \frac{1}{q^3} < \int_{[P^{1-\nu}]}^{\infty} q^{-3} dq \ll P^{-2+2\nu}.$$

Adding this to the sum in (38), and bringing in the notation introduced in (25), we finally obtain the asymptotic formula

$$W_N^{(I)} = \frac{[\Gamma(1+\nu)]^r}{\Gamma(r\nu)} \mathfrak{S}(N,r,n) N^{r\nu-1} + O(P^{r-n-\nu}). \tag{40}$$

Estimation of a certain kind of trigonometric sum. Before proceeding we prove two lemmas.

LEMMA 6. *Let $U \geq 1$, and α be a real number given in the form*

$$\alpha = \frac{a}{q} + \frac{\theta}{q\tau}, \ (a,q) = 1, \ 0 < q \leq \tau.$$

Then the following statements are valid:

1) *If $m > 0, f, q'$ are integers satisfying*

$$(m,q) = \delta, \quad m = m_1\delta, \quad q = q_1\delta, \quad 0 < q' \leq q_1,$$

[232] From 1 to N? *Trans.*

[233] E. Landau, "Über die neue Winogradoffsche Behandlung des Waringschen Problems. Anhang No. 2", *Math. Zeitschr.*, 31, 1929, pp. 319–338.

then

$$\Omega = \sum_{h=f+1}^{f+q'} \min\left(U, \frac{1}{2(\alpha m h)}\right) \ll \left(\frac{m_1 q_1}{\tau} + 1\right) U + q_1 \ln q_1.$$[234]

2) *If in addition we have* $-V < f+1 < f+q' < V$, *then*

$$\Omega \ll \left(\frac{m_1 V}{\tau} + 1\right) U + q_1 \ln q_1.$$

3) *Finally, if* $m = 1,\ f = 0,\ q' \le \frac{\tau}{2}$, *then*

$$\Omega \ll q \ln q.$$

PROOF. To prove the first statement, in the sum Ω we set $h = f + h_1,\ h_1 = 1, 2, \ldots, q'$. Clearly $\alpha m f = \frac{b}{q_1} + \frac{\theta'}{q_1}$, where b is an integer. Defining ρ by

$$am_1 h_1 + b \equiv \pm\rho \pmod{q_1}, \quad 0 \le \rho \le \left[\frac{q_1}{2}\right], \tag{41}$$

and noting that the function (x) (the distance from x to the nearest integer) is even and of period 1, we may write

$$(\alpha m h) = (\alpha m h_1 + \alpha m f) = \left(\frac{am_1 h_1 + b}{q_1} + \frac{\theta m_1 h_1}{q_1 \tau} + \frac{\theta'}{q_1}\right) = \left(\frac{\rho}{q_1} + \frac{\theta'' \sigma}{q_1}\right),$$

where $\sigma = \frac{m_1 q_1}{\tau} + 1$. Since $(am_1, q_1) = 1$, it follows from (41) that as h_1 ranges over the values $1, 2, \ldots, q'$ $(q' \le q_1)$, the number ρ takes on each of its allowed values

$$\rho = 0, 1, 2, \ldots, \left[\frac{q_1}{2}\right],$$

at most twice. Hence

$$\Omega \le 2 \sum_{\rho} \min\left(U, \frac{1}{2\left(\frac{\rho + \theta''\sigma}{q_1}\right)}\right). \tag{42}$$

Consider first the case $q_1 \ge 4(\sigma + 2)$; in this case the terms of the sum (42) for which

$$\rho = 0, 1, \ldots, [\sigma + 1], \text{ or } \rho \ge \frac{q_1}{2} - \sigma,$$

are equal to U, and obviously the number of such terms is $O(\sigma)$. For the remaining terms in (42) we set

$$\rho = [\sigma + 1] + s, \qquad 0 < s < \frac{q_1}{2} - 2\sigma < \frac{q_1}{2};$$

then

$$\frac{\rho + \theta''\sigma}{q_1} \le \frac{\rho + \sigma}{q_1} \le \frac{1}{2}, \quad \frac{\rho + \theta''\sigma}{q_1} \ge \frac{\rho - \sigma}{q_1} > \frac{s}{q_1}, \text{ whence } \left(\frac{\rho + \theta''\sigma}{q_1}\right) > \frac{s}{q_1}.$$

Hence

$$\Omega \ll \sigma U + \sum_{s=1}^{\frac{q_1}{2}} \frac{q_1}{2s} \ll \sigma U + q_1 \ln q_1 \ll \left(\frac{m_1 q_1}{\tau} + 1\right) U + q_1 \ln q_1.$$

In the case $q_1 < 4(\sigma + 2)$, the assertion 1) follows immediately since then

$$\Omega \ll q_1 U \ll \sigma U.$$

[234] If $(\alpha m h) = 0$ we take $\min\left(U, \frac{1}{0}\right) = U$.

This completes the proof of the first statement of the lemma. The proofs of the second and third statements are similar. □

LEMMA 7. *Let*

$$\alpha = \frac{a}{q} + \frac{\theta}{q^2}, \ (a, q) = 1, \ q > 0,$$

$M_1, \dots, M_s$ *be arbitrary integers,* $Z_1, \dots, Z_s$ *arbitrary positive integers,* $z_1, \dots, z_s$ *variables taking on the respective values*

$$z_j = M_j + 1, \dots, M_j + Z_j, \quad j = 1, 2, \dots, s,$$

and y *a variable ranging over* Y *distinct positive integers not exceeding* q. *Write*

$$T = \sum_{z_1} \cdots \sum_{z_s} \psi(z_1, \dots, z_s) \left| \sum_{y} \phi(y) e^{2\pi i (\chi(z_1, \dots, z_{s-1}, y) + \alpha y z_s)} \right|,$$

where $\psi(z_1, \dots, z_s)$ *satisfies the conditions*

$$\psi(z_1, \dots, z_s) \geq 0, \qquad \sum_{z_1} \cdots \sum_{z_s} [\psi(z_1, \dots, z_s)]^2 \leq W,$$

$\phi(y)$ *satisfies the condition*

$$0 \leq \phi(y) \leq \eta,$$

and, finally, $\chi(z_1, \dots, z_{s-1}, y)$ *is real-valued over the indicated range of* $z_1, \dots, z_{s-1}$, y. *We then have*

$$T \ll \eta \sqrt{W Z_1 \cdots Z_s Y (1 + Z_s^{-1} q \ln q)}.$$

PROOF. Applying to the sum T the Cauchy–Schwarz inequality for real x_i, y_i:

$$(x_1 y_1 + \cdots + x_k y_k)^2 \leq (x_1^2 + \cdots + x_k^2)(y_1^2 + \cdots + y_k^2),$$

we obtain

$$T^2 \leq W \sum_{z_1} \cdots \sum_{z_s} \left| \sum_{y} \phi(y) e^{2\pi i (\chi(z_1, \dots, z_{s-1}, y) + \alpha y z_s)} \right|^2. \tag{43}$$

Using $|\omega|^2 = \omega \bar{\omega}$, where $\bar{\omega}$ denotes the complex conjugate of the complex number ω, and introducing a new variable y_1 taking the same values as y, we can rewrite the multiple summation in (43) as follows:

(44)

$$\begin{aligned} &\sum_{z_1} \cdots \sum_{z_s} \left| \sum_{y} \phi(y) e^{2\pi i (\chi(z_1, \dots, z_{s-1}, y) + \alpha y z_s)} \right|^2 \\ &\quad = \sum_{z_1, \dots, z_s} \sum_{y} \sum_{y_1} \phi(y) \phi(y_1) e^{2\pi i [\chi(z_1, \dots, z_{s-1}, y) - \chi(z_1, \dots, z_{s-1}, y_1) + \alpha (y - y_1) z_s]} \\ &\quad = \sum_{z_1, \dots, z_{s-1}} \sum_{y, y_1} \phi(y) \phi(y_1) e^{2\pi i (\chi(z_1, \dots, z_{s-1}, y) - \chi(z_1, \dots, z_{s-1}, y_1))} \sum_{z_s} e^{2\pi i \alpha (y - y_1) z_s} \\ &\quad \leq Z_1 \cdots Z_{s-1} \sum_{y} \sum_{y_1} \phi(y) \phi(y_1) \left| \sum_{z_s} e^{2\pi i \alpha (y - y_1) z_s} \right|. \end{aligned}$$

The sum between the modulus signs in the last line of (44) has the form

$$S = \sum_{Z = M+1}^{N} e^{2\pi i \beta Z},$$

where β is real and M, N are integers. One can bound this sum in two ways: On the one hand clearly $|S| \le N - M$, while on the other

$$|S| = \left| \frac{e^{2\pi i\beta(N+1)} - e^{2\pi i\beta(M+1)}}{2ie^{\pi i\beta} \sin \pi\beta} \right| \le \frac{1}{|\sin \pi\beta|} = \frac{1}{\sin \pi(\beta)} \le \frac{1}{2(\beta)},$$

so that

$$|S| \le \min\left(N - M, \frac{1}{2(\beta)}\right). \tag{45}$$

Hence for the summations in (44) we infer that

$$\sum_y \sum_{y_1} \phi(y)\phi(y_1) \left| \sum_{z_s} e^{2\pi i\alpha(y-y_1)z_s} \right| \le \eta^2 Y \cdot \sum_{h=-q+1}^{q-1} \min\left(Z_s, \frac{1}{2(\alpha h)}\right).$$

By the first statement of Lemma 6 (with $m = 1, q_1 = q, \tau = q, U = Z_s$) the last summation in this inequality is $O(Z_s + q \ln q)$, whence

$$\sum_y \sum_{y_1} \phi(y)\phi(y_1) \left| \sum_{z_s} e^{2\pi i\alpha(y-y_1)z_s} \right| \ll \eta^2 Y Z_s (1 + Z_s^{-1} q \ln q). \tag{46}$$

The lemma is now immediate from (43), (44) and (46). □

The integral I_N. Let N be a sufficiently large integer (for which we wish to solve Waring's problem). We shall now assume that

$$r = 4n,$$

and write

$$P = [N^\nu], \; Y = [P^{\nu-\nu^2}].$$

Consider the integral

$$I_N = \int_0^1 L_\alpha^r U_\alpha^2 V_\alpha e^{-2\pi i\alpha N} d\alpha, \tag{47}$$

where

$$L_\alpha = \sum_{x=1}^{P} e^{2\pi i\alpha x^n}, \quad U_\alpha = \sum_u e^{2\pi i\alpha u}, \quad V_\alpha = \sum_v \sum_{t=1}^{Y} e^{2\pi i\alpha vt^n},$$

with u and v ranging over sequences of integers to be specified below. This integral may be rewritten as follows:

$$I_N = \sum_{x_1=1}^{P} \cdots \sum_{x_r=1}^{P} \sum_u \sum_{u_1} \sum_v \sum_{t=1}^{Y} \int_0^1 e^{2\pi i\alpha(x_1^n + \cdots + x_r^n + u + u_1 + vt^n - N)} d\alpha,$$

where u_1 ranges independently over the same sequence of values as u. In view of (27), we see from this that I_N is just the number of different representations of N in the form

$$N = x_1^n + \cdots + x_r^n + u + u_1 + vt^n.$$

The numbers u and v. Let $k = k(n)$ be a positive integer to be specified below, and

$$(48) \qquad \sigma = (1-\nu)^k.$$

Define the sequence of integers $P_1, \ldots, P_k$ successively by

$$(49) \qquad P_1 = \left[\frac{1}{4}P\right], \ P_2 = \left[\frac{1}{2}P_1^{1-\nu}\right], \ P_3 = \left[\frac{1}{2}P_2^{1-\nu}\right], \ldots, P_k = \left[\frac{1}{2}P_{k-1}^{1-\nu}\right].$$

Clearly, these numbers can all be made as large as we please by taking N sufficiently large. Now let ξ_j denote variables taking the following values:

$$\xi_j = P_j, P_j+1, \ldots, 2P_j - 1, \quad j = 1, 2, \ldots, k.$$

For each $i = 1, 2, \ldots, k$, we write

$$u_i = \xi_1^n + \cdots + \xi_i^n,$$

and denote by $\mathcal{E}_i$ the set of all[235] u_i together with $(2P_1)^n$. It is easy to see that all elements of $\mathcal{E}_i$ lie between P_1^n and $(2P_1)^n$, are distinct, and that the difference between any two numbers u', u'' from $\mathcal{E}_i$ that are neighbouring in size, satisfies the condition

$$u'' - u' > nP_i^{n-1}.$$

The inductive verification of the last of these claims is as follows: Taking $i = 1$, we have that the set $\mathcal{E}_1$ has as its elements

$$P_1^n, \ (P_1+1)^n, \ldots, (2P_1-1)^n, \ (2P_1)^n,$$

and the difference between neighbouring numbers of this sequence satisfies

$$u'' - u' \geq (P_1+1)^n - P_1^n > nP_1^{n-1}.$$

Suppose inductively that the claim in question holds for some $\mathcal{E}_i$ with $i < k$; we shall show that then it holds for $\mathcal{E}_{i+1}$. Let u', u'' be two numbers neighbouring in size, of the set $\mathcal{E}_i$ (so that $u' < u'' \leq (2P_i)^n$); to obtain $\mathcal{E}_{i+1}$ from $\mathcal{E}_i$ we need to add to each number in $\mathcal{E}_i$ all values taken by ξ_{i+1}^n. We add these values to u', obtaining

$$(50) \qquad u' + P_{i+1}^n, u' + (P_{i+1}+1)^n, \ldots, u' + (2P_{i+1}-1)^n,$$

and compare these numbers with u''; using the inductive assumption $u'' - u' > nP_i^{n-1}$ together with the definition of the number P_i, we obtain

$$u'' - (u' + (2P_{i+1}-1)^n) > nP_i^{n-1} - P_i^{(1-\nu)n} = (n-1)P_i^{n-1} > nP_{i+1}^{n-1},$$

since by taking N sufficiently large the ratio $\frac{P_{i+1}}{P_i}$ can be made as small as necessary. The difference between any two neighbouring numbers in the list (50) is also greater than nP_{i+1}^{n-1}. Hence there being no numbers from $\mathcal{E}_{i+1}$ in the interval (u', u''), other than those of the form $u' + \xi_{i+1}^n$, the desired conclusion follows for $\mathcal{E}_{i+1}$.

The values of u that we had yet to specify, we now define to be the numbers in the set $\mathcal{E}_k$. It follows that these are distinct from one another, lie between P_1^n and $(2P_1)^n \leq \left(\frac{P}{2}\right)^n$, and that each of them is a sum of k terms of the form ξ^n where ξ is a positive integer. Denoting by X the number of values taken by u,[236] and recalling (48), we have

$$(51) \qquad X = P_1 P_2 \cdots P_k \gg P_1^{1+(1-\nu)+(1-\nu)^2+\cdots+(1-\nu)^{k-1}} \gg P^{n-n\sigma}.$$

[235] As $\xi_1, \xi_2, \ldots, \xi_i$ range over their prescribed values. *Trans.*

[236] That is, the size of the set $\mathcal{E}_k$. *Trans.*

Analogously, writing

$$R = [P^{1-\nu}], \quad R_1 = \left[\frac{1}{2}R\right],$$

we generate a sequence of values for v lying between R_1^n and $(2R_1)^n \leq R^n$, distinct from one another, and each having the form of a sum of k nth powers η^n for some positive integer η. As before we shall have that the number X_1 of these values of v satisfies

$$X_1 \gg R^{n-n\sigma}. \tag{52}$$

Estimation of $I_N^{(II)}$. Set

$$\tau = 2nP^{n-1}$$

as before, and in the formula (47) replace the interval of integration by the interval

$$-\frac{1}{\tau} \leq \alpha \leq 1 - \frac{1}{\tau}.$$

Corresponding to the partition of this interval into class I and class II subintervals, the integral I_N decomposes as a sum $I_N = I_N^{(I)} + I_N^{(II)}$. We wish now to estimate $I_N^{(II)}$. Thus suppose α lies in a class II subinterval; this means that

$$\alpha = \frac{a}{q} + z, \quad (a, q) = 1, \quad P^{1-\nu} < q \leq \tau, \quad |z| \leq \frac{1}{q\tau}.$$

For the sum

$$V_\alpha = \sum_v \sum_{t=1}^{Y} e^{2\pi i \alpha v t^n}$$

we have

$$V_\alpha \ll \sum_v \left| \sum_{t=1}^{Y} e^{2\pi i \alpha v t^n} \right| = T.$$

Here T is a sum of the form considered in Lemma 7, namely

$$T = \sum_{z=1}^{R^n} \psi(z) \left| \sum_y e^{2\pi i \alpha z y} \right|,$$

where $\psi(z) = 1$ whenever z takes on one of the values v, and is otherwise zero, and y ranges over the numbers t^n, $t = 1, 2, \ldots, Y$, with Y restricted by $Y^n \leq P^{1-\nu} < q$. Applying Lemma 7 with $s = 1,\ Z_1 = R^n,\ \phi(y) = 1,\ W = X_1,\ \chi = 0$, we obtain

$$V_\alpha \ll \sqrt{X_1 R^n Y (1 + R^{-n} q \ln q)}.$$

Invoking the estimates $q \ll R^{n-1} \ll R^n$, and (52), we obtain

$$\begin{aligned} V_\alpha &\ll \sqrt{X_1 R^n Y \ln P} \\ &= X_1 Y \sqrt{X_1^{-1} R^n Y^{-1} \ln P} \ll X_1 Y R^{\frac{1}{2}n\sigma} Y^{-\frac{1}{2}} \ln P. \end{aligned}$$

Together with the obvious estimate

$$L_\alpha = \sum_{x=1}^{P} e^{2\pi i \alpha x^n} \ll P,$$

this yields

$$I_N^{(II)} \ll P^r X_1 Y R^{\frac{n\sigma}{2}} Y^{-\frac{1}{2}} \ln P \int_0^1 |U_\alpha|^2 d\alpha.$$

Note that here the integral is taken over the whole of the interval $0 \leq \alpha \leq 1$, rather than just the class II subintervals, but that this can result only in a larger value for that integral. This integral is of a type well-studied in Fourier analysis. From $|\omega|^2 = \omega\bar{\omega}$, and the formulae (27) and (51), we obtain

$$\begin{aligned}\int_0^1 |U_\alpha|^2 d\alpha &= \int_0^1 \left(\sum_u e^{2\pi i \alpha u} \sum_{u_1} e^{-2\pi i \alpha u_1}\right) d\alpha \\ &= \sum_u \sum_{u_1} \int_0^1 e^{2\pi i \alpha (u-u_1)} d\alpha = X = X^2 \cdot X^{-1} \ll X^2 \cdot P^{-n+n\sigma}.\end{aligned}$$

Hence since $R \leq P^{1-\nu}, Y \leq P^{\nu-\nu^2}$, we have

$$\begin{aligned}I_N^{(II)} &\ll P^{r-n} X^2 X_1 Y R^{\frac{n\sigma}{2}} P^{n\sigma} Y^{-\frac{1}{2}} \ln P \\ &\ll P^{r-n} X^2 X_1 Y P^{\left(\frac{3}{2}n - \frac{1}{2}\right)\sigma - \frac{1}{2}(\nu-\nu^2)} \ln P.\end{aligned} \tag{53}$$

Choosing

$$k = [2n \ln n] + 2n + 1, \tag{54}$$

whence

$$k > 2n \ln n + 2n > 2n \ln n + n \ln 4,$$

and invoking the well-known inequality

$$\left(1 - \frac{1}{n}\right)^n < e^{-1},$$

we obtain with this choice of k

$$\begin{aligned}\left(\frac{3}{2}n - \frac{1}{2}\right)\sigma - \frac{1}{2}(\nu - \nu^2) &< \left(\frac{3}{2}n - \frac{1}{2}\right)(1-\nu)^{2n\ln n + n \ln 4} - \frac{1}{2}(\nu - \nu^2) \\ &< \left(\frac{3}{2}n - \frac{1}{2}\right)\frac{1}{4n^2} - \frac{1}{2n} + \frac{1}{2n^2} = -\frac{1}{8n} + \frac{3}{8n^2}.\end{aligned}$$

Hence there exists a positive $\delta = \delta(n)$ such that

$$\left(\frac{3}{2}n - \frac{1}{2}\right)\sigma - \frac{1}{2}(\nu - \nu^2) < -\delta,$$

and since of course this then holds with any smaller number replacing δ, we infer that for all sufficiently large N,

$$P^{\left(\frac{3}{2}n - \frac{1}{2}\right)\sigma - \frac{1}{2}(\nu-\nu^2)} \ln P \ll P^{-\delta}.$$

Combining this with (53), we obtain finally (with $r = 4n$)

$$I_N^{(II)} \ll X^2 X_1 Y P^{3n-\delta}. \tag{55}$$

Estimation of I_N. We now turn to the calculation of the first part of the integral I_N, namely the sum

$$I_N^{(I)} = \sum_{\Delta} \int_{\Delta} L_\alpha^r U_\alpha^2 V_\alpha e^{-2\pi i\alpha N} d\alpha,$$

over the totality Δ of intervals of class I. Replacing U_α and V_α by their expressions as sums, and interchanging the order of summation and integration, we obtain

$$I_N^{(I)} = \sum_{u} \sum_{u_1} \sum_{v} \sum_{t=1}^{Y} \sum_{\Delta} \int_{\Delta} L_\alpha^r e^{-2\pi i\alpha(N-u-u_1-vt^n)} d\alpha.$$

Here the sum

$$\sum_{\Delta} \int_{\Delta} L_\alpha^r e^{-2\pi i\alpha(N-u-u_1-vt^n)} d\alpha$$

is just the first part $W_N^{(I)}$ say, of the integral W_N (see (32)) with

$$N_0 = N - u - u_1 - vt^n \tag{56}$$

in place of N; hence we have

$$I_N^{(I)} = \sum_{N_0} W_{N_0}^{(I)},$$

where the summation is taken over all N_0 given by (56), with u, u_1, v, t ranging over their above-indicated values. The conditions satisfied by u and v entail that for sufficiently large N we have

$$N_0 > N - 2\left(\frac{P}{2}\right)^n - R^n Y^n \geq N\left(1 - \frac{1}{2^{n-1}} - \frac{1}{P^\nu}\right) > \frac{N}{2}.$$

Hence we may apply to $W_{N_0}^{(I)}$ the asymptotic formula (40) (with $r = 4n$), to obtain

$$W_{N_0}^{(I)} = K_n N_0^3 \mathfrak{S}(N_0, 4n, n) + O(P^{3n-\nu}), \quad K_n = \frac{[\Gamma(1+\nu)]^{4n}}{6}. \tag{57}$$

Since the sizes of the domains of the variables u, v, t are respectively X, X_1, Y, the number of values taken by N_0 (see (56)) is

$$T = X^2 X_1 Y.$$

Summing the equation (57) over all N_0, we obtain

$$I_N^{(I)} = K_n \sum_{N_0} N_0^3 \mathfrak{S}(N_0, 4n, n) + O(TP^{3n-\nu}).$$

Combining this with (55), which may now be rewritten as

$$I_N^{(II)} = O(TP^{3n-\delta}),$$

and choosing $\delta < \nu$, we obtain finally

$$I_N = K_n \sum_{N_0} N_0^3 \mathfrak{S}(N_0, 4n, n) + O(TN^{3-\delta\nu}). \tag{58}$$

Estimate of $G(n)$. In view of the properties of the series $\mathfrak{S}$ indicated earlier (see the formula (26)), we have for all N_0

$$\mathfrak{S}(N_0, 4n, n) > c = c(n) > 0.$$

This together with the inequality $N_0 > \frac{N}{2}$, gives us that the main term on the right-hand side of (58) exceeds

$$K_n \left(\frac{N}{2}\right)^3 cT = K' N^3 T,$$

where $K' = K'(n)$ is constant. On the other hand the error term in (58) is in absolute value less than $LN^3T \cdot N^{-\delta\nu}$ for some positive constant $L = L(n)$. Hence from (58) we infer that for all sufficiently large N we have $I_N > 0$. Recalling now that I_N is the number of different representations of N in the form

$$N = x_1^n + \cdots + x_{4n}^n + u + u_1 + vt^n,$$

where each of u, u_1, v is a sum of k terms of the form ξ^n with ξ a positive integer, we see that we have shown that every sufficiently large integer N is expressible as a sum of $4n+3k$ nth powers of positive integers, i.e., have solved Waring's problem in the affirmative, and have furthermore obtained the following bound for the number of nth power summands sufficing:

$$G(n) \leq 4n + 3k < 6n \ln n + 11n.$$

The Goldbach Problem

In 1742 the St. Petersburg academician Ch. Goldbach, in a letter to L. Euler,[237] made the conjecture that every even number is the sum of two primes. This conjecture, supplemented in the obvious way for odd numbers, and widely known in our time as "Goldbach's conjecture", may be precisely formulated as follows:

Every even number ≥ 6 *is the sum of two, and every odd number* ≥ 9 *the sum of three, odd prime numbers.*

By the simplicity of its formulation together with the difficulty of finding a proof, this conjecture has attracted the attention of many mathematicians. There is now an extensive and interesting literature devoted to it.[238] Several mathematicians (Sylvester, Stäckel, Landau, Brun) attempted to find by heuristic means a formula for the number of representations of an even number as a sum of two primes, while others (Cunningham (1906), Maillet (1911)) invented various individual methods for verifying the conjecture numerically for very large numbers; however, none of these approaches was successful in settling the conjecture, and as relatively recently as 1912 there was an opinion abroad that the conjecture was beyond the reach of contemporary mathematical means.[239]

In 1923 there appeared the third[240] in the series of six memoirs of Hardy and Littlewood mentioned above. In this memoir they apply their method to the question of the decomposability of an odd number as a sum of three primes. However, their arguments depend on certain assumptions concerning the roots of so-called "Dirichlet L-series", i.e., infinite series of the form

$$L(s,\chi)=\sum_{n=1}^{\infty}\frac{\chi(n)}{n^s},$$

where $\chi(n)$ is an integer-valued function with the property $\chi(mn) = \chi(m)\chi(n)$, taking the same value at arguments congruent modulo a certain integer k.[241] The series $L(s,\chi)$ represents an analytic function of the complex variable $s=\sigma+it$. The assumption made by Hardy and Littlewood may in its simplest form be formulated as follows: no function of the form $L(s,\chi)$ has roots, i.e., zeroes, in the half-plane $\sigma > \frac{1}{2}$. This assumption is reminiscent of the Riemann hypothesis, mentioned

[237] "Corresp. Math. et Phys. de célèbres géomètres du 18-ième siècle" (1843), I, pp. 127 et 135. (Letter dated 7th June, 1742.)

[238] Details of the history of Goldbach's conjecture may be found in: L. E. Dickson, "History of the theory of numbers", vol. I, pp. 421–425. See also: N. G. Chudakov, "On Goldbach's problem", *Uspekhi Mat. Nauk*, IV, pp. 14–33.

[239] Expressed in Landau's lecture at the International Congress of Mathematicians, Cambridge, 1912.

[240] G. H. Hardy and J. E. Littlewood, "Some problems of partitio numerorum III: On the expression of a number as a sum of primes", *Acta Math.*, t. 44, pp. 1–70.

[241] The function $\chi(n)$ is called a *character* mod k.

earlier,[242] and like it remains to this day unproven. Under this assumption Hardy and Littlewood show that every sufficiently large odd number can be represented as a sum of three primes. Thus the Hardy–Littlewood proof is conditional; it in effect merely transfers the difficulty of the problem to a different area of number theory, namely to that concerned with the distribution of zeroes of functions related to the Riemann ζ-function $\zeta(s)$.

As far as the original conjecture (that every even number is representable as a sum of two primes) is concerned, Hardy and Littlewood were unable to prove it fully even under the assumption about the zeroes of L-functions; they were able to show only that "almost all" even numbers are so representable,[243] in the sense that if $f(m)$ denotes the number of even numbers in the sequence $1, 2, \ldots, m$, not representable as a sum of two primes, then

$$\lim_{m\to\infty} \frac{f(m)}{m} = 0. \tag{59}$$

In 1930 there appeared an article by the Soviet mathematician Shnirelman,[244] in which he proposed a new approach to solving problems of additive number theory. Consider an infinite sequence

$$n_1, n_2, n_3, \ldots, \tag{60}$$

of natural numbers, and let $N(x)$, $x > 0$, be the number of terms not exceeding x. We call the sequence (60) *almost dense* if $\frac{N(x)}{x} \geq \alpha$ (for some positive constant α) for all sufficiently large x. Shnirelman proved that if the sequence (60) is almost dense, then every positive integer N can be expressed in the form

$$N = n_{\alpha_1} + n_{\alpha_2} + \cdots + n_{\alpha_k} + \nu, \tag{61}$$

where k and ν are bounded above independently of N. The sequence $p_1, p_2, \ldots,$ of primes is *not* almost dense, since the number $\pi(x)$ of primes not exceeding x satisfies the formula[245]

$$\pi(x) = O\left(\frac{x}{\ln x}\right). \tag{62}$$

However, Shnirelman proved that the sequence of numbers of the form $p_i + p_j$ *is* almost dense, so that the formula (61) applies to it; thus since in that formula ν is the sum of a bounded number of primes, we infer the existence of an absolute constant r such that every integer N can be expressed as the sum of at most r primes. However, the number r afforded by Shnirelman's method turns out to be very large; minute analysis of that method, carried out by various mathematicians (Romanov, Landau, Heilbronn, Ricci), succeeded in reducing that number only to 67.

These then are the two chief approaches to the Goldbach problem antedating the work of Vinogradov. In both cases, although as a result new and fruitful ideas were introduced into additive number theory, there was little advance towards a solution of the problem itself: the work of Hardy and Littlewood reduced it to a different unsolved problem, while that of Shnirelman, although it contains a fully

---

[242] See the chapter on Chebyshev.

[243] "Some problems of partitio numerorum", V, 1924. See also: Landau, "Vorlesungen", Bd. I, V. Teil.

[244] "On additive properties of numbers", *Izv. Donsk. Politekhn. Inst.*, Novocherkassk, 1930; see also: L. Schnirelmann, "Über additive Eigenschaften von Zahlen", *Math. Ann.*, 1933.

[245] See the chapter on Chebyshev.

proven result (i.e., that involving 67 prime summands), can hardly be reckoned as a satisfactory approximation to a solution of Goldbach's problem as originally stated, since its chief merit, namely its simplicity, has been lost.

In 1937 there appeared the work of Vinogradov[246] in which he proved that every odd number from some point on is indeed expressible as a sum of three primes. His method, exceptional for its depth and elegance, is in essence the same as that employed earlier in his solution of Waring's problem, i.e., the number of representations of a number in the desired form is expressed as an integral over an interval which is subdivided into smaller subintervals. Thus Vinogradov's method is ultimately related to that of Hardy and Littlewood, and not to Shnirelman's.

Vinogradov's proof is divided into two parts. The first is devoted to the derivation of the dominant term in the formula for the number of representations of an odd number as a sum of three primes. Here Vinogradov uses known laws governing the distribution of primes in arithmetic progressions; such laws have been the subject of close investigation in western-European science in the work of a whole series of mathematicians (Dirichlet, Riemann, Hadamard, Siegel), and the results obtained by them are of essential help here. The second part of the proof is concerned with estimating the sum

$$\sum_{p\leq N} e^{2\pi i\alpha p},$$

taken over all primes p not exceeding a given bound N. (Here α is a certain real number.) The final result is achieved in part by means of methods invented by Vinogradov for estimating trigonometric sums, and in part by applying new ideas based ultimately on the well-known "sieve of Eratosthenes" of ancient times. In a comparison of the methods of Hardy–Littlewood and Vinogradov, this second part would correspond to the assumption concerning the zeroes of L-series that Hardy and Littlewood were compelled to make in their proof. Vinogradov's argumentation in this part of his proof is so elementary, that it would be difficult to find in all of the number-theoretic literature a comparable example of research leading to such a deep result by such simple means.

Thus, as for Waring's problem, so also as far as another long-unsolved problem of number theory is concerned—the decomposability of odd numbers as sums of three primes—we may regard Vinogradov's work as the final word.

The original conjecture that every even number is a sum of two primes continues to await solution. In this direction N. G. Chudakov, on the basis of Vinogradov's work, showed,[247] sharpening the conditional formula (59), that

$$f(m) = O\left(\frac{m}{(\ln m)^a}\right),$$

for every positive number a.

Estimating auxiliary trigonometric sums. To begin our exposition of Vinogradov's approach to the solution of the Goldbach problem,[248] we prove certain lemmas on trigonometric sums.

[246] I. M. Vinogradov, "The representation of an odd number as a sum of three prime numbers", *DAN SSSR*, vol. XV, Nos. 6, 7.

[247] N. G. Chudakov, "On the density of the set of even numbers not representable as a sum of two odd primes", *IAN SSSR*, 1938.

[248] "A new method in analytic number theory", 1937, Chapter IX.

LEMMA 8. *Let $\tau > 1$, $U > 1$, and α be real numbers, W, m positive integers, and, as in Lemma 6,*

$$\alpha = \frac{a}{q} + \frac{\theta}{q\tau}, \quad (a,q) = 1, \quad 0 < q \le \tau,$$
$$\delta = (m,q), \quad m = m_1\delta, \quad q = q_1\delta.$$

Then the sum

$$H = \sum_{h=-W}^{W} \min\left(U, \frac{1}{2(\alpha m h)}\right)$$

satisfies

(63) $$H \ll UW\left(\frac{\ln q_1}{U} + \frac{1}{W} + \frac{1}{q_1} + \frac{q_1 \ln q_1}{UW} + \frac{m_1}{\tau}\right).$$

PROOF. This lemma follows easily from Lemma 6. If $W \ge q_1$, the sum H decomposes into summands of the form

$$\Omega = \sum_{h=f+1}^{f+q'} \min\left(U, \frac{1}{2(\alpha m h)}\right),$$

in all of which, with the possible exception of the last, where $0 < q' \le q_1$, we have $q' = q_1$. The number of such summands is clearly $O\left(\frac{W}{q_1}\right)$. Applying the first statement of Lemma 6 to each of these summands, we obtain

$$H \ll \frac{W}{q_1}\left(\left(\frac{m_1 q_1}{\tau} + 1\right)U + q_1 \ln q_1\right)$$
$$= UW\left(\frac{m_1}{\tau} + \frac{1}{q_1} + \frac{\ln q_1}{U}\right),$$

whence (63). If on the other hand $W < q_1$, then we apply the second statement of Lemma 6, obtaining

$$H \ll \sum_{h=1}^{W} \min\left(U, \frac{1}{2(\alpha m h)}\right) \ll \left(\frac{m_1 W}{\tau} + 1\right)U + q_1 \ln q_1$$
$$= UW\left(\frac{m_1}{\tau} + \frac{1}{W} + \frac{q_1 \ln q_1}{UW}\right),$$

whence again (63) follows, completing the proof of the lemma. □

LEMMA 9. *Suppose given two increasing sequences $(x), (y)$ of positive integers, a positive integer m, and*

$$0 \le N' < N_1, \quad 1 < U \le N_1, \quad U_1 \le 2U,$$
$$\alpha = \frac{a}{q} + \frac{\theta}{q\tau}, \quad (a,q) = 1, \quad 0 < q \le \tau,$$
$$\delta = (m,q), \quad m = m_1\delta, \quad q = q_1\delta.$$

Write

$$T = \sum_x \sum_y e^{2\pi i \alpha m x y},$$

where x ranges over the terms of the sequence (x) satisfying the condition

$$U < x \le U_1,$$

and for each such x the index y ranges over all terms of the sequence (y) satisfying

$$\frac{N'}{x} < y \le \frac{N_1}{x}.$$

Then

$$T \ll N_1 \sqrt{\frac{\ln q_1}{U} + \frac{U}{N_1} + \frac{1}{q_1} + \frac{q_1 \ln q_1}{N_1} + \frac{m_1}{\tau}}. \tag{64}$$

PROOF. By the Cauchy–Schwarz inequality we have

$$T^2 \ll U \sum_x \left| \sum_y e^{2\pi i \alpha m x y} \right|^2 \le U \sum_{\xi > U}^{U_1} \left| \sum_y e^{2\pi i \alpha m \xi y} \right|^2,$$

where in the last sum the variable ξ ranges now over all integers in the interval $(U, U_1]$, and for each such value of ξ the variable y ranges as before over all terms of the sequence (y) satisfying

$$\frac{N'}{\xi} < y \le \frac{N_1}{\xi}.$$

Hence

$$T^2 \ll U \sum_\xi \sum_y \sum_{y_1} e^{2\pi i \alpha m \xi (y - y_1)},$$

where y_1 ranges over the same set of values as y.

We now change the order of summation. We shall sum over all y, y_1 satisfying the conditions

$$0 < y \le W, \quad 0 < y_1 \le W, \quad \text{where } W = \left[\frac{N_1}{U}\right],$$

and for each such pair of values of y, y_1, over all ξ satisfying the conditions

$$\frac{N'}{y} < \xi \le \frac{N_1}{y}, \quad \frac{N'}{y_1} < \xi \le \frac{N_1}{y_1}, \quad U < \xi \le U_1;$$

thus ξ will range over all integers (if any) in the interval

$$\eta < \xi \le \eta_1, \text{ where } \eta = \max\left(\frac{N'}{y}, \frac{N'}{y_1}, U\right), \ \eta_1 = \min\left(\frac{N_1}{y}, \frac{N_1}{y_1}, U_1\right).$$

The sum

$$\sum_\xi e^{2\pi i \alpha m (y - y_1) \xi},$$

where ξ ranges over an interval of consecutive integers, is of the type

$$\sum_\xi e^{2\pi i \beta \xi},$$

for whose modulus we earlier obtained an upper bound (see (45)); in the present situation, this yields

$$\begin{aligned} T^2 &\ll U \sum_y \sum_{y_1} \sum_{\xi > \eta}^{\eta_1} e^{2\pi i \alpha m (y - y_1) \xi} \\ &\ll U \sum_y \sum_{y_1} \min\left(U, \frac{1}{2(\alpha m (y - y_1))}\right). \end{aligned}$$

Since $-W < y - y_1 < W$, and the number of pairs y, y_1 satisfying $y - y_1 = h$ for each admissible h is $O(W)$, it follows that

$$T^2 \ll UW \sum_{h=-W}^{W} \min\left(U, \frac{1}{2(\alpha m h)}\right).$$

The desired estimate (64) is now immediate from Lemma 8 and the fact that $N_1 \ll UW \ll N_1$. □

LEMMA 10. *Suppose given two increasing sequences $(x), (y)$ of positive integers, a positive integer m, a sufficiently large number N_1, and*

$$r_1 = \ln N_1, \quad 1 < N' \leq N_1, \quad 1 < U_0 < U_1 \leq N_1, \quad 1 < \tau < N_1,$$

$$\alpha = \frac{a}{q} + \frac{\theta}{q\tau}, \quad (a, q) = 1, \quad 0 < q \leq \tau,$$

$$\delta = (m, q), \quad m = m_1\delta, \quad q = q_1\delta,$$

$$T = \sum_x \sum_y e^{2\pi i \alpha m x y},$$

where x ranges over all terms of (x) satisfying the condition

$$U_0 < x \leq U_1,$$

and for each such x the index y ranges over all terms of (y) satisfying

$$\frac{N'}{x} < y \leq \frac{N_1}{x}.$$

Then we have

$$T \ll N_1 \sqrt{\frac{r_1^2}{U_0} + \frac{U_1 r_1}{N_1} + \frac{q_1 r_1^3}{N_1} + \frac{r_1^2}{q_1} + \frac{m_1 r_1^2}{\tau}}. \tag{65}$$

PROOF. If we subdivide the interval $U_0 < x \leq U_1$ into the $k + 1 \ll r_1$ subintervals

$$U_0 < x \leq 2U_0, \ 2U_0 < x \leq 4U_0, \ldots, 2^k U_0 < x \leq U_1,$$

where $2^k U_0 < U_1 \leq 2^{k+1} U_0$, then the sum T will decompose into corresponding summands:

$$T = T_0 + T_1 + \cdots + T_k,$$

whence, by the Cauchy–Schwarz inequality,

$$T^2 \ll r_1 \sum_{s=0}^{k} |T_s|^2.$$

Applying Lemma 9 to each sum T_s, we obtain

$$T_s^2 \ll N_1^2 \left(\frac{\ln q_1}{2^s U_0} + \frac{2^s U_0}{N_1} + \frac{1}{q_1} + \frac{q_1 \ln q_1}{N_1} + \frac{m_1}{\tau}\right).$$

Summing this from $s = 0$ to $s = k$, and noting that

$$U_0 \sum_{s=0}^{k} 2^s \ll 2^k U_0 \ll U_1, \quad \ln q_1 \leq \ln \tau < r_1,$$

we obtain the desired result. □

Estimate for the basic trigonometric sum. We begin our exposition of Vinogradov's method by estimating for sufficiently large N the basic trigonometric sum

$$S = \sum_{p \le N} e^{2\pi i \alpha p},$$

where the summation is taken over the primes p in the indicated range. Write D for the product of all primes less than or equal to $\sqrt{N}$, and (d) for the sequence whose terms are in order the divisors of D. Further, write

$$r = \ln N,$$

let $h > 3$ be a constant, and let H be such as to satisfy the inequalities

$$r^h \le H \le e^{\frac{1}{3}\sqrt{r}}. \tag{66}$$

Note that any given power of H will be less than N provided N is sufficiently large, since

$$H^k \le e^{\frac{k}{3}\sqrt{r}} < e^r = N.$$

We write also

$$\tau_1 = \frac{N}{H}, \quad \tau = \frac{N}{H^3},$$

and express the number α figuring in the sum S in the form

$$\alpha = \frac{a}{q} + \frac{\theta}{q\tau}, \ (a, q) = 1, \ H^3 \le q \le \tau.$$

We shall need the following identity, valid for any function $f(z)$ of the positive integers:

$$\sum_{(z,D)=1} f(z) = \sum_{d \le N} \left(\mu(d) \sum_{z|d} f(z) \right), \tag{67}$$

where the sum

$$\sum_{(z,D)=1} f(z)$$

is taken over all positive integers $z \le N$ relatively prime to D, and the sum

$$\sum_{z|d} f(z)$$

over all positive integers $z \le N$ dividing d, with d ranging in turn over all terms of the sequence (d) not exceeding N; finally, $\mu(d)$ is the "Möbius function", defined for all positive integers by:

$\mu(1) = 1$;

$\mu(d) = 0$ if d is not square-free;

$\mu(d) = (-1)^k$ if $d = p_1 p_2 \cdots p_k$ for distinct primes p_i.

The identity (67) is easy to prove using well-known properties of the function $\mu(d)$. We now apply that identity with $f(z) = e^{2\pi i \alpha z}$. Write

$$S' = \sum_{(z,D)=1} e^{2\pi i \alpha z}, \quad S_d = \sum_{m=1}^{\frac{N}{d}} e^{2\pi i \alpha m d}.$$

If N' is any positive composite integer, then it has at least one prime divisor $\leq \sqrt{N'}$; hence the values of z over which the sum S' is taken will be just the primes p in the interval

$$\sqrt{N} < p \leq N,$$

together with 1. (Observe that it is at this juncture that the sieve of Eratosthenes is in effect being applied.) Hence we have

$$S' = e^{2\pi i\alpha} + \sum_{\sqrt{N}<p\leq N} e^{2\pi i\alpha p} = S + O(\sqrt{N}),$$

whence by (67)

$$S = \sum_{d\leq N} \mu(d) S_d + O(\sqrt{N}). \tag{68}$$

We now estimate the sum

$$\sum_{d\leq\tau_1} |S_d|.$$

Since

$$S_d \leq \min\left(\frac{N}{d}, \frac{1}{2(\alpha d)}\right),$$

we have

$$\sum_{d\leq\tau_1} |S_d| \leq \sum_{d\leq\frac{q}{2}} \min\left(\frac{N}{d}, \frac{1}{2(\alpha d)}\right) + \sum_{\frac{q}{2}<d\leq\tau_1} \min\left(\frac{N}{d}, \frac{1}{2(\alpha d)}\right). \tag{69}$$

Applying statement 3) of Lemma 6 to the first sum on the right-hand side in (69), we obtain

$$\sum_{d\leq\frac{q}{2}} \min\left(\frac{N}{d}, \frac{1}{2(\alpha d)}\right) \leq \sum_{h=1}^{\frac{q}{2}} \min\left(N, \frac{1}{2(\alpha h)}\right)$$
$$\ll q\ln q \ll NrH^{-3}.$$

The second sum on the right-hand side of (69) can be decomposed into several summands of the form

$$\Omega = \sum_{d>d_0}^{d_0+q'} \min\left(\frac{N}{d}, \frac{1}{2(\alpha d)}\right) \leq \sum_{h>d_0}^{d_0+q'} \min\left(\frac{N}{d_0}, \frac{1}{2(\alpha h)}\right),$$

where, as earlier, q' is equal to q except possibly in the last such summand, where $0 < q' \leq q$. Here the number d_0 ranges over the values

$$d_0 = \left[\frac{q}{2}\right], \left[\frac{q}{2}\right] + q, \ldots, \left[\frac{q}{2}\right] + s_1 q,$$

where s_1 is the largest integer satisfying the condition

$$\left[\frac{q}{2}\right] + s_1 q < \tau_1.$$

Noting that the number of summands Ω is $s_1+1 \ll \frac{\tau_1}{q}$, and applying statement 1) of Lemma 6 to each such summand, we obtain

$$\begin{aligned}\sum_{\frac{q}{2}<d\leq\tau_1} \min\left(\frac{N}{d}, \frac{1}{2(\alpha d)}\right) &\ll \sum_{s=0}^{s_1}\left(\left(\frac{q}{\tau}+1\right)\frac{N}{\left[\frac{q}{2}\right]+sq}+q\ln q\right)\\ &\ll \frac{N}{\left[\frac{q}{2}\right]}+\frac{N}{q}\sum_{s=1}^{s_1}\frac{1}{s}+\frac{\tau_1}{q}q\ln q\\ &\ll \frac{N}{q}\ln\frac{\tau_1}{q}+\tau_1\ln\tau \ll \frac{Nr}{H^3}+\frac{Nr}{H} \ll NrH^{-1}.\end{aligned}$$

Taking this into account in (69) gives us

$$\sum_{d\leq\tau_1}|S_d| \ll NrH^{-1},$$

and on substituting this in (68) we obtain the crucial estimate

$$S=\sum_{\tau_1<d\leq N}\mu(d)S_d+O(NrH^{-1}). \tag{70}$$

Transformation of the main term in (70). We first partition the sequence (d) into two subsequences (d_0) and (d_1), the first containing all terms of (d) with an even number of prime factors, and the second those d with an odd number of prime factors. The main term in (70) may then be rewritten as

$$\sum_{\tau_1<d\leq N}\mu(d)S_d=T_0-T_1, \text{ where } T_0=\sum_{(d_0)}S_d,\ T_1=\sum_{(d_1)}S_d, \tag{71}$$

where in both sums T_0 and T_1 the index d is in addition restricted to the interval $\tau_1<d\leq N$. It suffices to estimate T_0 since T_1 can be estimated similarly. We have

$$T_0=\sum_d\sum_m e^{2\pi i\alpha dm},$$

where d ranges over the terms of the sequence (d_0) satisfying

$$\tau_1<d\leq N,$$

and for each such value of d, the index m takes on the values

$$m=1,2,3,\dots,\left[\frac{N}{d}\right].$$

We now change the order of summation. Since $\left[\frac{N}{d}\right]<\frac{N}{\tau_1}=H$, the sum T_0 may be rewritten as

$$T_0=\sum_{m=1}^{H}T(m), \qquad T(m)=\sum_d e^{2\pi i\alpha md}, \tag{72}$$

where in the sum $T(m)$ the index d now ranges over the terms of the sequence (d_0) subject to

$$\tau_1<d_0\leq\frac{N}{m}.$$

The number of terms of the sequence (d') lying in the above interval. We now consider a different partition of the sequence (d) into two subsequences: Let (d') be the subsequence of those terms of (d) all of whose prime divisors are less than or equal to H^3, and (d'') the subsequence made up of the rest, i.e., with those terms d that have at least one prime divisor greater than H^3. We wish to estimate the number F of terms of (d') lying in the interval

$$\tau_1 < d' \leq \frac{N}{m} = N_1.$$

Consider a typical such term d', and suppose it has k prime divisors; then

$$\tau_1 < H^{3k}, \quad (3k+1)\ln H > r, \quad k > \sqrt{r} - \frac{1}{3}.$$

Hence the number $\tau(d')$ of distinct divisors of d' will satisfy (for sufficiently large N)

$$\tau(d') = 2^k > 2^{\sqrt{r}-\frac{1}{3}} > e^{0.69\left(\sqrt{r}-\frac{1}{3}\right)} > H.$$

On the other hand we have

$$\tau(1) + \tau(2) + \cdots + \tau([N_1]) = \left[\frac{N_1}{1}\right] + \left[\frac{N_1}{2}\right] + \cdots + \left[\frac{N_1}{N_1}\right] \ll N_1 r.$$

Hence

$$FH \ll N_1 r, \text{ whence } F \ll \frac{N}{m} r H^{-1}.$$

Estimate of the sum S. We begin by estimating the sum $T(m)$ (see (72)) for any given value of m. As before we write

$$N_1 = \frac{N}{m}.$$

We partition the sequence (d_0) into two subsequences (d_0') and (d_0'') made up of the terms of (d_0) in the sequences (d'), and (d'') respectively. The sum $T(m)$ then decomposes as a sum of corresponding summands:

$$T(m) = T'(m) + T''(m).$$

Since (d_0') is a subsequence of (d'), obviously the quantity $|T'(m)|$ is less than or equal to F, the number of terms of (d') contained in the interval $\tau_1 < d' \leq N_1$. Hence the above estimate for F gives immediately

$$T'(m) \ll \frac{N}{m} r H^{-1},$$

so that

$$T(m) = T''(m) + O\left(\frac{N}{m} r H^{-1}\right). \tag{73}$$

The values of d over which the other sum $T''(m)$ is taken are the terms of the sequence (d_0'') contained in the interval $\tau_1 < d \leq N_1$; let k be the number of prime factors $> H^3$ of a typical such value. Then $0 < k \leq r$, and the sum $T''(m)$ can be decomposed as

$$T''(m) = \sum_k T_k(m), \tag{74}$$

where $T_k(m)$ denotes the sum

$$T_k(m) = \sum e^{2\pi i \alpha m d},$$

with summation taken over those terms d of the sequence (d_0'') contained in the interval

$$\tau_1 < d \le N_1,$$

and having exactly k prime factors exceeding H^3.

In order to estimate $T''(m)$ we derive an estimate for $T_k(m)$. To this end we consider the more general sum

$$T_{k_0} = \sum_u \sum_v e^{2\pi i \alpha m u v},$$

where u ranges over all primes $> H^3$ contained in the sequence (d), and for each such u the index v ranges over all terms of the sequence (d_1) contained in the interval

$$\frac{\tau_1}{u} < v \le \frac{N_1}{u},$$

and with exactly $k-1$ prime factors $> H^3$. Thus all terms $e^{2\pi i\alpha m d}$ of the sum $T_k(m)$ are contained among those of $T_{k_0}(m)$; in fact each such term can be written in the form

$$e^{2\pi i \alpha m u v}$$

in k different ways (since one can choose u to be any of the k prime factors $> H^3$ of d). In the sum T_{k_0} there will also be terms of the form

$$e^{2\pi i\alpha m p^2 v_1}, \ \frac{\tau_1}{p^2} < v_1 \le \frac{N_1}{p^2},$$

where p is a prime $> H^3$, and v_1 ranges over the terms of the sequence (d_0) with exactly $k-2$ prime factors $> H^3$, none of which is p. Each term of this form appears once in $T_{k_0}(m)$. For each such p the number of integers of the form $p^2 v_1$ satisfying

$$\frac{\tau_1}{p^2} < v_1 \le \frac{N_1}{p^2},$$

is at most $\frac{N_1}{p^2}$. Hence

$$T_{k_0}(m) = kT_k(m) + O\left(\sum_{p>H^3}^{\sqrt{N}} \frac{N_1}{p^2}\right) = kT_k(m) + O\left(\frac{N}{m}H^{-3}\right). \tag{75}$$

We now apply Lemma 10 to the sum $T_{k_0}(m)$. In that lemma we take $N' = \tau_1$; then since $\tau_1 = \frac{N}{H}, m \le H$, we shall have $1 < N' \le N_1$. We take also $\tau = \frac{N}{H^3} < N_1$, and for the numbers U_0, U_1 set

$$U_0 = H^3, \ U_1 = \sqrt{N},$$

whence

$$1 < U_0 < U_1 < N_1,$$

fulfilling a further requirement of that lemma. For the sequence (x) in the lemma we take the sequence of values of u, and for the sequence (y) the sequence of values of v. Hence the conclusion of Lemma 10, together with the fact that $H^3 \le q \le \tau$, yields

$$T_{k_0}(m) \ll N_1\sqrt{\frac{r^2}{H^3}+\frac{r\sqrt{N}}{N_1}+\frac{Nr^3H^{-3}}{N_1}+\frac{mr^2}{H^3}+\frac{mr^2H^3}{N}}$$
$$\ll N_1\sqrt{\frac{mr^2}{H^3}+\frac{Nr^3H^{-3}}{N_1}+\frac{r^2H^3}{N_1}}$$
$$\ll N_1\sqrt{\frac{mr^3}{H^3}} \ll N\frac{r^{\frac{3}{2}}H^{-\frac{3}{2}}}{\sqrt{m}}.$$

This and (75) now give

$$T_k(m) \ll N\frac{r^{\frac{3}{2}}H^{-\frac{3}{2}}}{k\sqrt{m}},$$

and then from (74)

$$T''(m) \ll \sum_k N\frac{r^{\frac{3}{2}}H^{-\frac{3}{2}}}{k\sqrt{m}} \ll N\frac{r^{\frac{3}{2}}H^{-\frac{3}{2}}}{\sqrt{m}}\rho,$$

where

$$\rho = \ln r.$$

Substituting this expression for $T''(m)$ in (73), we obtain

$$T(m) \ll Nr^{\frac{3}{2}}H^{-\frac{3}{2}}\rho\frac{1}{\sqrt{m}} + NrH^{-1}\frac{1}{m}.$$

Summation over $m = 1, 2, \dots, [H]$ (see (72)), yields finally

$$T_0 \ll Nr^{\frac{3}{2}}H^{-\frac{3}{2}}\rho\sum_{m=1}^{H} m^{-\frac{1}{2}} + NrH^{-1}\sum_{m=1}^{H}\frac{1}{m}$$
$$\ll Nr^{\frac{3}{2}}H^{-1}\rho + Nr^{\frac{3}{2}}H^{-1} \ll Nr^2H^{-1}.$$

Exactly the same estimate being valid for T_1, we infer at last via (70) and (71) that

$$S \ll Nr^2H^{-1}.$$

We formulate this as a

THEOREM. *Let $h > 3$ be a constant, H satisfy the inequalities*

$$r^h \le H \le e^{\frac{1}{3}\sqrt{r}},$$

and suppose

$$\alpha = \frac{a}{q} + \frac{\theta}{q\tau}, \quad (a,q) = 1, \quad H^3 \le q \le \tau, \quad \tau = NH^{-3}.$$

Then

$$\sum_{p \le N} e^{2\pi i\alpha p} \ll Nr^2H^{-1}.$$

The number of representations. Using the integral in (27) we may express the number I_N of representations of a positive integer N as a sum

$$N = p_1 + p_2 + p_3$$

of three primes, by the formula

$$I_N = \int_0^1 S_\alpha^3 e^{-2\pi i \alpha N} d\alpha, \text{ where } S_\alpha = \sum_{p \le N} e^{2\pi i \alpha p}.$$

In order to estimate this integral we shall use the technique applied earlier in solving Waring's problem. Thus we first replace the indicated limits of integration by the following:

$$-\frac{1}{\tau} \le \alpha \le 1 - \frac{1}{\tau},$$

and express each number α in this interval in the form

$$\alpha = \frac{a}{q} + \frac{\theta}{q\tau}, \quad (a, q) = 1, \quad 0 < q \le \tau.$$

We consider, centred on each value

$$\alpha = \frac{a}{q}, \quad (a, q) = 1, \quad 0 \le a < q, \quad q = 1, 2, \ldots, [r^{3h}],$$

the subinterval

$$\alpha = \frac{a}{q} + z, \ -\frac{1}{\tau} \le z \le \frac{1}{\tau},$$

and call these *intervals of class I.* It is easy to see that they are nonoverlapping. The subintervals remaining after class I intervals are removed will be called *intervals of class II.* Thus each α lying in some class II interval may be written in the form

$$\alpha = \frac{a}{q} + \frac{\theta}{q\tau}, \quad (a, q) = 1, \quad r^{3h} < q \le \tau.$$

Corresponding to this partition of the interval of integration into the two types of subinterval, the integral I_N decomposes into two summands:

$$I_N = I_N^{(I)} + I_N^{(II)}.$$

The second summand, $I_N^{(II)}$, can be immediately estimated using the above theorem: Setting

$$H = r^h, \text{ so that } \tau = \frac{N}{r^{3h}},$$

we have by that theorem that for those α contained in a class II interval,

$$S_\alpha \ll N r^{2-h},$$

whence

$$I_N^{(II)} \ll N r^{2-h} \int_0^1 \left| \sum_{p \le N} e^{2\pi i \alpha p} \right|^2 d\alpha.$$

Here the integral is easily identified (we encountered a similar such integral in the solution of Waring's problem): it is equal to the number of terms in the sum S_α, i.e., the number $\pi(N)$ of primes not exceeding N. Hence by (62) we have

$$I_N^{(II)} \ll N^2 r^{1-h}. \tag{76}$$

Estimate of S_α for α in a class I interval. In order to estimate S_α in the case where α lies in a class I interval, we need to use the laws governing the distribution of primes in arithmetic progressions. We formulate the precise result needed as a lemma, a proof of which may be found in an article by Walfisz.[249]

LEMMA 11. *Let h_1 and h_2 be arbitrarily large positive constants, $N > 2$, and set $r = \ln N$. Then the number $\pi(N, q, l)$ of primes not greater than N and occurring in the arithmetic progression*

$$qx + l, \text{ where } 0 < q \le r^{h_1}, \quad (q, l) = 1, \quad 0 \le l < q,$$

satisfies

$$\pi(N, q, l) = \frac{1}{q_1} \int_2^N \frac{dx}{\ln x} + O\left(\frac{N}{q_1 r^{h_2}}\right), \tag{77}$$

where $q_1 = \phi(q)$, the value of Euler's ϕ-function at q.

Before applying this lemma to the estimation of S_α when

$$\alpha = \frac{a}{q} + z, \quad |z| \le \frac{1}{\tau}, \quad 0 < q \le r^{3h},$$

we need to prepare the ground somewhat. The sum S_α can be decomposed as a sum of $O(r^{9h})$ terms of the form

$$S_{\alpha M} = \sum_{p>M}^{M_1} e^{2\pi i(\frac{a}{q}+z)p}, \quad 0 < M_1 - M \le Nr^{-9h}.$$

Each such sum can in turn be written in the form

$$S_{\alpha M} = \sum_l S_{\alpha M l},$$

where $S_{\alpha M l}$ denotes the sum of those terms of the sum $S_{\alpha M}$ indexed by values of p of the form

$$p = p_l = qx + l,$$

and the summation $\sum_l$ is taken over all numbers l relatively prime to q among $0, 1, \ldots, q-1$. For the p_l in the range $M < p_l \le M_1$, we have

$$zp_l = zM + O(|z|Nr^{-9h}),$$

and since

$$O(|z|Nr^{-9h}) \ll \frac{Nr^{-9h}}{\tau} = r^{-6h}$$

is arbitrarily small for sufficiently large N, it follows that

$$e^{2\pi i(\frac{a}{q}+z)p_l} = e^{2\pi i\frac{a}{q}l + 2\pi i z M} + O(|z|Nr^{-9h}).$$

Denoting by π_l the number of primes p_l satisfying the condition $M < p_l \le M_1$, we then infer that

$$S_{\alpha M l} = \sum_{p_l>M}^{M_1} e^{2\pi i(\frac{a}{q}+z)p_l} = e^{2\pi i\frac{a}{q}l+2\pi i z M}\pi_l + O(|z|Nr^{-9r}\pi_l).$$

[249]A. Walfisz, "Zur additiven Zahlentheorie, II", *Math. Zeitschr.*, Bd. 40, 1935, pp. 592–607.

We now apply Lemma 11 to the intervals $(2, M), (2, M_1)$, with $h_1 = 3h, h_2 = 15h + 1$. Noting that $M < M_1 \le N$, and that the function $\frac{M}{(\ln M)^{h_2}}$ is from some value of M on always increasing with increasing M, we obtain from Lemma 11 that

$$\pi_l = \frac{1}{q_1} J_M + O\left(\frac{N}{q_1 r^{15h+1}}\right), \text{ where } J_M = \int_M^{M_1} \frac{dx}{\ln x}, \; q_1 = \phi(q).$$

Hence

$$\begin{aligned} S_{\alpha M l} &= \frac{1}{q_1} e^{2\pi i \frac{a}{q} l + 2\pi i z M} J_M + O\left(\frac{N}{q_1 r^{15h+1}}\right) \\ &\quad + O\left(|z| N r^{-9h} \frac{1}{q_1} J_M\right) + O\left(|z| N r^{-9h} \frac{N}{q_1 r^{15h+1}}\right) \\ &= \frac{1}{q_1} e^{2\pi i \frac{a}{q} l + 2\pi i z M} J_M + O\left(\frac{N}{q_1 r^{15r+1}} + \frac{1}{q_1} J_M |z| N r^{-9h}\right). \end{aligned} \tag{78}$$

Since

$$|z|(x - M) \le |z| N r^{-9h} \text{ for } M \le x \le M_1,$$

we have

$$e^{2\pi i z M} J_M = \int_M^{M_1} \frac{e^{2\pi i z M}}{\ln x} dx = \int_M^{M_1} \frac{e^{2\pi i z x}}{\ln x} dx + O(J_M |z| N r^{-9h}),$$

whence (78) gives

$$S_{\alpha M l} = \frac{1}{q_1} e^{2\pi i \frac{a}{q} l} \int_M^{M_1} \frac{e^{2\pi i z x}}{\ln x} dx + O\left(\frac{N}{q_1 r^{15h+1}} + \frac{1}{q_1} J_M |z| N r^{-9h}\right).$$

Summing this over all $l = 0, 1, \ldots, q - 1$ relatively prime to q, and invoking the easily proved identity

$$\sum_l e^{2\pi i \frac{a}{q} l} = \mu(q)$$

for the Möbius function, we obtain

$$S_{\alpha M} = \frac{\mu(q)}{q_1} \int_M^{M_1} \frac{e^{2\pi i x z}}{\ln x} dx + O(N r^{-15h-1} + J_M |z| N r^{-9h}).$$

We now sum these equalities in turn over all M (which number $O(r^{9h})$), obtaining

$$S_\alpha = \frac{\mu(q)}{q_1} J(z) + O(N r^{-6h-1}) + O\left(|z| N r^{-9h} \int_2^N \frac{dx}{\ln x}\right),$$

where

$$J(z) = \int_2^N \frac{e^{2\pi i x z}}{\ln x} dx.$$

Hence, via the easy estimate

$$\int_2^N \frac{dx}{\ln x} \ll \frac{N}{r},$$

we obtain finally

$$S_\alpha = \frac{\mu(q)}{q_1} J(z) + O(N r^{-6h-1}). \tag{79}$$

The integral $J(z)$. The integral

$$J(z)=\int_2^N \frac{e^{2\pi izx}}{\ln x}dx$$

is similar to that considered in Lemma 5. An argument along the same lines as the proof of Lemma 16 in Chapter I of Vinogradov's book "A new method in analytic number theory", yields the estimate

$$J(z)\ll Z,$$

where

$$Z=\frac{N}{r}\quad\text{for } |z|\le\frac{1}{N};\qquad Z=\frac{1}{|z|r}\quad\text{for } \frac{1}{N}<|z|\le\frac{1}{\tau}. \tag{80}$$

Estimate of Q_{aq}. We next estimate that portion Q_{aq} of the integral $I_N^{(I)}$ corresponding to an arbitrarily given fraction $\frac{a}{q}$. Note first that the main term on the right-hand side of (79) is $O\left(\frac{1}{q_1}Z\right)$. Since we always have $Z\ge\frac{\tau}{r}=Nr^{-3h-1}$, it follows that

$$\frac{Nr^{-6h-1}}{\frac{1}{q_1}Z}\ll\frac{q_1Nr^{-6h-1}}{Nr^{-3h-1}}\ll\frac{q}{r^{3h}}\ll 1.$$

Hence on cubing (79) we obtain

$$S_\alpha^3=\frac{\mu(q)}{q_1^3}J(z)^3+O(q_1^{-2}Z^2Nr^{-6h-1}).$$

Multiplying this by

$$e^{-2\pi i\alpha N}d\alpha=e^{-2\pi i\frac{a}{q}N-2\pi izN}dz,$$

and integrating from $z=-\frac{1}{\tau}$ to $z=\frac{1}{\tau}$, we obtain

$$Q_{aq}=\frac{\mu(q)}{q_1^3}e^{-2\pi i\frac{a}{q}N}R+H,\ \text{where } R=\int_{-\frac{1}{\tau}}^{\frac{1}{\tau}}J(z)^3e^{-2\pi izN}dz,$$

and

$$H\ll\int_0^{\frac{1}{\tau}}q_1^{-2}Z^2Nr^{-6h-1}dz=q_1^{-2}Nr^{-6h-1}\left[\int_0^{\frac{1}{N}}\frac{N^2}{r^2}dz+\int_{\frac{1}{N}}^{\frac{1}{\tau}}\frac{1}{z^2r^2}dz\right]$$
$$\ll q_1^{-2}N^2r^{-6h-3}.$$

Asymptotic formula for I_N. Summing the above expression for Q_{aq} over all a relatively prime to q in the interval $0\le a<q$, we obtain

$$\sum_a Q_{aq}=\frac{\mu(q)}{q_1^3}G(q)R+O(q_1^{-1}N^2r^{-6h-3}),\ \text{where } G(q)=\sum_a e^{-2\pi i\frac{a}{q}N}.$$

Summing this equation over all

$$q=1,2,\dots,[r^{3h}],$$

and using the trivial estimate

$$\sum_{q=1}^{3h}q_1^{-1}\ll r^{3h},$$

we now obtain

$$I_N^{(I)} = R\sum_{q=1}^{r^{3h}} \frac{\mu(q)}{q_1^3} G(q) + O(N^2 r^{-3h-3}).$$

Combining this with the estimate (76) for $I_N^{(II)}$, yields

$$I_N = R\sum_{q=1}^{r^{3h}} \frac{\mu(q)}{q_1^3} G(q) + O(N^2 r^{1-h}). \tag{81}$$

We now find a suitable estimate for the integral

$$R = \int_{-\frac{1}{\tau}}^{\frac{1}{\tau}} J(z)^3 e^{-2\pi i z N} dz.$$

We have

$$R - \int_{-\frac{\sqrt{r}}{N}}^{\frac{\sqrt{r}}{N}} J(z)^3 e^{-2\pi i z N} dz \ll \int_{\frac{\sqrt{r}}{N}}^{\frac{1}{\tau}} Z^3 dz = \int_{\frac{\sqrt{r}}{N}}^{\frac{1}{\tau}} \frac{1}{z^3 r^3} dz \ll \frac{N^2}{r^4},$$

whence

$$R = \int_{-\frac{\sqrt{r}}{N}}^{\frac{\sqrt{r}}{N}} J(z)^3 e^{-2\pi i z N} dz + O\left(\frac{N^2}{r^4}\right). \tag{82}$$

Write

$$R_0 = \frac{1}{r^3} \int_{-\frac{1}{2}}^{\frac{1}{2}} I(z)^3 e^{-2\pi i z N} dz, \text{ where } I(z) = \int_0^N e^{2\pi i z x} dx.$$

Direct calculation of the integral $I(z)$ yields the immediate estimate

$$I(z) \ll rZ,$$

where Z is as in (80). Hence

$$R_0 - \frac{1}{r^3} \int_{-\frac{\sqrt{r}}{N}}^{\frac{\sqrt{r}}{N}} I(z)^3 e^{-2\pi i z N} dz \ll \int_{\frac{\sqrt{r}}{N}}^{\frac{1}{2}} Z^3 dz = \int_{\frac{\sqrt{r}}{N}}^{\frac{1}{2}} \frac{1}{z^3 r^3} dz \ll \frac{N^2}{r^4},$$

and therefore

$$R_0 = \frac{1}{r^3} \int_{-\frac{\sqrt{r}}{N}}^{\frac{\sqrt{r}}{N}} I(z)^3 e^{-2\pi i z N} dz + O\left(\frac{N^2}{r^4}\right).$$

Subtracting this from (82), we obtain

$$R - R_0 = \int_{-\frac{\sqrt{r}}{N}}^{\frac{\sqrt{r}}{N}} \left\{ J(z)^3 - \left(\frac{1}{r} I(z)\right)^3 \right\} e^{-2\pi i z N} dz + O\left(\frac{N^2}{r^4}\right). \tag{83}$$

Now

$$\begin{aligned} J(z) - \frac{1}{r} I(z) &\ll \left| \int_2^{\frac{N}{r}} \frac{e^{2\pi i x z}}{\ln x} dx \right| + \frac{1}{r} \left| \int_0^{\frac{N}{r}} e^{2\pi i x z} dx \right| + \left| \int_{\frac{N}{r}}^N \left(\frac{1}{\ln x} - \frac{1}{\ln N} \right) e^{2\pi i z x} dx \right| \\ &\ll \frac{N}{r^2} + \frac{N}{r^2} + \left(\frac{1}{r-\rho} - \frac{1}{r} \right) N \ll \frac{N\rho}{r^2}, \end{aligned}$$

where $\rho = \ln r$. Hence

$$J(z)^3 - \left(\frac{1}{r}I(z)\right)^3 \ll \frac{N\rho}{r^2}Z^2,$$

and substitution of this in (83) yields

$$R - R_0 \ll \frac{N\rho}{r^2}\int_{-\frac{\sqrt{r}}{N}}^{\frac{\sqrt{r}}{N}} Z^2 dz + O\left(\frac{N^2}{r^4}\right)$$

$$\ll \frac{N\rho}{r^2}\left\{\int_0^{\frac{1}{N}} \frac{N^2}{r^2}dz + \int_{\frac{1}{N}}^{\frac{\sqrt{r}}{N}} \frac{1}{z^2 r^2}dz\right\} + O\left(\frac{N^2}{r^4}\right) \ll \frac{N^2\rho}{r^4},$$

whence

$$R = R_0 + O\left(\frac{N^2\rho}{r^4}\right). \tag{84}$$

We now turn to the relatively easy estimation of the integral R_0. Let K_N denote the number of representations of N in the form

$$N = x_1 + x_2 + x_3,$$

where x_1, x_2, x_3 are positive integers. In the by now familiar way using (27), K_N may be expressed as

$$K_N = \int_{-\frac{1}{2}}^{\frac{1}{2}} \left(\sum_{x=1}^{N} e^{2\pi i z x}\right)^3 e^{-2\pi i z N} dz.$$

From the easily proved estimate

$$\sum_{x=1}^{N} e^{2\pi i z x} = I(z) + O(1),$$

it follows that

$$K_N = \int_{-\frac{1}{2}}^{\frac{1}{2}} \{I(z) + O(1)\}^3 e^{-2\pi i z N} dz = r^3 R_0 + H,$$

where

$$H \ll \int_0^{\frac{1}{2}} r^2 Z^2 dz \ll \int_0^{\frac{1}{N}} N^2 dz + \int_{\frac{1}{N}}^{\frac{1}{2}} \frac{1}{z^2}dz \ll N,$$

whence

$$R_0 = \frac{1}{r^3}K_N + O\left(\frac{N}{r^3}\right).$$

To estimate K_N, observe that for each $k = 1, 2, \ldots, N-1$, the equality

$$x_1 + x_2 = k$$

is fulfilled in $k-1$ ways, whence

$$K_N = \sum_{k=1}^{N-1}(k-1) = \frac{(N-1)(N-2)}{2} = \frac{N^2}{2} + O(N).$$

Hence

$$R_0 = \frac{N^2}{2r^3} + O\left(\frac{N}{r^3}\right).$$

Substitution of this in (84) yields

$$R = \frac{N^2}{2r^3} + O\left(\frac{N^2\rho}{r^4}\right). \tag{85}$$

Returning to the formula (81), we note that

$$G(q) = \sum_a e^{-2\pi i \frac{a}{q} N} \ll q_1,$$

whence

$$\sum_{q=1}^{r^{3h}} \frac{\mu(q)}{q_1^3} G(q) \ll \sum_{q=1}^{\infty} \frac{1}{q_1^2} \ll 1.$$

From this and (85) we infer that (81) (with $h = 5$) may be rewritten as

$$\begin{aligned} I_N &= R \sum_{q=1}^{r^{15}} \frac{\mu(q)}{q_1^3} G(q) + O\left(\frac{N^2}{r^4}\right) \\ &= \frac{N^2}{2r^3} \sum_{q=1}^{r^{15}} \frac{\mu(q)}{q_1^3} G(q) + O\left(\frac{N^2\rho}{r^4}\right). \end{aligned} \tag{86}$$

Consider now the infinite series

$$S(N) = \sum_{q=1}^{\infty} \frac{\mu(q)}{q_1^3} G(q). \tag{87}$$

In this series nonzero terms occur only for square-free values of the index q; let $q = p_1 p_2 \cdots p_s$ be a typical such value of q, where $p_1, \ldots, p_s$ denote distinct primes, and as usual denote by $\tau(q)$ the number of divisors of q. Then

$$\frac{q}{q_1} = \prod_{i=1}^{s} \left(1 + \frac{1}{p_i - 1}\right) \le 2^s = \tau(q).$$

It is a known theorem[250] that

$$\tau(q) < K q^{\epsilon},$$

where $K = K(\epsilon)$ is a constant. Hence taking $\epsilon = \frac{1}{3}$, we infer that $\frac{1}{q_1} < K q^{-\frac{2}{3}}$, whence

$$\sum_{q > r^{15}}^{\infty} \frac{\mu(q)}{q_1^3} G(q) \ll \sum_{q > r^{15}} \frac{1}{q_1^2} \ll \sum_{q > r^{15}} q^{-\frac{4}{3}} \ll r^{-5}.$$

Taking this into account in (86), we finally obtain the desired estimate for I_N:

$$I_N = \frac{N^2}{2r^3} S(N) + O\left(\frac{N^2\rho}{r^4}\right).$$

We note that this estimate of the number of representations of N as a sum of three primes is the same as that derived conditionally by Hardy and Littlewood.

[250]See for instance: Landau, "Vorlesungen", Bd. I, p. 188.

The Goldbach theorem. The series $S(N)$ is the Hardy–Littlewood "singular series" for the problem we are considering. Obtaining a lower bound for $S(N)$ does not present great difficulties.[251] It is easy to show that

$$W(q) = \frac{\mu(q)}{q_1^3} G(q)$$

satisfies

$$W(qq') = W(q)W(q')$$

for relatively prime q, q'. It is also relatively easy to calculate $W(p^k)$ for prime powers p^k. One thus obtains

$$S(N) = \prod \left(1 + \frac{1}{(p-1)^3}\right) \prod{}' \left(1 - \frac{1}{p^2 - 3p + 3}\right),$$

where the first product is taken over all primes, and the second over all primes dividing N. For *odd* N it turns out that one always has

$$S(N) > \frac{1}{2}.$$

Thus the final result is as follows:

THEOREM. *The number of representations of an odd positive integer N as a sum of three primes satisfies the formula*

$$I_N = \frac{N^2}{2(\ln N)^3} S(N) + O\left(\frac{N^2 \ln \ln N}{(\ln N)^4}\right),$$

and furthermore

$$S(N) > \frac{1}{2}.$$

This theorem implies the existence of a constant N_0 such that $I_N > 0$ for all odd $N \geq N_0$, i.e., that all sufficiently large odd integers can in fact be expressed as the sum of three primes (the Goldbach theorem).

In 1937, when Vinogradov produced this brilliant result, his proof depended on Lemma 11 (concerning primes in arithmetic progressions), whose proof was not effective, i.e., although it established the existence of a certain constant (namely the one in the error term in the formula (77)), it was not possible to estimate that constant. Such proofs are frequently encountered in modern number theory, arising from the necessity of carrying out proofs by *reductio ad absurdum*. However, somewhat earlier several foreign mathematicians[252] had proposed in place of Lemma 11 other similar lemmas of effective character, on the basis of which a constant N_0 (as above) can be actually computed. For the full theorem, in addition to computing N_0 one has of course to verify its validity for all odd positive integers up to N_0; however, this presents no theoretical difficulties.

---

[251]See: Landau, "Vorlesungen", Bd. I, Teil V, Kap. 4; Vinogradov, "A new method in analytic number theory", Chapter IX.

[252]See: Estermann, "Proof that every large integer is the sum of two primes and a square", *Proc. London Math. Soc.* (2) **42** (1937), pp. 511–516.

Estimation of Weyl Sums and the Problem of the Fractional Parts of a Polynomial

We conclude our discussion of Vinogradov's work with a description of his contributions to two further important questions of contemporary number theory taking up a considerable amount of his research activity: the estimation of so-called "Weyl sums", and the problem of the fractional parts of a polynomial.

In our exposition of Vinogradov's solution of Waring's problem we encountered a sum of the form

$$S_{aq} = \sum_{x=1}^{q} e^{2\pi i \frac{a}{q} x^n}. \tag{88}$$

This is a particular case of the more general sum

$$S = \sum_{x=1}^{P} e^{2\pi i f(x)}, \tag{89}$$

where

$$f(x) = a_0 x^n + a_1 x^{n-1} + \cdots + a_n \tag{90}$$

is a given polynomial with real coefficients and P is a large positive integer. A sum of the form (89) is called a *Weyl sum*, since Hermann Weyl[253] was the first, in 1916, to give a general method for estimating such sums—with the aim of using them to solve certain number-theoretical problems. In order to bring home to the reader more clearly the degree of accuracy of the estimates to be given below, we note, to begin with, the trivial estimate

$$|S| \le P.$$

The estimate of the sum S depends crucially on the first coefficient a_0 of the polynomial $f(x)$, or, more precisely, on the approximation of a_0 by fractions $\frac{a}{q}$. If

$$a_0 = \frac{a}{q} + \frac{\theta}{q^2}, \quad \text{where } (a, q) = 1, \; q > 0, \tag{91}$$

then Weyl's method yields for any arbitrarily small $\epsilon > 0$, a constant $c(n, \epsilon)$ satisfying

$$|S| \le P\gamma, \quad \text{where } \gamma = cP^{\epsilon}(P^{-1} + q^{-1} + qP^{-n})^{\rho}, \; \rho = \frac{1}{2^{n-1}}. \tag{92}$$

In order to better appreciate the significance of this bound, suppose for definiteness that $q \le P$, and that q increases to infinity together with P according to some rule. Then one easily sees that the inequality (92) yields one of the form

$$|S| \le cPq^{\epsilon-\rho}. \tag{93}$$

[253] H. Weyl, *Math. Ann.*, 77, 1916, pp. 313–352.

Since the quantity $\rho = \frac{1}{2^{n-1}}$ tends to zero rapidly as $n \to \infty$, for large n the improvement afforded by (93) over the trivial bound is very substantial. Of those mathematicians who have attempted to improve on Weyl's bound, it is appropriate to mention van der Corput;[254] however, even he was unable to achieve a significant increase in the exponent ρ.

However, by means of his new method Vinogradov was able to obtain an essential improvement of the bound (at least for large n); for instance for $q \leq P$, in place of (93) Vinogradov proves that[255]

$$|S| \leq c_1 P q^{-\rho_1},$$

where $c_1 = c_1(n)$ is a constant, and

$$\rho_1 = \frac{2}{n^3(\ln n + 2.5 \ln\ln n)}.$$

The exponent ρ_1, being of order

$$\frac{1}{n^3 \ln n},$$

tends to zero much more slowly than $\frac{1}{2^{n-1}}$. Note also that the sum S_{aq} featuring in the solution of Waring's problem (see (88)), can in certain cases be calculated exactly. For instance it can easily be shown[256] that for $P = q = p^n$, where p is a prime not dividing n, one has $S_{aq} = p^{n-1}$. Expressing this in the form of the right-hand side of (93), we have

$$\sum_{x=1}^{P} e^{2\pi i \frac{a}{q} x^n} = P q^{-\frac{1}{n}} \quad \text{for } P = q = p^n,$$

whence it is clear that the exponent ρ in (93) cannot in general be replaced by any quantity of order less than $\frac{1}{n}$, so that Vinogradov's exponent ρ_1 is already close to being best possible.

A problem closely related to that of estimating Weyl sums is that of the fractional parts of a polynomial (or other functions). The question of interest here is as follows: Given a real polynomial (90) and a large positive integer P, how are the numbers

$$\{f(x)\}, \quad x = 1, 2, \ldots, P, \tag{94}$$

distributed in the interval $(0, 1)$? The first results of significance in this direction were obtained by Hardy and Littlewood in 1914. In 1916 Weyl proved in the above-cited work that the fractional parts (94) are distributed uniformly in the interval $(0, 1)$, i.e., for any fixed number δ, $0 < \delta < 1$, the number T_δ of fractional parts (94) contained in the interval $(0, \delta)$ satisfies

$$\lim_{P \to \infty} \frac{T_\delta}{P} = \delta. \tag{95}$$

Weyl's proof involves in an essential way his method for estimating sums of the form (89). Further developments relating to the question of the distribution of the fractional parts of polynomials and other functions are mainly due to the efforts of

---

[254] See, for instance: J. G. van der Corput, "Neue zahlentheoretische Abschätzungen", *Math. Ann.*, 89, H. 3/4, 1923, p. 215.

[255] "A new method in analytic number theory", Chapter VI.

[256] Loc. cit., Chapter II, Lemma 4.

the Dutch mathematicians van der Corput and Koksma, who in connection with this general theme posed and solved several important general problems.

The work of Vinogradov on this question has a different emphasis. He aimed at obtaining an asymptotic formula for T_δ (which by (95) must have δP as its main term) with an estimate of the error term as accurate as possible, in order to draw conclusions concerning, for instance, the approximation of a given real number by fractions with denominator an exact power. Vinogradov's first work in this direction dates from 1926,[257] and is distinguished by the unusual elegance of the purely arithmetic means it employs. (This work involves only whole numbers, to the exclusion even of exponential functions and trigonometric sums.) In his next article on this problem Vinogradov used a new analytic method (involving Fourier analysis), together with Weyl's method for estimating sums of the form (89), to establish the formula

$$T_\delta = \delta P + O(P \cdot q^{\epsilon - \rho}), \quad \rho = \frac{1}{2^{n-1}},$$

where it is assumed that the leading coefficient a_0 of the given polynomial is expressed in the form (91) with $q \leq P$. This then yields the following theorem concerning approximation by means of a fraction with denominator an exact power:

One can find a constant $L = L(n, \epsilon)$ with the property that for any integer $P > 1$ and any real number α, there exist integers x, y satisfying the conditions

$$|\alpha x^n - y| < \frac{L}{P^\rho}, \quad \text{where } \rho = \frac{1}{2^{n-1} + n^{-1}} - \epsilon, \ 0 < x < P.$$

Thus for example for $n = 2$, there exist integers x, y such that

$$|\alpha x^2 - y| < \frac{L}{P^{\frac{2}{5} - \epsilon}}, \quad 0 < x < P.$$

Finally, the cardinal improvement of the bound for Weyl sums for large n obtained by Vinogradov in 1937 using his "new method", yields a corresponding improvement in results relating to the present question; thus for instance in the above theorem the exponent ρ may be replaced by a quantity of order $\frac{1}{n^2 \ln n}$.[258]

This completes our description of the chief applications of the new method in analytic number theory invented by Vinogradov. However, it should be noted that the potential of his method has by no means yet been exhausted, and that Vinogradov himself has over the past several years[259] persistently worked to deepen and broaden it. In this connection we note the remarkable articles of 1940 and 1943 concerned with certain sums depending on the primes in an arithmetic progression.[260]

We conclude this summary with a brief description of the work of Vinogradov's chief collaborators and students.

[257] "On the fractional parts of an integer polynomial", *IAN SSSR*, 1926.

[258] "A new method in analytic number theory", Chapter V.

[259] Prior to 1947. *Trans.*

[260] I. M. Vinogradov, "The distribution relative to a given modulus of the primes in an arithmetic progression", *IAN*, 4, 1940; "A sharpening of a method for estimating sums involving prime numbers", *IAN*, 7, 1943.

Nikolaĭ Grigoryevich Chudakov (Professor at Saratov University). The work of Chudakov is chiefly concerned with the analytic theory of the distribution of primes (along the lines of the classical investigations of Riemann, Hadamard, de la Vallée Poussin, etc.). In the chapter on Chebyshev we mentioned that in 1896 Hadamard and de la Vallée Poussin proved independently of one another the "Prime Number Theorem" giving the rule for the the distribution of primes, namely that the ratio

$$\frac{\pi(x)}{\int_2^x \frac{dx}{\ln x}}$$

tends to 1 as $x \to \infty$, i.e., that the integral

$$\int_2^x \frac{dx}{\ln x}$$

must be the main term in any asymptotic formula for the number $\pi(x)$ of primes not exceeding x. There naturally arises the question as to the form of the error term in such an asymptotic formula. (Such a formula was used above; see the formula for $\pi(x, q, l)$ in Lemma 11.) The known results in this direction up to 1936 were as follows:

$$\pi(x) = \int_2^x \frac{dx}{\ln x} + O\left(xe^{-c(\ln x)^{\frac{1}{2}}}\right) \quad \text{(de la Vallée Poussin, 1899)},$$

$$\pi(x) = \int_2^x \frac{dx}{\ln x} + O\left(xe^{-c_1(\ln x \ln\ln x)^{\frac{1}{2}}}\right) \quad \text{(Littlewood, 1924)},$$

where $c, c_1 > 0$ are constants. In 1936 Chudakov proved[261] that

$$\pi(x) = \int_2^x \frac{dx}{\ln x} + O\left(xe^{-c_2(\ln x)^{\mu}}\right), \quad \text{where } \mu = \frac{1}{2} + \frac{1}{42} - \epsilon.$$

Note that from the truth of the Riemann hypothesis concerning the zeroes of $\zeta(s)$, it would follow that

$$\pi(x) = \int_2^x \frac{dx}{\ln x} + O\left(x^{\frac{1}{2}+\epsilon}\right). \tag{96}$$

It is a question of great interest as to how the difference between two neighbouring primes changes. Let

$$p_1 = 2, p_2, p_3, \ldots, p_n, \ldots,$$

denote the sequence of primes in increasing order, and write $\delta_n = p_{n+1} - p_n$. It follows from the Prime Number Theorem that

$$\lim_{n\to\infty} \frac{\delta_n}{p_n} = 0,$$

and again there arises the question of estimating the rate of growth of δ_n. Up to 1936 the following estimates had been obtained:

$$\delta_n = O\left(p_n^{\frac{33{,}000}{33{,}001}}\right) \quad \text{(G. Hoheisel, 1930)};$$

$$\delta_n = O\left(p_n^{\frac{249}{250}}\right) \quad \text{(Hans Heilbronn, 1933)}.$$

[261] N. G. Chudakov, "On the zeroes of the function $\zeta(s)$", *DAN SSSR*, 1936, No. 5, pp. 197–200.

In his 1936 article cited above, Chudakov, using the recently obtained new estimates of Vinogradov for Weyl sums, also proved that

$$\delta_n = O\left(p_n^{\frac{3}{4}+\epsilon}\right). \tag{97}$$

This result of Chudakov may be reformulated in a striking, though not completely accurate, way, as follows: The primes are distributed in the sequence of natural numbers with the same density as the fourth powers $1^4, 2^4, 3^4, \ldots$ (more precisely, as the powers of the form $(a+kd)^{4+\epsilon}$, $k = 1, 2, 3, \ldots$, of the terms of a certain arithmetic progression).

Chudakov deduces from the formula (97) a curious consequence in connection with Waring's problem, namely that any sufficiently large integer n can be expressed as a sum of 8 cubes, each differing from $\frac{n}{8}$ by a quantity of order $O\left(n^{\frac{77}{78}+\epsilon}\right)$ ("almost equal cubes").[262]

Earlier we defined the very important integer function $h(-m)$ giving the number of equivalence classes of properly primitive binary quadratic forms of determinant $-m < 0$. Gauss conjectured that $h(-m)$ grows to infinity with m. This was proved only in 1935, by Heilbronn, whose result was then sharpened and generalized by Siegel.[263] Siegel's inequality for the function $h(-m)$ may be written as

$$h(-m) \gg m^{\frac{1}{2}-\epsilon}. \tag{98}$$

In 1942 Chudakov gave a different proof of Siegel's inequality, avoiding the use of the theory of algebraic numbers.[264]

Konstantin Konstantinovich Mardzhanishvili (Researcher at the Steklov Mathematical Institute, Academy of Sciences of the USSR, Moscow). In a 1941 article[265] Mardzhanishvili introduced a variation into the derivation of the asymptotic formula for the number I_N of representations of an odd number N as a sum of three primes (in the Goldbach problem). He started from the same integral representation of I_N as Vinogradov, but his treatment of the "class I intervals", given by

$$\alpha = \frac{a}{q} + z, \quad |z| \leq \frac{1}{\tau}, \quad \tau = \frac{N}{\xi}, \quad 0 < q \leq \xi = (\ln N)^{18}$$

(similar to the "major arcs" of the Hardy–Littlewood approach) was different from Vinogradov's. Mardzhanishvili's approach in dealing with these intervals is based instead on results of the English mathematician A. Page[266] concerning characters. We mention just one of these results: *There exists a constant $c > 0$ such that at most one of the series $L(s, \chi)$, $s = \sigma + it$, corresponding to primitive characters χ modulo q, $0 < q \leq \xi$, has a real root $\sigma > 1 - \frac{c}{\ln \xi}$.* (It can be shown easily that no series $L(s, \chi)$ has a zero in the half-plane $\sigma > 1$.) For class II intervals ("minor

---

[262] N. Tchudakoff, "On the difference between two neighbouring prime numbers", *Mat. Sbornik*, 1 (43): 6, 1936, pp. 799–814.

[263] *Acta Arithmetica*, 1936, I, pp. 83–86.

[264] N. G. Chudakov, "On Siegel's theorem", *IAN*, 6, 1942, pp. 135–142.

[265] C. Mardjanichvili, "Sur la démonstration du théorème de Goldbach–Winogradoff", *Comptes Rendus (Doklady) de l'Acad. des Sci. de l'URSS*, 1941, vol. XXX, No. 8, pp. 687–689.

[266] A. Page, "On the number of primes in an arithmetic progression", *Proc. London Math. Soc.*, Ser. 2, 39, 1935, pp. 116–141.

arcs") Mardzhanishvili stays with Vinogradov's method. The upshot is a somewhat more accurate formula than Vinogradov's, namely

$$I_N = R \cdot S(N) + O\left(\frac{N^2}{(\ln N)^{5-\epsilon}}\right),$$

where R and $S(N)$ are as before given by (85) and (87).

In other works[267] Mardzhanishvili investigated, using Vinogradov's method, the question of the simultaneous representation of finitely many given integers as sums of powers of integers or, more particularly, primes, i.e., the question as to whether a system of equations of the form

$$N_k = x_1^k + x_2^k + \cdots + x_s^k, \quad k = 1, 2, \ldots, n,$$

has a solution in integers $x_i \geq 0$ or primes $x_i = p_i$. Here n, s, and $N_1 < N_2 < \cdots < N_n$, are given positive integers.

Mardzhanishvili also worked on the problem of estimating sums of the form

$$\sum_{m=1}^{n} (\tau_k(m))^l,$$

where $\tau_k(m)$ is an integer function generalizing the function $\tau(m)$ giving the number of divisors of m, namely that giving the number of solutions $(x_1, x_2, \ldots, x_k)$ in positive integers x_i, of the equation[268]

$$m = x_1 x_2 \cdots x_k.$$

Yuriĭ Vladimirovich Linnik (Professor at Leningrad University). An outstanding Leningrad mathematician, Linnik's work was devoted mainly to the analytic theory of the distribution of the primes.

In a long article of 1944,[269] Linnik considers the question of the least prime in an arithmetic progression

$$Dx + l, \quad 0 < l < D, \quad (D, l) = 1, \quad x = 1, 2, 3, \ldots . \tag{99}$$

By the classical result of Dirichlet the progression (99) contains infinitely many primes; however, Dirichlet's proof affords no way of estimating for instance the least prime $p_{\min}(l, D)$ occurring in the progression (99). Of course estimates can be inferred from more recent results on the distribution of primes in arithmetic progressions, but these turn out to be high. Linnik, on the other hand, shows that

$$p_{\min}(l, D) < D^c, \tag{100}$$

where c is an absolute constant. Linnik's idea may be described roughly but vividly as follows: Significant reductions in various estimates relating to the primes (for example as in the formula (96)) would follow from the knowledge that the function $\zeta(s)$, $s = \sigma + it$, has no zeroes in the critical strip $\frac{1}{2} < \sigma < 1$; however, this, the Riemann hypothesis, remains up till now, as we well know, unproven. But might

[267]K. K. Mardzhanishvili, "On simultaneous representation of n natural numbers as sums of full first, second, ..., nth powers", *IAN*, ser. mat., 1937, pp. 609–631; "On a problem of additive number theory", *IAN*, 4, 1940, pp. 193–214.

[268]C. Mardjianichvili, "Estimation d'une somme arithmétique", *C. R. (Doklady) de l'Acad. des Sci. de l'URSS*, 1939, vol. XXII, No. 7, pp. 387–389.

[269]U. V. Linnik, "On the least prime in an arithmetic progression. I. The basic theorem", *Mat. Sb.*, 15 (57):2 (1944), pp. 139–178; "II. The Deuring–Heilbronn phenomenon", *Mat. Sb.*, 15 (57):3 (1944), pp. 347–368.

one not obtain results of interest using only the fact that the zeroes of $\zeta(s)$ or the functions $L(s,\chi)$ are thinly distributed in certain portions of the critical strip? Such rare occurrences of zeroes have been established in certain cases (see for example the result of Page cited above). In its precise formulation the inequality (100) rests on two basic theorems proved by Linnik in Parts I and II of the above-cited work.

The first of these is as follows:

Let[270]

$$L(s,\chi_1), L(s,\chi_2), \ldots, L(s,\chi_{\phi(D)}), \tag{101}$$

be all L-series with mod *D characters, $\psi(D)$ some number in the interval $2 \le \psi(D) \le \frac{1}{3}\ln D$, and Q the number of series in (101) having at least one zero in the rectangle*

$$1 \ge \sigma \ge 1 - \frac{\psi(D)}{\ln D}, \quad |t| \le \min\{(\psi(D))^{100}, \ln^3 D\}.$$

Then

$$Q < e^{A\psi(D)},$$

where $A > 0$ is an absolute constant.

The second theorem, called by Linnik "the Deuring–Heilbronn phenomenon", is concerned with the exceptional real zero $\beta_0 > 1 - \frac{c_0}{\ln D}$, which at most one of the functions $L(s,\chi)$ modulo D can have; it is as follows:

Setting $\sigma_0 = 1 - \beta_0$, $\mu_0 = \frac{1}{\sigma_0 \ln D}$, we have that the rectangle defined by

$$1 \ge \sigma \ge 1 - c_1\frac{\ln(c_0\mu_0)}{\ln D}, \quad |t| \le \ln^3 D,$$

contains no zeroes of any $L(s,\chi)$ for all moduli $\le D^2$ (so in particular for the modulus D), except for β_0.

In 1945 Linnik gave a different proof of the Goldbach–Vinogradov theorem concerning three primes,[271] in which, so he asserts, Vinogradov's estimate of the sum

$$\sum_{p \le N} e^{2\pi i\alpha p},$$

may be replaced by results concerning thin distribution of the zeroes of $\zeta(s)$ and the functions $L(s,\chi)$. Let N be a large odd integer (which we seek to express as a sum of three primes), ϑ a real number, and write

$$\Lambda(n) = \begin{cases} \ln p & \text{if } n = p^k,\ k > 0,\ p \text{ prime}, \\ 0 & \text{otherwise.} \end{cases}$$

Linnik begins by considering the function

$$S(N,\vartheta) = \sum_{n=1}^{\infty} \Lambda(n) e^{-\frac{n}{N}} e^{-2\pi i\vartheta n},$$

[270] As is well known, there exist $\phi(D)$ characters for a given modulus D.

[271] Yu. V. Linnik, "On the possibility of a unified approach to certain questions of the 'additive' and 'distributive' theories of prime numbers", *DAN SSSR*, nov. ser., t. XLIX, No. 1, 1945, pp. 3–7.

close in form to the initial function of Hardy and Littlewood. It is not difficult to show that $Q(N)$, defined by

$$Q(N) = \sum_{N=p+p'+p''} \ln p \ln p' \ln p'',$$

satisfies

$$Q(N) = e \int_0^1 S(N, \vartheta)^3 e^{2\pi i N\vartheta} d\vartheta + O(N^{\frac{3}{2}+\epsilon}).$$

The Goldbach theorem will be established if it can be shown that $Q(N) > 0$ for all N from some point on. To this end the interval $(0, 1)$ is again subdivided into "major and minor arcs". For the "major arcs" the argument is based on results like those of Page (see above), while for the "minor arcs", $S(N, \vartheta)$ is estimated with the aid of Littlewood's formula[272]

$$\sum_{n=1}^{\infty} \chi(n)\Lambda(n)e^{-nx} = E(\chi)x^{-1} - \sum_{\rho} x^{-\rho}\Gamma(\rho) - \frac{L'(0,\chi)}{L(0,\chi)} + \frac{1}{2\pi i}\int_{-\frac{1}{2}-i\infty}^{-\frac{1}{2}+i\infty} x^{-s}\left(-\frac{L'(s,\chi)}{L(s,\chi)}\Gamma(s)\right) ds,$$

where x has positive real part, ρ ranges over all critical zeroes of the function $L(s, \chi)$ (i.e., the zeroes in the strip $0 < \sigma < 1$), and

$$E(\chi) = \begin{cases} 1 & \text{if } \chi \text{ is a principal character,} \\ 0 & \text{otherwise.} \end{cases}$$

The following lemma of Carlson and Titchmarsh from 1930 on the zeroes of $\zeta(s)$ also plays an important role here: *For $0 < \nu < \frac{1}{2}$, the number Q of zeroes of $\zeta(s)$ in the rectangle defined by $1 \geq \sigma \geq \frac{1}{2} + \nu$, $|t| \leq T$, satisfies*

$$Q < \min\{c_2(\epsilon)T^{1-\frac{\nu}{1-\nu}+\epsilon},\ c_3 T^{1-4\nu^2}\ln^7 T\}.$$

It is appropriate however to note that at the time of writing (1947) the article in question (in *Doklady Akad. Nauk*) is in the form of an announcement, and detailed proofs have not appeared, so that we shall refrain from giving here a more detailed assessment of this work of Linnik.

In conclusion we note that Linnik has obtained via Shnirelman's approach yet another solution of Waring's problem.[273]

Rodion Osievich Kuz'min (Professor at Leningrad University). In one of his earliest articles (of 1916) Vinogradov gives an estimate for the modulus of the sum

$$\sum_n e^{2\pi i f(n)}, \tag{102}$$

where $f(n)$ is a real-valued function with the property that the difference $f(n) - f(n+1)$ varies monotonically while remaining always in an interval $(\theta, 1-\theta)$, $0 < \theta < \frac{1}{2}$, and the sum is taken over the successive integer values of n for which

[272] See: E. C. Titchmarsh, "The Zeta-function of Riemann", *Cambridge Univ. Press*, 1930.

[273] Yu. V. Linnik, "An elementary solution of Waring's problem by means of Shnirel'man's method", *Mat. Sb.*, 12 (54):2, 1943, pp. 225–230.

$f(n)$ satisfies this condition.[274] In his quite short article "On certain trigonometric inequalities",[275] Kuz'min gives an elegant proof of Vinogradov's result, obtaining

$$\left|\sum_n e^{2\pi i f(n)}\right| < \frac{2}{\sin \pi\theta}.$$

In another article[276] Kuz'min extends Vinogradov's theorem on the fractional parts of a polynomial to the case of several polynomials in a single variable.

Mikhail Aleksandrovich Gel'bke (Docent at the Leningrad Polytechnic Institute; perished during the blockade of Leningrad). As we have seen, in a 1917 article Vinogradov gave an asymptotic formula for the sum

$$\sum_{x>Q}^{R} \{f(x)\}$$

of the fractional parts of values of a function f of a single variable. The first articles of Gel'bke[277] are devoted to generalizing this formula to functions of two variables; his main result is the following:

Suppose we are given a region Ω of the (x, y)-plane, bounded by a closed curve, with the property that any straight line parallel to the coordinate axes divides the region into at most two parts. Let $F(x, y)$ be a function defined in the region Ω and having all first and second partial derivatives, where in addition each of the first partial derivatives takes on the same value at only finitely many points (or on continuous portions of the boundary) and the second derivatives satisfy the following conditions:

$$\frac{1}{A} \le F''_{xx}(x, y) \le \frac{k}{A}, \qquad \frac{1}{B} \le F''_{yy}(x, y) \le \frac{l}{B},$$
$$|F''_{xy}(x, y)| \le s\sqrt{F''_{xx}(x, y)F''_{yy}(x, y)},$$

where k, l, s are constants with $0 < s < 1$, and

$$A \ge T > 1, \ B \ge T,$$

for some constant T. Finally, assume that the diameters of the region Ω parallel to the coordinate axes[278] *are less than or equal to gA and hB for some constants g, h. Under these conditions we have*

$$\sum_{\Omega} \{F(x, y)\} = \frac{\Pi}{2} + O\left((AB)^{\frac{3}{4}}\sqrt{\ln(AB)}\right),$$

where the summation is over all integer points (x, y) of the region Ω, and Π denotes the number of such points.

The rest of Gel'bke's work was devoted to Waring's problem.

---

[274] Presumably the largest sequence of successive integers on which f is defined. *Trans.*

[275] *Zh. Leningrad. Fiz.-Mat. Obshch. (Journal of the Leningrad Physics and Mathematics Society)*, I, 2 (1927), pp. 233–239.

[276] R. O. Kuz'min, "On the theory of simultaneous Diophantine approximations", ibid., II, 2 (1929), pp. 1–12.

[277] M. A. Gel'bke, "On an asymptotic expression for the sum of the fractional parts of a function of two variables", ibid., I, 2 (1927), pp. 281–299; "On an asymptotic expression for the sum of the fractional parts of a function of two variables", *IAN, IOFM*, 1930, pp. 409–423.

[278] That is, the suprema of the vertical and horizontal distances between pairs of points of Ω. *Trans.*

A List of Works in Number Theory by Chebyshev, Korkin, Zolotarev, Markov, Voronoĭ, and Vinogradov

P. L. Chebyshev

[1] *The theory of congruences*, book; Russian editions: first, St. Petersburg, 1849; second, St. Petersburg, 1879; third, St. Petersburg, 1901; Translations: *Theorie der Congruenzen*, Berlin, 1888; *Theoria delle congruenze*, Roma, 1895. Also reprinted in the Collected Works of P. L. Chebyshev, published by the Academy of Sciences of the USSR in 1944 (Russian).

[2] *On the determination of the number of primes not exceeding a given number*, first published in 1849 as a third appendix to the book *The theory of congruences* (Russian), and also as a separate article under the title *Sur la fonction qui détermine la totalité des nombres premiers inférieures à une limite donnée*, Mémoires des savants étrangers de l'Acad. Imp. Sci. de St.-Pétersbourg, VI, 1848. Published also under the same title in the Journal de math. pures et appl., I série, XVIII, 1852.

[3] *Mémoire sur les nombres premiers*, Mémoires de l'Acad. Imp. Sci. de St.-Pétersbourg, VII, 1850, and under the same title in the Journal de math. pures et appl., I série, XVII, 1852.

[4] *Sur les formes quadratiques*, Journal de math. pures et appliquées, I série, XVI, 1851.

[5] *Note sur différentes séries*, Journal de math. pures et appl., I série, XVI, 1851.

[6] *On an arithmetical question*, An appendix to Notes of the Imp. Akad. of Sciences, X, No. 4, 1868. (Russian)

[7] *Lettre à M. Fuss sur un nouveau théorème relatif aux nombres premiers contenus dans les formes* $4n+1$ *and* $4n+3$, Bull. de la classe phys.-math. de l'Akad. Imp. Sci. de St.-Pétersbourg, XI, 1853.

[8] *Sur la généralisation de la formule de M. Catalan et sur une formule arithmétique qui en résulte*, C. R. de l'Assoc. franç. pour l'avancement des sciences, 1876.

[9] *Sur une transformation des séries numériques*, Nouv. corr. math. rédigé par M. Catalan, IV, 1879.

[10] *Proof of a theorem of Chebyshev* (Russian), note written up by A. A. Markov from papers of Chebyshev remaining after his death, C. R. de l'Acad. Sci. Paris, CXX, 1895.

(Except for the last, all of these works were reproduced in both Russian and French in the first edition of Chebyshev's collected works (Akad. Nauk, 1899–1907), and in Russian in vol. 1 of a new edition, AN SSSR, 1944).)

A. N. Korkin

[1] *Sur les formes quadratiques positive quaternaires* (joint with E. I. Zolotarev), Math. Ann., Bd. 5, 1872.

[2] *Sur les formes quadratiques* (joint with E. I. Zolotarev), Math. Ann., Bd. 6, 1873.

[3] *Sur les formes quadratiques positives* (joint with E. I. Zolotarev), Math. Ann., Bd. 11, 1877.

[4] *Sur l'impossibilité de résoudre l'équation* $X^n + Y^n + Z^n = 0$ *en fonctions entières*, C. R. de l'Acad. des Sci., Paris, 1880.

A paper ready for publication with the following title was found among Korkin's remains; it was subsequently incorporated in part in D. A. Grave's course in number theory:

[5] *Sur la distribution des nombres entiers suivant le module premier, et les congruences binômes, avec une table des racines primitives et des caractères qui s'y rapportent pour les nombres premiers inférieurs à 4000.*

E. I. Zolotarev

[1] *On an indeterminate equation of the third degree*, dissertation for the degree of Master of Mathematics, St. Petersburg, 1869. (Russian)
[2] *Sur les formes quadratiques positives quaternaires* (joint with A. Korkin), Math. Ann., Bd. 5, 1872.
[3] *Nouvelle démonstration de la loi de réciprocité de Legendre*, Nouv. Ann. de Math., 2-e série, t. 11, 1872.
[4] *Sur l'équation* $Y^2 - (-1)^{\frac{p-1}{2}} pZ^2 = 4X$, ibid.
[5] *Extrait d'une lettre de M. G. Zolotareff*, Nouv. Ann. de Math., 2-e série, t. 12, 1873.
[6] *Sur les formes quadratiques* (joint with A. Korkin), Math. Ann. Bd. 6, 1873.
[7] *The theory of integral complex numbers with an application to the integral calculus*, Doctoral Dissertation, St. Petersburg, 1874. (Russian)
[8] *Sur les nombres complexes*, Bull. de l'Acad. Sci. de St.-Pétersbourg, 3-e série, 24, 1878.
[9] *Sur les formes quadratiques positives* (joint with A. Korkin), Math. Ann., Bd. 11, 1877.
[10] *Sur la théorie des nombres complexes*, Journal de math. pures et appliquées, (3), t. 6, 1885.

A. A. Markov

[1] *On indefinite ternary quadratic forms*, Izv. AN, vol. XIV, 1901. (Russian)
[2] *On indefinite quadratic forms in four variables*, Izv. AN, vol. XVI, 1902. (Russian)
[3] *On three indefinite ternary quadratic forms*, Izv. AN, vol. XVII, 1902. (Russian)
[4] *G. F. Voronoĭ. 1868–1908.* Obituary, Izv. AN, vol. II, 1908. (Russian)
[5] *Table des formes quadratiques ternaires indéfinies ne répresentant pas zéro, pour tous les déterminants positifs* $D \leq 50$, Notes of A. N. in the physics and mathematics department, series VIII, vol. XXII, No. 7, 1909.
[6] *On binary quadratic forms of positive determinant*, Master's Dissertation, St. Petersburg, 1880. (Russian)
[7] *Sur les nombres entiers dépendants d'une racine cubique d'un nombre entier ordinaire*, Mém. de l'Acad. Sci. de St.-Pétersbourg, VII série, t. 38, 1892.

G. F. Voronoĭ (A complete list of his works)

[1] *On the Bernoulli numbers*, Soobshch. Khar'kov. Mat. Obshch., 1890. (Russian)
[2] *On integral algebraic numbers depending on a root of an irreducible equation of the third degree*, Master's dissertation, 1893. (Russian)
[3] *On a generalization of the continued fraction algorithm*, Doctoral dissertation, Warsaw, 1896. (Russian)
[4] *On the number of roots of a congruence of the third degree with respect to a prime modulus*, Notes of the Xth conference of scientific professions, 1897. (Russian)
[5] *On the sum of the quadratic residues of the form* $4m+3$ *modulo a prime number* p, Protokoly St.-Peterburg. Mat. Obshch., 1899. (Russian)
[6] *A generalization of the concept of the sum of a series*, Notes of the XIth conference of scientific professions, 1901. (Russian)
[7] *Sur un problème du calcul des fonctions asymptotiques*, Crelle, Bd. 126, 1903.
[8] *Sur une fonction transcendente et ses applications à la sommation de quelques séries*, Ann. de l'École Norm. Sup., t. 20, III série, 1903.
[9] *Propriété du discriminant des fonctions entières*, Ver. des III internationalen math. Kongr., 1904.
[10] *Sur le développement à l'aide des fonctions cylindriques des sommes doubles* $\sum f(pm^2 + 2qmn + rn^2)$, ibid.
[11] *Sur quelques propriétés des formes quadratiques positives parfaites*, Crelle, Bd. 133, 1907.
[12] *Recherches sur les paralleloèdres primitifs*, Crelle, Bde. 134, 136, 1908, 1909.

I. M. Vinogradov[279]

[1] *A new method for obtaining asymptotic expressions for arithmetic functions*, Izv. Ros. Akad. Nauk, vol. II, No. 16, 1917, pp. 1347–1378. (Russian)

[2] *On an asymptotic equivalence in the theory of quadratic forms*, Zh. Fiziko-Matematicheskogo Obshch. Perm. Univ., vol. 1, 1918, pp. 18–28. (Russian; French summary)

[3] *On the average value for the number of classes of purely fundamental forms of negative determinant*, Soobshch. Khar'kov. Mat. Obshch., vol. 16, Nos. 1/2, 1918, pp. 10–38. (Russian)

[4] *Sur la distribution des résidus et des non-résidus des puissances*, Zh. Fiziko-Matematicheskogo Obshch. Perm. Univ., issue 1, 1918, pp. 94–98. (Russian)

[5] *On the distribution of quadratic residues and non-residues*, Zh. Fiziko-Matematicheskogo Obshch. Perm. Univ., vol. 2, 1919, pp. 1-16. (Russian; French summary)

[6] *On asymptotic equivalences in number theory*, Izv. Ros. Akad. Nauk, vol. 5, 1921, pp. 158–160. (Supplement to the proceedings of the XVI meeting of the Phys.-Math. Section of the Acad. of Sciences on November 23, 1921.) (Russian)

[7] *Sur un théorème général de Waring*, Matem. Sb., vol. 31, 1924, pp. 490–507. (French; Russian summary)

[8] *An elementary proof of a general theorem of analytic number theory*, Izv. Akad. Nauk SSSR, vol. 19, Nos. 16/17, 1925, pp. 785–796. (Russian; French summary)

[9] *An elementary proof of a general proposition of analytic number theory*, Izv. Leningrad. Politekhn. Inst., vol. 29, 1925, pp. 3–12. (Russian; French summary)

[10] *On the distribution of indices*, Dokl. Akad. Nauk SSSR, April 1926, pp. 73–76. (Russian)

[11] *On a bound for the least non-residue of degree n*, Izv. Akad. Nauk SSSR, vol. 20, Nos. 1/2, 1926, pp. 47–58. (Russian)

[12] *On the fractional parts of an integer polynomial*, Izv. Akad. Nauk SSSR, vol. 20, No. 9, 1926, pp. 585–600. (Russian)

[13] *On the question of the distribution of the fractional parts of the values of a function of one variable*, Zh. Leningrad. Fiz.-Mat. Obshch., vol. 1, 1926, pp. 56-65. (Russian; French summary)

[14] *On a general theorem concerning the distribution of residues and nonresidues of powers*, Bull. Amer. Math. Soc., vol. 32, No. 6, 1926, p. 596.

[15] *Analytic proof of a theorem on the distribution of the fractional parts of an integer polynomial*, Izv. Akad. Nauk SSSR, vol. 21, Nos. 7/8, 1927, pp. 567–578. (Russian)

[16] *On the distribution of the fractional parts of the values of a polynomial in two variables*, Izv. Leningrad. Politekhn. Inst., vol. 30, 1927, pp. 31–52. (Russian; French summary)

[17] *Démonstration élémentaire d'un théorème de Gauss*, Zh. Leningrad. Fiz.-Mat. Obshch., vol. 1, 1927, pp. 187–193. (French; Russian summary)

[18] *On a general theorem concerning the distribution of the residues and non-residues of powers*, Trans. Amer. Math. Soc., vol. 29, 1927, pp. 209–217.

[19] *On the bound of the least non-residues of n-th powers*, Trans. Amer. Math. Soc., vol. 29, 1927, pp. 218–226.

[20] *On Waring's theorem*, Izv. Akad. Nauk SSSR (Otd. Fiz.-Mat. Nauk), Nos. 4/5, 1928, pp. 393–400. (Russian)

[21] *On the representation of a number by an integer polynomial in several variables*, Izv. Akad. Nauk SSSR (Otd. Fiz.-Mat. Nauk), Nos. 4/5, 1928, pp. 401–414. (Russian)

[22] *On a class of simultaneous Diophantine equations*, Izv. Akad. Nauk SSSR (Otd. Fiz.-Mat. Nauk), No. 4, 1929, pp. 355–376. (Russian)

[23] *On the least primitive root*, Dokl. Akad. Nauk SSSR, No. 1, 1930, pp. 7–11. (Russian)

[24] *The elements of higher mathematics*, Leningrad., Izd. KUBUCH, Part 1, *Analytic geometry*, 1932, 248 pages. (Russian)

[279]The original list of Vinogradov's publications includes appointments at various times to editorial positions on the following journals and series: Zhurnal Leningradskogo Fiziko-Matematicheskogo Obshchestva; Mathematics in Monographs. Survey series; Matematicheckiĭ Sbornik. New series; Izv. Akad. Nauk SSSR, OMEN, Ser. Mat.; Mathematics in Monographs. Basic series.

[25] *Problems of analytic number theory. Summary of the lecture.* In the book: *Lectures on the occasion of the Jubilee session of the Acad. of Sciences of the USSR, devoted to the XV-th anniversary of the October Revolution*, Leningrad, Izd. Akad. Nauk SSSR, 1932, p. 13. (Russian)

[26] *On the number of integer points inside a circle*, Izv. Akad. Nauk SSSR, No. 3, 1932, pp. 313–336. (Russian)

[27] *The elements of higher mathematics*, Leningrad, Izd. KUBUCH, Part 2, *Differential calculus*, 1933, 176 pages. (Russian)

[28] *On the problems of analytic number theory.* In the book: *Proc. of the session devoted to the fifteenth anniversary of the October revolution, November 12–19, 1932*, Leningrad, Izd. Akad. Nauk SSSR, 1933, pp. 27–37; see also a separate brochure, 1933, pp. 1–11. (Russian)

[29] *An application of finite trigonometric sums to the question of the distribution of the fractional parts of an integer polynomial.* In *Trudy Fiz.-Mat. Inst. Steklova*, vol. 4, 1933, pp. 5–8. (Russian)

[30] *On a trigonometric sum and its application to number theory*, Dokl. Akad. Nauk SSSR, No. 3, 1938, pp. 195–204. (Russian; English summary)

[31] *On certain trigonometric sums and their applications*, Dokl. Akad. Nauk SSSR, No. 6, 1933, pp. 249–255. (Russian; English summary)

[32] *New applications of trigonometric sums*, Dokl. Akad. Nauk SSSR, vol. 1, No. 1, 1934, pp. 10–14. (Russian; English summary)

[33] *New asymptotic expressions*, Dokl. Akad. Nauk SSSR, vol. 1, No. 2, 1934, pp. 49–51. (Russian; English summary)

[34] *Trigonometric sums depending on a composite modulus*, Dokl. Akad. Nauk SSSR, vol. 1, No. 5, 1934, pp. 225–229. (Russian; English summary)

[35] *New theorems on the distribution of quadratic residues*, Dokl. Akad. Nauk SSSR, vol. 1, No. 6, 1934, pp. 289–290. (Russian; French summary)

[36] *New theorems on the distribution of primitive roots*, Dokl. Akad. Nauk SSSR, vol. 1, No. 7, 1934, pp. 366–369. (Russian; English summary)

[37] *A new solution of Waring's problem*, Dokl. Akad. Nauk SSSR, vol. 2, No. 6, 1934, pp. 337–341. (Russian; English summary)

[38] *On certain new problems of number theory*, Dokl. Akad. Nauk SSSR, vol. 3, No. 1, 1934, pp. 1–3. (Russian; English version, pp. 4–6.)

[39] *Certain theorems of analytic number theory*, Dokl. Akad. Nauk SSSR, vol. 4, No. 4, 1934, pp. 185–187. (Russian; English summary)

[40] *A new estimate for $G(n)$ in Waring's problem*, Dokl. Akad. Nauk SSSR, vol. 4, Nos. 5/6, 1934, pp. 249-251. (Russian; English version, pp. 251–253)

[41] *On an upper bound for $G(n)$ in Waring's problem*, Izv. Akad. Nauk SSSR, No.10, 1934, pp. 1455–1469. (Russian; English summary)

[42] *Certain theorems on the distribution of indices and primitive roots.* In *Trudy Fiz.-Mat. Inst. Steklova*, vol. 5, 1934, pp. 87–93. (Russian)

[43] *Sur quelques nouveaux résultats en théorie analytique des nombres*, C. R. Acad. Sci. Paris, vol. 199, No. 3, 1934, pp. 171–175.

[44] *On approximations by means of rational fractions with denominator an exact power*, Dokl. Akad. Nauk SSSR, vol. 2, No. 1, 1935, pp. 1–5. (Russian; French summary)

[45] *On certain rational approximations*, Dokl. Akad. Nauk SSSR, vol. 3, No. 1, 1935, pp. 3–6. (Russian)

[46] *On the fractional parts of polynomials and other functions*, Dokl. Akad. Nauk SSSR, vol. 3, No. 3, 1935, pp. 99–100. (Russian)

[47] *New estimates for Weyl sums*, Dokl. Akad. Nauk SSSR, vol. 3, No. 5, 1935, pp. 195–198. (Russian)

[48] *A new variant of the derivation of Waring's theorem*, Trudy Matem. Inst. Steklova, vol. 9, 1935, pp. 5–15. (Russian)

[49] *The number of integer points in a sphere*, Trudy Matem. Inst. Steklova, vol. 9, 1935, pp. 17–38. (Russian)

[50] *Une nouvelle variante de la démonstration du théorème de Waring*, C. R. Acad. Sci. Paris, vol. 200, 1935, pp. 182–184.

[51] *Sur les sommes de M. H. Weyl*, C. R. Acad. Sci. Paris, vol. 201, No. 13, 1935, pp. 514–516.

[52] *On Waring's problem*, Ann. of Math., vol. 36, No. 2, 1935, pp. 395–405.

[53] *On approximation to zero with the help of numbers of a certain general form*, Matem. Sb., vol. 42, No. 2, 1935, pp. 149–156.

[54] *On Weyl's sums*, Matem. Sb., vol. 42, No. 5, 1935, pp. 521–530.

[55] *An asymptotic formula for the number of representations in Waring's problem*, Matem. Sb., vol. 42, No. 5, 1935, pp. 531–534.

[56] *Fundamentals of number theory*, Moscow-Leningrad, ONTI, 1936. (Russian)

[57] *A new improvement of the estimates of trigonometrical sums*, Dokl. Akad. Nauk SSSR, vol. 1, No. 5, 1936, p. 195. (Russian; English version, pp. 199–200)

[58] *New results concerning the distribution of the fractional parts of a polynomial*, Dokl. Akad. Nauk SSSR, vol. 2, No. 9, 1936, pp. 355–357. (Russian; English version, pp. 361–364)

[59] *Approximation with the help of certain fractions*, Ann. of Math., vol. 37, No. 1, 1936, pp. 101–106.

[60] *On fractional parts of certain functions*, Ann. of Math., vol. 37, No. 2, 1936, pp. 448–455.

[61] *Sur les nouveaux résultats de la théorie analytique des nombres*, C. R. Acad. Sci. Paris, vol. 202, 1936, pp. 179–180.

[62] *Sur quelques inégalités nouvelles de la théorie des nombres*, C. R. Acad. Sci. Paris, vol. 202, 1936, pp. 1361–1362.

[63] *On the number of fractional parts of a polynomial lying in a given interval*, Matem. Sb., vol. 1 (43), No. 1, 1936, pp. 3–8.

[64] *A new method of resolving certain general questions of the theory of numbers*, Matem. Sb., vol. 1 (43), No. 1, 1936, pp. 9–20.

[65] *Approximation by means of fractional parts of a polynomial*, Matem. Sb., vol. 1 (43), No. 1, 1936, pp. 21–27.

[66] *On an asymptotic formula in Waring's problem*, Matem. Sb., vol. 1 (43), No. 2, 1936, pp. 169–174.

[67] *A new method of estimation of trigonometrical sums*, Matem. Sb., vol. 1 (43), No. 2, 1936, pp. 175–178.

[68] *Supplement to the paper "On the number of fractional parts of a polynomial lying in a given interval"*, Matem. Sb., vol. 1 (43), No. 3, 1936, pp. 405–407.

[69] *A new method in analytic number theory*, Trudy Matem. Inst. Steklova, vol. 10, 1937. (Russian)

[70] *The distribution of the fractional parts of the values of a polynomial under the condition that the argument ranges over the primes in an arithmetic progression*, Izv. Akad. Nauk SSSR,, No. 4, 1937, pp. 505–514. (Russian; English summary)

[71] *Represention of an odd number as a sum of three primes*, Dokl. Akad. Nauk SSSR, vol. 15, Nos. 6/7, 1937, pp. 291–294. (Russian)

[72] *Some new problems of the theory of prime numbers*, Dokl. Akad. Nauk SSSR, vol. 16, No. 3, 1937, pp. 139–141. (Russian; English version, pp. 131–132)

[73] *New estimations of trigonometrical sums containing primes*, Dokl. Akad. Nauk SSSR, vol. 2 (44), No. 4, 1937, pp. 165–166.

[74] *A new estimate of a sum containing primes*, Matem. Sb., vol. 2 (44), No. 5, 1937, pp. 783–792. (Russian; English summary)

[75] *Some theorems concerning the theory of primes*, Matem. Sb., vol. 2 (44), No. 2, pp. 179–195. (English; Russian summary)

[76] *Some general theorems in the theory of primes*, Trudy Tbilis. Mat. Inst., vol. 3, 1937, pp. 35–67. (Russian)

[77] *Einige allgemeine Primzahlsätze*, Trudy Tbilis. Mat. Inst., vol. 3, 1937, pp. 35–67.

[78] *Fundamentals of number theory*, Second revised edition, Moscow-Leningrad, ONTI, 1938. (Russian)

[79] *A new estimate of a certain trigonometric sum containing primes*, Izv. Akad. Nauk SSSR, No. 1, 1938, pp. 3–14. (Russian; English summary)

[80] *Improvement in the estimate of a certain trigonometric sum containing primes*, Izv. Akad. Nauk SSSR, Ser. Mat., No. 1, 1938, pp. 15–24. (Russian; English summary)

[81] *Some new estimates in analytic number theory*, Dokl. Akad. Nauk SSSR, vol. 19, No. 5, 1938, pp. 339–340. (Russian)

[82] *The distribution of quadratic residues and non-residues of the form $p + k$ with respect to a prime modulus*, Matem. Sb., vol. 3 (45), No. 2, 1938, pp. 311–319. (Russian; English summary)

[83] *Elementary estimates of a certain trigonometric sum involving primes*, Izv. Akad. Nauk SSSR, Ser. Mat., 1939, pp. 111–122. (Russian; English summary)

[84] *Estimates of certain of the simplest trigonometric sums involving primes*, Izv. Akad. Nauk SSSR, Ser. Mat., 1939, pp. 371–398. (Russian; English summary)

[85] *New perfection of a method for estimating trigonometric sums involving primes*, Dokl. Akad. Nauk SSSR, vol. 22, No. 2, 1939. (Russian)

[86] *The simplest trigonometric sums involving primes*, Dokl. Akad. Nauk SSSR, vol. 23, 1939. (Russian)

[87] *A certain general property of the distribution of primes*, Matem. Sb., vol. 7 (49), 1940, pp. 365–372. (Russian; English summary)

[88] *The distribution relative to a given modulus of the primes in an arithmetic progression*, Izv. Akad. Nauk SSSR, Ser. Mat., vol. 4, 1940, pp. 27–36. (Russian; English summary)

[89] *Two theorems relating to the theory of the distribution of prime numbers*, Dokl. Akad. Nauk SSSR, vol. 30, No. 4, 1941, pp. 287–288.

[90] *Improvement of some theorems of the theory of primes*, Dokl. Akad. Nauk SSSR, vol. 37, 1942, pp. 115–117.

[91] *General theorems on estimations of trigonometric sums*, Dokl. Akad. Nauk SSSR, vol. 43, No. 2, 1944, pp. 47–48.

[92] *The analytic theory of numbers*, Izv. Akad. Nauk SSSR, Ser. Mat., vol. 9, No. 3; *The works of P. L. Chebyshev in the theory of numbers* (joint with B. N. Delone). In the collection: *Scientific heritage of P. L. Chebyshev*, vol. I, Izd. Akad. Nauk SSSR, 1945. (Russian)

[93] *A general distribution law for the fractional parts of the values of a polynomial when the variable ranges over the primes*, Dokl. Akad. Nauk SSSR, vol. 1, No. 7, 1946, pp. 491–492.

Titles in This Series

26 **B. N. Delone,** The St. Petersburg school of number theory, 2005
25 **J. M. Plotkin, Editor,** Hausdorff on ordered sets, 2005
24 **Hans Niels Jahnke, Editor,** A history of analysis, 2003
23 **Karen Hunger Parshall and Adrain C. Rice, Editors,** Mathematics unbound: The evolution of an international mathematical research community, 1800–1945, 2002
22 **Bruce C. Berndt and Robert A. Rankin, Editors,** Ramanujan: Essays and surveys, 2001
21 **Armand Borel,** Essays in the history of Lie groups and algebraic groups, 2001
20 Kolmogorov in perspective, 2000
19 **Hermann Grassmann,** Extension theory, 2000
18 **Joe Albree, David C. Arney, and V. Frederick Rickey,** A station favorable to the pursuits of science: Primary materials in the history of mathematics at the United States Military Academy, 2000
17 **Jacques Hadamard (Jeremy J. Gray and Abe Shenitzer, Editors),** Non-Euclidean geometry in the theory of automorphic functions, 1999
16 **P. G. L. Dirichlet (with Supplements by R. Dedekind),** Lectures on number theory, 1999
15 **Charles W. Curtis,** Pioneers of representation theory: Frobenius, Burnside, Schur, and Brauer, 1999
14 **Vladimir Maz′ya and Tatyana Shaposhnikova,** Jacques Hadamard, a universal mathematician, 1998
13 **Lars Gårding,** Mathematics and mathematicians: Mathematics in Sweden before 1950, 1998
12 **Walter Rudin,** The way I remember it, 1997
11 **June Barrow-Green,** Poincaré and the three body problem, 1997
10 **John Stillwell,** Sources of hyperbolic geometry, 1996
9 **Bruce C. Berndt and Robert A. Rankin,** Ramanujan: Letters and commentary, 1995
8 **Karen Hunger Parshall and David E. Rowe,** The emergence of the American mathematical research community, 1876–1900: J. J. Sylvester, Felix Klein, and E. H. Moore, 1994
7 **Henk J. M. Bos,** Lectures in the history of mathematics, 1993
6 **Smilka Zdravkovska and Peter L. Duren, Editors,** Golden years of Moscow mathematics, 1993
5 **George W. Mackey,** The scope and history of commutative and noncommutative harmonic analysis, 1992
4 **Charles W. McArthur,** Operations analysis in the U.S. Army Eighth Air Force in World War II, 1990
3 **Peter L. Duren et al., Editors,** A century of mathematics in America, part III, 1989
2 **Peter L. Duren et al., Editors,** A century of mathematics in America, part II, 1989
1 **Peter L. Duren et al., Editors,** A century of mathematics in America, part I, 1988